D0363198

Green Britain
or
Industrial Wasteland?

. . . this sceptred isle, this earth of majesty . . . this other Eden, demi-paradise, this fortress built by Nature for herself . . . this precious stone set in the silver sea . . . this blessed plot, this earth, this realm, this England.

Green Britain
or
Industrial Wasteland?

EDITED BY

Edward Goldsmith and Nicholas Hildyard

Polity Press

© Edward Goldsmith and Nicholas Hildyard, 1986.

First published 1986 by Polity Press, Cambridge, in association with Basil Blackwell, Oxford

Editorial Office:
Polity Press, Dales Brewery, Gwydir Street, Cambridge CB1 2LJ, UK.

Basil Blackwell Ltd, 108 Cowley Road, Oxford OX4 1JF, UK.

Basil Blackwell Inc., 432 Park Avenue South, Suite 1503, New York, NY 10016, USA.

All rights reserved. Except for the quotation of short passages for the purposes of criticism and review, no part of this publication may be reproduced, stored in a retrieval system, or transmitted, in any form or by any means, electronic, mechanical, photocopying, recording or otherwise, without the prior permission of the publisher.

Except in the United States of America this book is sold subject to the condition that it shall not by way of trade or otherwise be lent, hired out or otherwise circulated without the publisher's prior consent in any form or binding or cover other than that in which it is published and without a similar condition including this condition being imposed on the subsequent purchaser.

British Library Cataloguing in Publication Data

Green Britain or industrial wasteland?:
    a critique of Britain's environmental record.
    1. Environmental policy—Great Britain—History
    I. Goldsmith, Edward II. Hildyard, Nicholas
    333.7′0941   HC260.E5

    ISBN 0-7456-0249-5
    ISBN 0-7456-0350-9 Pbk

Library of Congress Cataloging in Publication Data applied for

Typeset by Alan Sutton Publishing Limited, Gloucester
Printed in Great Britain by T J Press Padstow, Cornwall

Contents

List of Contributors

Edward Goldsmith is editor of *The Ecologist*. He has authored or co-authored four previous books on the environment, including *A Blueprint for Survival*, *The Stable Society* and *The Social and Environmental Effects of Large Dams*. His articles have appeared in major environmental magazines around the world.

Nicholas Hildyard is a co-editor of *The Ecologist*. He is the author of *Cover-Up*, and co-author of *The Toxic Time Bomb* and *The Social and Environmental Effects of Large Dams*.

Ken Powell is Secretary of Save Britain's Heritage, 68 Battersea High Street, London SW11.

Alice Coleman is Director of the Land Use Research Unit, Kings College, London. She advises local authorities on design improvement of problem estates. She is author of *Utopia on Trial*.

Robert Waller is a poet, novelist, biographer and well-known writer and broadcaster on agriculture.

Chris Rose is Campaigns and Publications Officer of the World Wildlife Fund International, 1196 Gland, Switzerland. He was previously acid rain and pesticides campaigner for Friends of the Earth.

R. P. C. Morgan is Reader in Applied Geomorphology at the Department of Agricultural Engineering, Silsoe College.

Colin Prince, Christine Cahalan and **Don Harding** are at the Department of Forestry and Wood Science, University College of North Wales, Bangor, Gwynedd.

Nigel Dudley is a consultant researcher at Earth Resources Research Ltd, 258 Pentonville Road, London N1 9JY. He is co-author of *The Acid Rain Controversy* published by Earth Resources Research.

David Harris is Member of Parliament for St Ives and former Member of the European Parliament for Cornwall and Plymouth.

Alan Long is an organic chemist. He is Honorary Research Adviser to the Vegetarian Society, 53 Marloes Road, London W86 LA.

Chris Kaufman is editor of the *T and G Record*, the monthly newspaper of the Transport and General Workers Union, and former editor of *Landworker*. He is co-author of *Portrait of a Poison: The 2,4,5-T Story*.

Professor A. H. Walters was scientific adviser to the Soil Association from 1969–71. He is author of many books and editor of several international symposia.

Erik Millstone is a lecturer at Sussex University in Science and Technology Policy. Trained in physics and philosophy, he has studied the food industry for the past 12 years. He is author of *Food Additives*.

Brian Price is a freelance pollution consultant and an adviser to Friends of the Earth.

Angela Singer is a former *Guardian* journalist who has written extensively on industrial hazards.

Fred Pearce is news editor of *New Scientist*. He is author of *Watershed: The Water Crisis in Britain*.

Hilary Bacon is co-author of *Power Corrupts*. She has testified as a health witness at five public inquiries into the health effects of overhead high-voltage electricity cables.

Peter Bunyard is co-editor of *The Ecologist* and editor of *Industry and the Environment*. He is author of *Nuclear Britain* and co-author of *The Politics of Self-Sufficiency*.

Jim Slater is General Secretary of the National Union of Seamen, Alembic House, 93 Albert Embankment, London SE1.

Charles Medawar is co-founder of Social Audit Ltd. He has worked for Social Audit on a variety of consumer/public interest issues, but has a particular interest in the use of medicines and medicine policy in developing countries.

Alan Irwin is a lecturer in Science and Technology Policy at the University of Manchester and was involved with the group of authors who wrote *Cancer in Britain: The Politics of Prevention*.

Doogie Russell is a Workers' Educational Association tutor-organizer based in Manchester and was involved with the group of authors who wrote *Cancer in Britain: The Politics of Prevention*.

John Madeley is editor of the journal *International Agricultural Development*. He also writes for *The Observer*, the *Guardian* and other newspapers, and broadcasts on the BBC's World Service North-South on economic and ecological issues.

Maurice Frankel is Director of the Campaign for Freedom of Information, 3 Endsleigh Street, London, WC1H 0DD. He is a contributor to *Citizen Action*.

Jonathon Porritt is Director of Friends of the Earth, 377 City Road, London EC1. He is author of *Seeing Green*.

Nick Gallie is Fundraiser at Greenpeace, 36 Graham Street, London N1 8LL.

Lindy Williams is Co-Chair of the Green Party, 36/38 Clapham Road, London SW9 0JQ.

Acknowledgements

We would like to thank the Ecological Foundation for its generous support in financing *Green Britain or Industrial Wasteland*.

We would also like to thank Hilary Datchens and Jane Embury for helping us in the co-ordination of this project and not least for typing the final manuscript; David Held for his invaluable criticism of the various drafts of the articles; and Pat Lawrence for copy-editing the manuscript. Finally we would like to thank Helen Pilgrim, Anna Oxbury and other staff of Polity Press and Basil Blackwell. It has been a pleasure working with them.

Finally, we wish to make it clear that the views expressed by the individual contributors to *Green Britain or Industrial Wasteland* are their own. Inclusion in this volume should not be taken to imply that individual authors subscribe to the views expressed by other contributors, or that they support the policies advocated by *The Ecologist*, Friends of the Earth, Greenpeace or the Green Party.

Edward Goldsmith and Nicholas Hildyard.

List of Abbreviations

ACP	Advisory Committee on Pesticides
ADAS	Agricultural Development and Advisory Service
AGRs	advanced gas-cooled reactors
ASLEF	Association of Steam Locomotive Engineers and Firemen
ASTMS	Association of Scientific, Technical and Managerial Staff
BAA	British Agrochemical Association
BCME	bischloromethylether
BIBRA	British Industrial Biological Research Association
BNFL	British Nuclear Fuels Ltd
BYDV	Barley Yellow Dwarf Virus
CA	Consumers' Association
CBI	Confederation of British Industry
CEGB	Central Electricity Generating Board
CFP	Common Fishery Policy
CHP	combined heat and power
CLA	Country Landowners' Association
CLEAR	Campaign for Lead Free Air
CND	Campaign for Nuclear Disarmament
COGEMA	Compagnie Générale des Matières Nucléaires
COPA	Control of Pollution Act
CPRE	Council for the Protection of Rural England
CRM	Committee on Review of Medicines
CSM	Committee on Safety of Medicines
DCF	discounted cash flow
DES	Diethylstilbestrol
DHSS	Department of Health and Social Security
DoE	Department of the Environment
DPWA	Deposit of Poisonous Waste Act

EdF	Électricité de France
EEC	European Economic Community
ELF	extremely low frequency
e/m	electromagnetic
EMAS	Employment Medical Advisory Service
EPA	Environmental Protection Agency
FOE	Friends of the Earth
FRL	Fisheries Radiobiological Laboratory
GLs	guide levels
GMBATU	General, Municipal, Boilermakers and Allied Trades Union
GMBU	General, Municipal and Boilermakers Union
HSC	Health and Safety Commission
HSE	Health and Safety Executive
HV	high voltage
HWI	Hazardous Waste Inspectorate
IAEA	International Atomic Energy Agency
ICD	Infestation Control Division
ICES	International Council for the Exploration of the Sea
ICRP	International Commission on Radiological Protection
IPM	Integrated Pest Management
LDC	London Dumping Convention
MACs	Maximum Admissible Concentrations
MAFF	Ministry of Agriculture, Fisheries and Food
MPE	Meat Promotion Executive
NC	Nature Conservancy
NCC	Nature Conservancy Council
NCI	National Cancer Institute
NEC	net effective cost
NFU	National Farmers' Union
NHS	National Health Service
NPC	National Parks Commission
NPT	Nuclear Non-Proliferation Treaty
NRPB	National Radiological Protection Board
NSCA	National Society for Clean Air
NUAAW	National Union of Agricultural and Allied Workers Union
NUF	National Union of Farmworkers
NUS	National Union of Seamen
PAN	Pesticides Action Network
PDO	Potentially Damaging Operations
PERG	Political Ecology Research Group
POM	prescription-only medicine
PSPS	Pesticide Safety Precautions Scheme

PWR	pressurised water reactor
RSPB	Royal Society for the Protection of Birds
SANA	Scientists against Nuclear Arms
SPAID	Society for the Prevention of Asbestosis and Industrial Disease
SSSIs	Sites of Special Scientific Interest
TAC	Total Allowable Catches
TCPA	Town and Country Planning Association
TDRI	Tropical Development and Research Institute
TGWU	Transport and General Workers Union
THORP	thermal oxide reprocessing plant
TICA	Thermal Insulation Contractors Association
UCATT	Union of Construction, Allied Trades and Technicians
UKAEA	United Kingdom Atomic Energy Authority
WDA	Waste Disposal Authority
WHO	World Health Organization

Introduction

Edward Goldsmith and Nicholas Hildyard

The Costs of Modernization

Several themes run through the essays in this book. The first is that there is a direct, historical link between the increasingly serious environmental problems we are experiencing today and the 'modernization' of our economic activities.

Such modernization – be it in the field of forestry, agriculture, fishing, food processing, manufacturing or power generation – inevitably sets in train a series of closely related changes with profound social, economic and ecological implications. Activities (bread-making, for example) which were previously a vocation or a way of life have become industrial undertakings. Rather than being carried out at the domestic level, or as small family businesses, they are now carried out by giant commercial concerns. The scale of operations increases correspondingly; the use of natural products gradually gives way to that of synthetics; and increasingly sophisticated machines take over more and more of the functions previously fulfilled by human labour.

Alongside these changes, there has been a critical shift in our attitude towards the management of our resources and the running of our economy. The accent is now on the short term, with little or no thought being shown for the future. The achievement of economic efficiency (with the aim of maximizing short-term gains) has seemingly become the be-all and end-all of human endeavour. All other considerations have been ruthlessly subordinated to that one overriding goal. No matter if a new 'development' destroys an ancient monument, flattens the historic centre of a town, or uproots a local community. No matter if our countryside is torn apart by bulldozers, or if the country's wildlife is steadily deprived of habitats in which to survive. No matter if the quality of the air we breathe,

the food we eat and the water we drink is eroded through pollution, or if our health is undermined as a result. Such considerations are simply brushed aside as the inevitable price we must pay for economic efficiency.

Forestry and Farming

The effects of 'modernization' are evident in the field of forestry. The Forestry Commission operates on a large scale, using machinery and methods that are entirely geared to the achievement of short-term results. There is no concern for the environmental costs. Vast spraying programmes are regularly undertaken using pesticides known to be health hazards. Ugly conifer monocultures are planted on unsuitable lowland soils which, as **Colin Price**, **Christine Cahalan** and **Don Harding** note in chapter 7, can 'cause podzolization, a process which results in the acidification of soils', in addition to giving rise to 'nitrogen deficiencies'.[1] Clearcutting also leads to erosion, especially on slopes. For these and other reasons, modern forestry cannot last for long. Indeed, the experience in Czechoslovakia suggests that growing seven generations of conifers on sandy soil leads to such environmental degradation that the soil is no longer capable of supporting commercial forestry.[2]

Robert Waller also notes how in farming, efficiency has become the catchword. 'Farming is a business now, not a way of life' has become 'the slogan nailed to the masthead of the ministry'.[3] Farms have grown bigger and bigger, putting the traditional small farmer out of business and destroying the very fabric of rural society. Ever more expensive equipment has been introduced with the result that debts have skyrocketed bringing many farmers to the brink of bankruptcy. At the same time, sound husbandry has been replaced by chemical farming,[4] our countryside has been ruthlessly destroyed to provide more and more agricultural land and ever bigger fields,[5] and our wildlife has been all but exterminated. Our best soils are suffering from accelerating erosion,[6] our groundwaters are increasingly polluted with nitrates (the result of fertilizer use) and our rivers with pesticides. Inevitably, our drinking water,[7] and our food, is increasingly contaminated – the latter with antibiotics and hormones,[8] in addition to nitrates,[9] and pesticide residues.[10]

Can we seriously regard such environmental costs as an acceptable price to pay for having a modern agricultural industry? Unquestionably not. As **Robert Waller** asks, 'Surely the "business of farming" includes the conservation of the environment? How can it be healthy "business" if it erodes the soil, contaminates our waterways and our groundwater and even the food it produces?'[11]

Fishing

The fishing industry is another industry which has caused widespread ecological damage as a result of modernization. In order to increase economic efficiency, we have introduced massive trawlers, equipped with the latest capital-intensive technology, that can harvest vast quantities of fish in a very short time. Fish catches have increased dramatically as a result, but at what cost and for how long?

As **David Harris** notes, small fishermen using traditional methods have been put out of business; fishing communities have been disrupted; and overfishing has brought the herring 'to the verge of commercial extinction'.[12] Now, says Harris, 'it is the turn of the mackerel.' In the future, other species are also likely to be fished into extinction. Indeed, 'nearly all species within easy reach are under intense pressure.'

Once the fish stocks are gone, the big trawler owners will in turn go out of business, as is already beginning to happen, leaving a new breed of untutored and possibly part-time fishermen to eke out a marginal livelihood from our polluted and depleted seas.

Even now that it is readily apparent that the fishing industry is rushing headlong towards destruction, the authorities are doing next to nothing to halt (let alone reverse) these disastrous trends. On the contrary, government policy actually militates against the conservation of fish stocks.

Food Processing

In the last 30 years, we have seen the development of the modern, capital-intensive, science-based food-processing industry which now churns out 75 per cent of the food we eat. According to **Erik Millstone**, some 3,500 or more additives are now used 'in millions of combinations' to make this industry economically efficient.[13] As a result, the average Briton eats four kilogrammes a year of a mixture of some 3,500 different chemical additives, most of which have never been tested for possible health effects. Yet, as **Alan Irwin** and **Doogie Russell** imply, such additives must make a significant (though as yet unquantified) contribution to the incidence of cancer in this country – a disease which may already be killing as many as 150,000 Britons a year.[14]

The food-processing industry argues that additives are essential if the price of food is to be kept down. In reality, however, many additives are used to transform fresh food which would otherwise be sold relatively cheaply into packaged foods which, with sufficient publicity, can be sold at a much higher price – even though they are laced with synthetic chemicals and thoroughly devitalized.

Potato crisps (which, today, play an important part in the diet of most of our children) are a case in point. 'When we spend 13p to buy a packet of crisps, we are buying 1p's worth of potatoes which have been peeled, sliced, fried, flavoured, preserved, packaged, distributed and advertised into a highly profitable product, instead of a simple but relatively unprofitable spud.'[15]

Such products cannot provide us with a sound and healthy diet. Thankfully, however, the food-processing industry is unlikely to survive in its present form for very much longer. Already consumer pressure is forcing manufacturers to change their ways. In January 1985, Sainsbury's began to eliminate additives from their own-brand goods: as from March 1986, shoppers have been able to identify which products contain additives and (perhaps more important) which do *not*, through a system of colour coding introduced by the company. Birds Eye has also announced that it intends to eliminate 'all artificial colours' (including tartrazine, the yellow dye suspected of causing hyperactivity in children) from all of its products. The company has also promised a reduction in the number of other additives it uses.

The Electricity Supply Industry

Over the last decade, we have also witnessed the modernization of the electricity supply industry – and, in particular, the substitution of nuclear for coal- and oil-fired power stations.

The insidious pollution to which nuclear power gives rise is well described by **Peter Bunyard**, who also examines the likely effects of Britain's nuclear programme on our health.[16] Britain's reprocessing plant at Sellafield in Cumbria is particularly polluting. As **Nick Gallie** notes in his contribution, the plant, which is owned by British Nuclear Fuels Ltd (BNFL), routinely releases 'two million gallons of radioactive waste-contaminated water directly into the Irish Sea every day. It has done so for over 20 years.'[17] Accidental releases to the environment add still more radioactivity to the sea. Such accidents occur with monotonous regularity. In February 1986 alone, there were four major accidents, one of which released half a tonne of reprocessed uranium into the Irish Sea.

The extent to which the Irish Sea is now contaminated and the effects of radioactive pollution from Sellafield are now coming to light, with clusters of childhood leukaemias appearing in villages on the Cumbrian coast – and, indeed, on the other side of the sea in Ireland.

Is the pollution really worth while? Is it a justifiable price to pay for the cheap electricity that nuclear power is said to provide? The answer is a resounding 'No'. The costs of generating nuclear electricity, as **Peter**

Bunyard explains only too clearly, are incomparably higher than they are made out to be by the Central Electricity Generating Board (CEGB) and Britain's other generating boards.[18] Significantly, in the USA, the nuclear industry is almost dead. There have been no new orders for nuclear power stations for several years, while over 100 orders have been cancelled since the late 1970s. This is not primarily because of opposition to nuclear power – though this is certainly a factor – but because *they are too expensive*. In the USA, power is generated by private companies which do not have a monopoly on electricity supply and which cannot *afford* the massive investment required for a single nuclear power station. By contrast, the CEGB is one of the largest monopolies in the world. With most electricity being generated by coal, it can afford, at this stage, to divert its resources to a build-up of nuclear power.

If a referendum were called tomorrow, it is likely that Britain's nuclear industry would be phased out in the near future.[19] Of those questioned in a recent public opinion poll (conducted prior to the Chernobyl accident), 70 per cent said that they thought Sellafield unsafe: 39 per cent wanted the plant closed, whilst 40 per cent felt it should only reprocess spent fuel from Britain's reactors.[20] Only 17 per cent of those interviewed believed ministerial assurances that the plant was safe – and a mere 11 per cent wished the government to continue building nuclear plants.[21]

Despite such public disquiet over nuclear power (disquiet which the government puts down to 'emotionalism' and 'irrationality') the government remains firmly committed to its nuclear programme. The reason, as **Peter Bunyard** makes clear, lies in the well-established (but much denied) connection between Britain's civil nuclear programme and the maintenance of her independent nuclear deterrent.[22] Put simply, Britain needs nuclear power to generate the plutonium for atom bombs.

Third World Relations

Our relationship with the Third World is also increasingly governed by hard-nosed business considerations. Since 1980, it has been official government policy to allocate aid according to 'political, industrial and commerical considerations' with the aim of 'helping the poorest people in the poorest countries'. As John Tanner notes in *The Times*, however, 'the second definition of aid seems to have been forgotten.'[23] Indeed, the government has actually *reduced* its aid to the people of Africa in the face of the worst famine of all time.

As **John Madeley** points out, this reduction in aid is intended to ensure that 'there is more in the kitty for better off countries such as Turkey and Mexico.'[24] The logic is that such countries are more likely to have the cash to spend on British goods and services than the poor nations of Africa. The 'goods and services' that are available from Britain include pesticides whose use is often prohibited in other Western countries, armaments (of which Britain now sells £1,200 million worth a year) and even instruments of torture.

Does the British public share the cynicism of its political leaders? Judging by its overwhelming support for Bob Geldof in his efforts to raise money for real aid to the world's starving masses, this is unlikely.

Indeed **John Madeley** suggests:

> There could yet be votes for politicians who show people that they understand why and how Britain's relationship with the Third World does matter – why it is not in our interests that Third World people are poisoned by our chemicals, go hungry because we refuse to pay a fair price for their products and give aid to build giant white elephants, like the Mahaweli project, which can only serve to impoverish still further the poor of the Third World.[25]

Rationalizing Inaction

The second theme to emerge from this book concerns the hand-in-glove relationship that has developed between industry, politicians and the civil service. Whenever efforts have been made to impose controls, however modest, on the activities of our most polluting industries, civil servants have done their level best to water down those controls – or, worse still, to stifle them at birth.

To choose an example at random: consider the support given to the highly polluting pesticide industry by successive British governments. As **Chris Rose** notes:[26]

> recent research in the Kew Public Records Office by Maurice Frankel of the Campaign for Freedom of Information, has revealed how early attempts to bring pesticides under a comprehensive system of legal controls in order to protect farmworkers, the public and the environment, were undermined and finally defeated by concerted lobbying from within the civil service on behalf (it would appear) of the pesticide industry. Official papers from the early 1950s show how civil servants deliberately manoeuvered and steered 'expert' committees away from imposing legal controls – and even rewrote and reversed their findings.

Rose goes on to comment:

> the ministry not only went out of its way to help manufacturers and commercial users of farm chemicals to fend off controls over their use, it also

played a considerable role in undermining controls over the sale of pesticides in shops.

It is worth considering some of the expedients resorted to by government, industry and the civil service to prevent the imposition of controls not only over the use of pesticides but over other destructive activities.

A Question of Evidence

The most obvious expedient is to insist that there is no 'scientific evidence' that a particular product or activity is in fact harmful, and hence no need to control it. Most people generally accept such an assurance at its face value, especially if it is provided by a well-known and highly qualified scientist. Few realize that such an assurance is often only 'true' because no one has ever bothered to look for the evidence – in other words, there is 'no evidence' because the necessary research has never been undertaken to find it.

The point is made by **Erik Millstone** with regard to food additives.[27] 'When industry says that there is no evidence of any chronic hazards from additives' he writes 'this does not mean that it has looked for such hazards.' The truth is that there has been very little research into the effects of food additives on our health.

Nor for that matter has there been much research on the environmental effects of agricultural chemicals. As **Harry Walters** recently noted in an article in *The Ecologist*, agricultural research in Britain has been 'concentrated on the most cost-effective ways of using the new machines, fertilizers, pesticides and herbicides'; 'environmental research' has been 'virtually ignored and has remained neglected to this day.'[28]

The fact that the environmental effects of the majority of chemicals have never been examined has not prevented successive British governments from encouraging their use. Thus the Agricultural Development and Advisory Service (ADAS) has encouraged the use of more and more nitrate fertilizers, though it has never conducted 'any research on the quality of food produced or on the health of those who eat it'.

Chris Rose notes that 'the complete absence of figures for the amounts of different pesticides used on farms makes the detailed study of pesticide related cancers, nervous disorders, or other potential effects extremely difficult, if not impossible.' In addition, 'baseline environmental monitoring has been studiously ignored or even reduced, so ministers can safely reply that *there is "no evidence" of* problems.' Nor does the government have much idea of the extent to which drinking water is contaminated since 'data on pesticides detected in rivers and groundwater are not held centrally.'[29]

Worse still, as David Wheeler of the University of Surrey points out, Britain's water authorities have been specifically asked by the Department of the Environment 'under direct instructions from ministers . . . effectively to ignore contamination of the public water supplies by pesticides'.[30] It would clearly be an embarrassment for the public to know the extent to which its drinking water is contaminated with such poisons. Even when environmental research is undertaken, it is often carried out with the apparent aim of *justifying* the continued use of a chemical or the continuation of a given policy. In 1982, for example, Sir Derek (now Lord) Rayner was appointed to conduct an audit of research undertaken by the Ministry of Agriculture, Fisheries and Food (MAFF). Commenting on the need to continue monitoring the biological effects of dumping waste at sea, Rayner listed the three reasons why such monitoring should be carried out:

(1) Because the (Oslo and London) Conventions require it;
(2) As a check on paper predictions of the effects of dumping;
(3) In order to demonstrate to national and international opinion that dumping is safe.[31]

Research to demonstrate that dumping is safe? Whatever happened to scientific objectivity?

Another tactic is to fund research which attempts to pin the blame for environmental damage or adverse health effects on factors which are either outside our control or whose regulation does not demand any drastic changes in policy. Thus, although many eminent epidemiologists now believe that between 50 and 80 per cent of human cancers are caused by exposure to radiation or to chemicals in the environment, little research is devoted to the environmental causes of cancer. Instead, the bulk of the funding goes on researching the mechanisms of carcinogenesis at the cellular level and the role that viruses play as possible causes of cancer.

Similarly, **Nigel Dudley** notes that according to Steve Elsworth of Friends of the Earth, 'The CEGB's scientific research is framed so that it does not ask the question "what causes acid rain . . ." but rather "what apart from sulphur oxide emissions could cause acid damage to the environment?"'[32]

What is Scientific Evidence?

Even when sufficient data have been acquired to justify the banning of a dangerous environmental pollutant, government scientists often insist that it does not constitute 'scientific evidence'. Thus, although the literature on the connection between nitrosamines and cancer is extensive, this does not prevent government medical advisers from declaring the link to be 'not proven'.

This brings us to the rarely discussed question of what actually constitutes 'scientific evidence'. One of its necessary features is that it must have been obtained by a qualified scientist. This is not a frivolous comment. Often evidence which incriminates a chemical or an activity as harmful is dismissed by the authorities because it has been gathered by a 'layman' rather than a scientist. For example, reports by farmers of abortions and birth defects among sheep following their exposure to the herbicide 2,4,5–T have invariably been classified as 'anecdotal evidence' – no matter how many cases are reported by farmers, or how well documented those reports might be. In effect, 'scientific evidence' is a commodity over which scientists have a virtual monopoly – which is of course very convenient, since most of them are employed directly or indirectly by government or industry. Those who are not are unlikely to obtain funding for the relevant research.

Another feature of scientific evidence is its claimed indubitability. But is it really possible to guarantee that a chemical is completely safe? It is certainly difficult, if not impossible, to prove the harmfulness of a chemical substance epidemiologically. **Erik Millstone** makes this point with regard to food additives: 'With 3,500 or more additives being used in millions of combinations, and often in minute quantities (some products contain up to 30 different additives), it is next to impossible for epidemiology to identify any long-term or chronic effects from using particular food additives.'[33]

Even under laboratory conditions, it is difficult to *prove* for certain that a chemical is 100 per cent safe. Consider, for example, the problems of testing for carcinogenicity:

*   Cancer tends to have a long 'latency' period – that is, it takes a considerable time (5 to 30 or more years) after exposure to a carcinogenic agent before the disease manifests itself. Clearly it is very difficult (if not impossible) to carry out experiments over such a period of time. Yet to expose laboratory animals to large doses over shorter periods (thus artificially reducing the latency period) may give misleading results.

*   Cancer can be caused by exposure to numerous chemicals. Since we are exposed to a vast number of possible carcinogens (it is reckoned that some 9,000 different synthetic organic chemicals are in current everyday use, and many of them are either established or suspected carcinogens), a slight carcinogenic effect on the part of an individual

chemical could be of significance. However, to test for slight carcinogenic effects is complicated and expensive, since it requires exposing large numbers of laboratory animals to very small doses. Not surprisingly, such tests are rarely carried out.

* Any comprehensive safety test would have to test for all the *additive* effects and (more difficult still) the *synergic* effects of all the different combinations of chemicals to which we are exposed.

* It is logistically impossible to carry out *all* the tests which would be required to establish that a chemical is safe beyond all doubt. To reveal *every* possible carcinogenic, let alone mutagenic or teratogenic, effect would necessitate testing *sufficient* samples of *every* possible combination of *all* the chemicals introduced into our environment, over a *long* enough period for the tests to yield meaningful results. Among other things, there are probably not enough technicians, laboratories or indeed laboratory animals available in the world, much less in Britain, to carry out the vast number of experiments which would be required to undertake such tests.

* The tests carried out on laboratory animals, living in totally artificial conditions, cannot be taken as providing hard and fast evidence as to the effects of a chemical on human beings living in the real world. Thalidomide, for instance, had hundreds of times less effect on laboratory hamsters than on human beings.

The Myth of 'Acceptable' Levels

If we take all these, and many other factors, into account, one cannot avoid coming to the conclusion that the 'scientific evidence' to which government scientists attach so much importance does not amount to very much in the way of a guarantee of safety. All we can hope to establish is the *probability* that a chemical is harmful, and that should be regarded as sufficient information to act upon. Were such an approach to be adopted, however, few of the chemicals we use today would be permitted.

Faced with that reality, the authorities have frequently reacted by simply burying their heads in the sands. Thus a spokesman for MAFF's Advisory Committee on Pesticides (ACP) actually told one of us, Edward Goldsmith, in a telephone conversation in 1977 that there was no evidence that synergic effects existed – this despite a large body of research to the contrary. Indeed, the scientific literature makes it clear that synergic effects are present more often than they are absent.

The refusal of MAFF scientists to face up to the existence (let alone the dangers) of synergy is reflected in the 'permissible' levels which the ministry sets for pesticide residues in food. Those levels are based on the implausible assumption that whilst it is unsafe to consume more than a given amount of a *single* pesticide, it is perfectly safe to consume a cocktail of many different pesticides – so long as the level of each pesticide in the cocktail does not exceed the permitted level. The truth is that it is impossible to set a safe level for a given pesticide when our food contains tens if not hundreds of pesticide residues – not to mention all the other chemicals purposefully introduced into it by the food industry.

In fact, few of the 'acceptable' levels which have been set for exposure to pollutants (be it pesticides or other dangerous substances) have any serious biological basis. More often than not the 'acceptable' exposure level to a pollutant is the lowest level acceptable to the industry which generates it.

Angela Singer shows how this is so with respect to exposure to asbestos.[34] Indeed the Health and Safety Executive (HSE) has itself admitted that the officially accepted level of asbestos particles in the air we breathe is 'of no biological significance'; the level being no more than 'an empirical level which we have some hope of enforcing' or 'merely what engineering controls achieve'.

Other examples abound. Consider the following extract, for instance, from the 1967 Report of the Food Additives and Contaminants Committee on Aldrin and Dieldrin Residues in Food.

> We should like to recommend that no aldrin and dieldrin be permitted in milk and baby foods but we are aware that with the great sensitivity of analytic methods it has become possible to detect very low residues of aldrin and dieldrin in food and also that at present *it would be impossible to produce milk or baby foods that were entirely free from aldrin and dieldrin.* For these reasons we reluctantly decide against a zero tolerance and recommend that a limit of 0.003 p.p.m. be placed on aldrin and dieldrin in liquid milk, this being the lowest practicable limit of analysis. We recommend a corresponding limit of 0.02 p.p.m. in baby foods (including dried milk) which would take account of the difference in residues likely to be found in liquid and dried products. We also recommend that all ingredients for baby foods should be chosen by manufacturers with a view to keeping the aldrin and dieldrin content to the lowest possible level. While these limits seem to us realistic, we do not accept them readily or with equanimity.[35] (Our emphasis.)

Secrecy

Research that conflicts with the interests of industry or which, if made public, would cast government policy in a poor light is frequently kept

"For God's sake, Parkinson, you've been telling the truth again."

secret. This is often achieved by pleading the Official Secrets Act. **Maurice Frankel** deals with this whole issue – and the vital need for a Freedom of Information Act – in some detail.[36]

Other authors give examples of how secrecy has prevented the public from knowing the real dangers it runs from exposure to specific pollutants:

* **Angela Singer** shows how a key Department of Health report on asbestos was kept secret for no less than eight years. 'When the BBC television programme, *Nationwide*, wanted a copy . . . it had to be obtained in the USA under the US Freedom of Information Act.'[37]

* **Brian Price** notes how Sir Henry Yellowlees, Chief Medical Officer at the Department of Health, wrote a letter in which he stated that 'the simplest and quickest way of reducing general population exposure to lead is by reducing sharply or by entirely eliminating lead in petrol.' This letter, however, was not made public until it was eventually leaked to CLEAR, the Campaign for Lead Free Air.[38]

* Secrecy also prevents the public from knowing anything about the chemical additives in processed foods. As **Erik Millstone** points out: 'all the deliberations of the Food Advisory Committee of MAFF, as well as all the technical data on which its gives its advice to ministers, are covered by the Official Secrets Act.'[39]

* So too, secrecy prevents us from knowing anything about the pesticides used to grow the food we eat, and hence about the pesticide residues which we inevitably consume. As **Chris Rose** notes, MAFF 'insists that pesticide safety and environmental clearance data should be kept secret'.[40]

Such secrecy over the health effects of pollutants is particularly irresponsible. Serious and objective students of carcinogenesis are agreed that somewhere between 50 and 80 per cent of human cancers are caused by exposure to radiation, or by the chemicals in the food we eat, the water we drink and the air we breathe. This is certainly the position of Professor Samuel Epstein – one of the leading authorities on the subject in the USA. **Alan Irwin** and **Doogie Russell**, though not venturing a figure themselves, point to research sponsored by the trade unions which shows that up to 30 per cent of cancers are caused in the work place by exposure to such pollutants as asbestos, 2,4,5–T, vinyl chloride, benzene, BCME, low-level radiation and aromatic amines – all of which are well-documented carcinogens.[41]

The chemical industry and the scientists whom it employs either directly or indirectly insist that industrial pollutants are not a major cause of cancer. Instead, they blame the majority of cancers on viruses, smoking and the consumption of alcohol and animal fats. At most, industry tells us, chemicals in the workplace cause 5 per cent of cancers. But even if this figure were true, it would mean, as Allan Irwin and Doogie Rusell note, that about 7,000 people are dying of cancer every year as a result of occupational exposure to carcinogens.[42] If Epstein is right, then the real figure is far higher – between 70,000 and 112,000 every year.

How can we possibly justify having hidden from so many people information which, if it had been acted upon, could have prevented them from dying a slow and painful death?

If we are to reverse the rising tide of cancer deaths in Britain, there is only one possible course open to us. As **Irwin** and **Russell** argue, our first priority must be to tackle carcinogenic substances in the environment.[43] To do this we must counteract 'the secrecy and the complacency that has characterized the state's approach to regulation' and indeed oppose 'the powerful lobbies in this country which act against good health'. We can no longer tolerate the stonewalling and secrecy of government and industry on this vital issue – and it is up to the environmental movement to make this clear.

Misleading the Public

Today, much of the information supplied by government and industry on key environmental issues is designed to *rationalize* current practices and policies. To that end, numerous public statements have been made which can only be described as *purposefully misleading*.

Alice Coleman, for example, considers that the land-use surveys provided by the civil service were designed, 'consciously or unconsciously', to conceal information rather than reveal it.[44] This, she says, may reflect the fact that the department responsible for advising on land-use policies (the Department of the Environment) also designs the surveys that might expose the adverse consequences of its advice. Elsewhere she refers to the official land-use surveys as 'smokescreens of disinformation'.

Likewise, the CEGB is shamelessly cooking the books in order to suggest that nuclear power stations are more economic than coal-fired ones. The Board has resorted to every possible accounting trick in order to delude the public on that score. Even the House of Commons Select Committee on Energy has concluded that 'the method used by the CEGB to justify past investments in Nuclear Power is highly misleading as a guide to past investment decisions' and 'entirely useless for appraising future ones'.[45]

The National Radiological Protection Board is also misleading the public when it states that 'no overriding reason connected with radiological protection considerations has been identified which would preclude the disposal of suitably conditioned high-level waste on the ocean floor.' The NRPB *knows* that there is no commercially obtainable material that can contain high-level wastes for the tens of thousands of years (in some cases hundreds of thousands) during which they are potentially dangerous. Certainly the stainless steel drums in which the wastes are normally encased are quite inadequate to the task in hand – not least because they will be corroded by the action of the salt water in a matter of decades. The NRPB also knows that it is impossible to predict the movement of radionuclides on the ocean floor, especially if one considers that violent storms are known to occur in the ocean's depths which could transport them just about anywhere. Besides, the Board's reassurances are based on the assumption that radionuclides are diluted in the sea water – 'the myth of dilution', as Professor Paul Ehrlich of Stanford University refers to it. In reality, radionuclides tend to concentrate in specific organisms, organs and tissues. Thus ruthenium concentrates in seaweed, strontium in the bones of living things, plutonium in bones and testicles (which might partly explain the current epidemic of testicular cancer among young men in the industrial world), iodine 129 and 131 in the thyroid gland and so on. Such concentrations can be anything up to 300,000 times the levels in the surrounding environment.

Patrick Jenkin MP also misled the public when he said that although the contamination of Cumbrian beaches (as the result of an accidental release of nuclear waste from Sellafield in 1983) was 'very unsatisfactory', there was 'no evidence it could cause significant damage to anyone's health'. He went on to say that the worst that anyone might suffer would be 'localised irritation of the skin from prolonged contact with one of a number of pieces which have been found with much higher than usual levels of radioactivity'. As **Peter Bunyard** describes in detail, even very small doses of radioactivity can cause cancer. Yet, in the case of Sellafield, we are dealing with the release of a substantial amount of radioactive waste, quite sufficient to effect the health of anyone exposed to it.

The government, as we now know, was also guilty of hoodwinking the public when it insisted that plutonium from civil nuclear reactors had never been exported to the USA for military purposes. Nigel Lawson, when Secretary of State for Energy, stated quite explicitly that 'There is no more connection between the generation of power in a nuclear power station and weapons than there is between a conventional power station and conventional weapons.' CEGB representatives took the same line at the Sizewell inquiry. Significantly, Lord Hinton, the first Chairman of the CEGB to commission nuclear power stations for generating electricity, made the following comment: 'I am absolutely certain the CEGB statement is incorrect . . . I don't know whether they should get permission for a PWR at Sizewell or not, but what is important is that they shouldn't tell bloody lies in their evidence.'

In March 1986, Lord Marshall, Chairman of the CEGB, admitted that some civil plutonium had indeed been diverted for military use.

Suppressing Information

All too often, government scientists who actually take it upon themselves to tell the truth on some vital environmental issue immediately get into trouble with the authorities, usually on the grounds that they have broken the Official Secrets Act and are thereby some sort of traitor.

Thus, when in 1982 Dr Matthews, a soil scientist working for the MAFF, dared warn Cumbrian mothers not to take their children onto the radioactive beaches in the area around Sellafield, he was immediately dismissed from his job. No one was supposed to know that the beaches were significantly radioactive. So, too, when Dr Ross Hesketh revealed in 1983 that the British government, contrary to all its assurances, was exporting plutonium to the USA for military purposes, he suffered the same fate.

When a government-appointed body decides to publish environmental information that is not consistent with the government line, it is also likely to get into trouble. Thus, the Standing Technical Advisory Commmittee on Water Quality warned the government in 1983 (in two leaked draft reports) of the alarming increases in the nitrate contents of most of our drinking water. It even dared show that some of our drinking water was so contaminated that we would be in breach of an EEC Directive on water quality, when it came into force in 1985. The government's response was simply to abolish the committee and to decree that, in future, reports on water quality would be drawn up only for the purpose of advising ministers, and would remain unpublished.

Scientists from the Soil Survey of England and Wales found their funds cut off after they reported on the true extent of soil erosion from arable land in Britain. The survey revealed that soil erosion is widespread, thus giving the lie to previous government assurances that erosion is not a problem in Britain. The results of the survey highlighted how destructive and unsustainable are the modern capital-intensive methods of farming which successive governments have been encouraging farmers to adopt. The Soil Survey is now likely to be discontinued.

Given such heavy-handed attempts to manipulate the flow of information to the public, it is hardly surprising that government announcements are treated with increasing scepticism by those within the environmental movement. Nor is that scepticism restricted to environmental activists. An informal poll conducted some years ago revealed that no more than 12 per cent of BNFL's employees at Sellafield actually believed what their management was telling them regarding the safety of the plant they worked in. How long will it be before the electorate, as a whole, reacts in a similar way to the pronouncements of its elected government?

Delaying Tactics

On those occasions when the government has been forced to 'do something' about a pollutant or a hazardous activity (usually as a result of a public outcry), its first reaction has invariably been to set up a scientific committee to 'look into the problem'. This delaying tactic (well known to fans of *Yes, Minister*) gains considerable time, since committees can take several years before they report. It also gives the public the impression that the government's response is even-handed, objective and well considered.

In recent years, no fewer than ten committees have been set up to examine the hazards of the herbicide 2,4,5–T. Each one has pronounced the poison 'safe' if used according to the manufacturers' instructions – a recommendation which, as **Chris Kaufman** points out, completely ignores the realities of spraying the chemical under farming conditions.[46] Similar committees in the USA and elsewhere have reviewed the same basic evidence and come to dramatically different conclusions. Indeed, 2,4,5–T is now banned in the USA (for most uses), Sweden, Norway, Denmark, Italy, Japan and Holland.

In 1976, the Labour government set up the Simpson Committee (known after its chairman, Bill Simpson) to look into the relationship between asbestos and lung cancer – a full 70 years after the critical link between the two had first been established. The committee took three years to come out with its final report, which contained 41 recommendations. By January 1986, only four of those recommendations had been implemented. According to **Angela Singer**, 'Many of the proposed reforms would have made no real difference to industrial practices anyway.'[47]

Similarly, in the mid-1970s, under pressure from environmental groups to ban the addition of lead to petrol, the government commissioned

Professor Patrick Lawther to look into the health effects of leaded petrol. 'In doing so,' notes **Brian Price**, 'it bought time during which it did not have to act and also hoped for a whitewash of sufficient opacity to enable it to rebuff the environmentalists for some considerable time.'[48] In the event, 'the ploy backfired'; the report was too deeply flawed to stand up to more than superficial analysis, and it was quickly discredited.

More recently, the British government announced that it would take no action to reduce sulphur dioxide emissions from coal-fired power stations until the Royal Society had completed a five year study into the connection between sulphur dioxide emissions and acid rain. The study is to be funded by the National Coal Board and the CEGB, the very industries which stand to lose most by any decision to tighten controls on emissions. Britain is the only major industrial country in Europe not to have joined the so-called '30 per cent Club' – a group of 20 countries which are now committed to reducing sulphur dioxide emissions by 30 per cent of 1980 levels by 1993. As **Nigel Dudley** notes: 'Britain's major role as a polluter is not open to doubt, however, and has produced strong European pressure for a reduction in emissions. This Britain has steadfastly resisted, blocking resolutions within the UN and European parliament wherever possible.'[49]

Non-Implementation of Environmental Legislation

When it was first introduced into parliament, Britain's main piece of environmental legislation, the 1974 Control of Pollution Act, was widely acclaimed as a responsible and well-meaning attempt to grapple with the problem of pollution. At the time, Mrs Thatcher, then opposition spokeswoman on the environment, welcomed the act as 'likely to have a greater, more lasting impact on the quality of life in many parts of Britain than most other measures'.

Once she had gained power, however, her interest in the quality of life noticeably diminished. Ten years after the act received royal assent, few of its clauses on water pollution had even been implemented. As **Fred Pearce** notes: 'Almost every polluting pipe or drain that the Act was intended to bring within the law has been granted an exemption.'[50] The act's clauses on waste disposal were similarly delayed. Indeed, had it not been for the need to comply with the EEC's Directive on Toxic and Dangerous Waste, many of the act's provisions on waste disposal might still not be implemented.

Commenting on the delays in implementing the Control of Pollution Act, the 1984 Royal Commission on Environmental Pollution made its view quite plain:

whilst we recognise that financial considerations will inevitably continue to be uppermost in Ministers' minds, we wish to stress the importance of tackling pollution problems in an order of priority which has been determined on merit, not on grounds of expediency or merely in response to the pressure of international obligations. We must sound a warning against the use of exemption orders as a mere device for postponing action on the nastier forms of uncontrolled discharge.[51]

It is not only 'home grown' legislation like the Control of Pollution Act which successive British governments have failed to implement or sought to delay. Britain, to her eternal shame, has fought tooth and nail to stymie numerous EEC Directives aimed at protecting the environment and improving public health. Thus:

* The government did everything it could to prevent the EEC from passing a directive to ban the use of growth-promoters in cattle. It argued that to pass such a law would simply encourage the development of a black market in hormones. Were one to accept that argument then no legislation would ever be passed against harmful products or activities unless they could be shown to be unprofitable – in which case, such legislation would be unnecessary since industry and consumers would have no reason to adopt them.

* The government has also failed to implement the EEC Directive on the Quality of Water Intended for Human Consumption, which became law in 1985. The Directive establishes guide levels (GLs) and maximum admissible concentrations (MACs) for a range of pollutants. The guide level for nitrate is set at 25mg per litre, while the maximum admissible concentration is 50mg per litre. As **Brian Price** points out, however: 'The British government is intending to ignore the new limit by issuing "derogations" (that is, exemptions) in the case of some 350 specified water sources.'[52]

* Britain also tried to side-step having to implement an EEC Directive requiring the cleaning up of bathing beaches in Europe. It did so by exploiting a loophole in the wording of the Directive, which allows member states to come up with their own definition of what constitutes a bathing beach. 'Britain decided on a definition so restrictive that it did not include Blackpool – by a long way the nation's favourite resort – as a bathing beach,' notes **Fred Pearce**.[53] As a result, Britain designated only 27 of her 600 beaches as 'bathing beaches' as against the 3,000 so designated by the French and the Italians. Although the EEC has set a scientifically testable

definition of what constitutes a 'safe' beach, England's water authorities have effectively ignored it. Instead, as Pearce points out, they define a beach as satisfactory 'if there are no signs of sewage, such as sewer slicks or solids, in areas that people might bathe in'.

* Perhaps the greatest battle between Britain and the EEC, however, has been over industrial discharges to rivers, estuaries and coastal waters. Britain' partners in the Community are adamant that the discharge of certain 'blacklisted' substances should be strictly controlled through a system of 'limit values' which would lay down the *maximum* permissible concentration of a given pollutant in any single discharge. The limit values would apply throughout the EEC. Britain, however, argues that such an approach is too 'inflexible' and would penalize British industry. Because our rivers are short and fast flowing, says the government, they can absorb more pollution than rivers on the continent. Britain therefore insists that emission standards should be allowed to vary from river to river. Directives which seek to impose limit values have been summarily rejected by the government.

Britain's frequent use of the veto to block EEC Directives and her *laissez-faire* attitude to environmental legislation has earned her the opprobrium of Europe. Our government is seen as petty minded, arrogant, and insular. Sadly our reputation outside Europe is little better. It sank to its lowest level when Britain attempted to steamroller the London Dumping Convention (the international body which polices the dumping of waste at sea) into lifting its moratorium on dumping nuclear waste at sea. Led by Spain, Australia and New Zealand, 25 countries voted for an indefinite ban on the practice. Only six countries (including Britain, France and the USA) voted in favour of nuclear dumping. When Britain lost the vote, the British delegation insisted that the resolution was not legally binding and that Britain (who is responsible for having dumped 90 per cent of the radioactive waste which has ever been dumped at sea) would continue to look upon ocean dumping as a possible waste-disposal option.

Public Inquiries: Rubber Stamping Decisions?

Under the 1971 Town and County Planning Act, the Secretary of State for the Environment has powers to 'call in' any planning application which, if granted, would have 'implications of more than local significance'. Once an application has been called in, a local planning inquiry is usually set up

"Having marshalled all the available scientific evidence we find the defendant not guilty."

to advise the Secretary of State whether or not to permit the proposed development. Any 'interested' parties may present evidence to the inquiry, which is presided over by an inspector and (if the evidence is likely to be highly technical) one or more assessors.

The Thatcher administration considers that too much time and money is being spent on public inquiries. To that end, it decided to limit their terms of reference so that basic issues (such as the *desirability* of the project) cannot be raised. Thus, the Dounreay inquiry, which opened in April 1986 to consider the Atomic Energy Authority's application to build a reprocessing plant for spent fuel from Britain's future fast-breeder reactor programme, had its terms of reference limited to a consideration of planning and purely technical issues. The government declared that it did not wish to see the Dounreay Inquiry become a 'trial for nuclear waste' in the same way as the Sizewell Inquiry has become a 'trial for nuclear power'.

For many people, the government's decision to limit the Dounreay inquiry merely confirmed a long-held suspicion that public inquiries are little more than PR exercises. Certainly, the government's refusal to provide funds to the objectors makes a mockery of the claim that inquiries are open, objective and democratic. At both the Windscale and Sizewell inquiries, the objectors had to raise for themselves the money with which to put forward their case, sometimes without the benefit of a lawyer. In sharp contrast, the expenses of both BNFL and the CEGB – amounting to several million pounds – were paid for by the taxpayer, since both are state-owned companies. If inquiries are really intended to be open, democratic and objective, surely the objectors should also be funded by the state?

The Windscale inquiry itself had all the hallmarks of a rubber stamp. The report of Mr Justice Parker, the Inspector at the inquiry, not only ignored the evidence of several distinguished objectors – notably Professor Radford, chairman of the US National Academy of Sciences' Committee on the Biological Effects of Ionizing Radiation – but also distorted the evidence of others. As *The Ecologist* commented at the time:

> Parker has a way of twisting the argument so that the objectors' case seems to support BNFL's. Thus he manages to argue that instead of increasing the chances of proliferation, reprocessing actually reduces them; instead of incurring a greater threat from radioactive waste, it diminishes it; and instead of leading to a greater drain of energy resources it actually increases them.

Changing Values

If our government refuses to listen to us, if any evidence we provide as to the harmfulness of its policies is summarily dismissed as 'hearsay', if the research required to reveal hazards is starved of funds, if vital information on pollution and other dangers is kept secret under the Official Secrets Act, and if public inquiries are little more than PR exercises, is there anything we can do to change government policy? Are there any grounds for hope?

The answer is a guarded 'Yes' – although we should never underestimate the likely resistance to change. The 1985 Social Attitudes Survey reveals that a significant proportion of the population is already beginning to regard a number of environmental problems as 'very serious' (see table 1). Public hostility to nuclear power is also growing daily, and unless Britain is transformed into a police state, it is difficult to see how our government can actually implement its present highly ambitious nuclear programme.

In the meantime, the environmental movement in Britain has never been more active. Friends of the Earth are producing extremely high-quality reports on key environmental issues, obtaining wider press coverage than ever before. Greenpeace, also, has shown the effectiveness of non-violent direct action which, as **Nick Gallie** points out, is one of the best means available for attracting the public's attention, arousing its sympathy and stimulating debate on the need to safeguard our environment against the depredations of industry.[54] Many other groups (such as Ecoropa, the Soil Association, the Henry Doubleday Research Association, the Farm and Food Society and the Conservation Society), together with countless local groups which have been set up to oppose irresponsible developments in their own areas, are also doing invaluable work.

The trade unions are increasingly taking up the green banner. Thus, the National Union of Seamen (NUS), the train drivers' union ASLEF, and the Transport and General Workers' Union (TGWU) have combined to

Table 1 How the public ranks environmental problems

	Very serious	Quite serious	Not very serious	Not at all serious
Nuclear power waste	69	18	9	2
Industrial effluents	67	25	6	1
Industrial air pollution	46	40	11	2
Lead from petrol	45	39	11	2
Traffic noise/dirt	20	45	29	4
Aircraft noise	7	24	50	17

Note: All figures in %. Don't knows omitted.

Source: The Social Attitudes Survey, 1985.

ban the movement of nuclear waste to be dumped at sea and have even persuaded the international Transport Workers' Federation to ban ocean-dumping worldwide.[55] The National Union of Agricultural and Allied Workers has instructed its members to refuse to handle 2,4,5–T and, as **Chris Kaufman** points out,[56] many other unions now support a total ban on the herbicide. Meanwhile, the Union of Construction, Allied Trades and Technicians (UCATT) has instructed its members to refuse to work with any form of asbestos.[57]

Britain's Green Party is now becoming better known and could eventually play an important role in the political life of Britain. Its activities attract considerable public attention at election time, publicizing both the ecological policies that it supports and the ecological value system which they reflect. This is of vital importance, for if *real* change is to be achieved, we must face up to the need for a radical shift in our values. There is, as **Jonathon Porritt** notes, a limit to what can be gained by acting on specific issues:

> The superficial sound and fury of many environmental battles conceals the fact that the real struggle is between . . . different value systems, a deeper confrontation which is only marginally influenced by the outcome of one specific issue. It has therefore been deeply frustrating to have to re-fight the same battle in different places at different times, as if nothing has been learnt from previous clashes.[58]

Underlying the destruction we are witnessing today lies a value system which is wedded to the belief that

> human needs can only be met through *permanent* expansion of the process of production and consumption – regardless of the damage done to the planet, to the rights of future generations, to the human spirit and to the living standards of all those who end up as the losers in this global, all-encompassing human race.[59]

Indeed, in terms of the value-system of industrialism (and, in particular, in terms of modern economic theory) the environmental damage we are inflicting on the planet is of no consequence whatsoever. Today's economists are trained to maximize 'benefits' and minimize 'costs'. By definition, a 'cost' must reflect a deprivation of some sort – more precisely, the deprivation of some 'benefit'. Yet Nature's benefits – the fresh, clean water that flows in unpolluted streams and rivers, the rain that naturally irrigates our crops, a stable and predictable climate, and the fertility of the soils upon which our agricultural system depends – are all taken for granted by economists and ascribed no economic value of any sort. As a result, the loss of Nature's benefits (which only accrue through the proper functioning of the ecosystem) is not considered a cost. It does not appear to have occurred to economists that if our activities interfere too radically with the workings of Nature, then Nature might no longer be capable of providing the benefits we now take for granted and upon which our very survival depends.

So long as we adhere to such a cock-eyed view of the world, we will continue to believe that if a project is 'economic' – that is, if it maximizes the short-term return on the resources it uses – it must be 'good' for the country, regardless of the environmental damage it causes. The environmental movement must reveal such thinking for the nonsense it clearly is. We must convince the public that economic growth – the 'permanent expansion of the process of production and consumption' – cannot solve the basic problems that confront us today: material goods cannot compensate for the breakdown of communities or the destruction of the environment; institutions, staffed by anonymous civil servants wearing (to use John McKnight's phrase) 'the mask of care', cannot replace a mother or, in a cohesive community, even a neighbour; and technology (however sophisticated) cannot solve the problems of alienation, alcoholism or drug abuse simply because these are not technological problems. The crisis we are facing today is not caused by a lack of material goods, nor yet by a lack of technology: it is caused by the social, biological and ecological disruption we have inflicted on the world in our relentless pursuit of material 'progress'. It can only be solved by re-establishing the social, biological and ecological systems we have disrupted. Only then can we hope to achieve a sustainable, just and self-reliant society.

The Chilean economist Manfred Von Neef notes how even Lord Keynes warned that 'the importance of economic problems should not be overestimated with the result that matters of higher and more permanent significance are sacrificed to its supposed necessities.'[60] The contributors to this book show how completely we have ignored that warning. We cannot afford to do so any longer. A fundamental change in the attitude of our political leaders is required – one which will lead to a veritable reversal of present policies. At stake is whether or not we and our children are to inhabit the industrial wasteland our politicians are busily creating for us, in what can still be 'a green and pleasant land'.

Notes

1. Colin Price, Christine Cahalan and Don Harding, chapter 7, 'The Environment in Forestry Policy', p. 88.

2. J. Pellisek, 'Conifers and Soil Deterioration', *The Ecologist*, vol. 5, no. 9, November 1975.
3. Robert Waller, chapter 3, 'Britain's Farm Policy', p. 47.
4. Alan Long, chapter 10, 'Down on the Pharm', pp. 120–35.
5. Chris Rose, chapter 5, 'The Destruction of the Countryside', pp. 66–78.
6. R. P. C. Morgan, chapter 6, 'Soil Erosion in Britain', pp. 79–84.
7. Brian Price, chapter 16, 'Lead Astray', pp. 189–97.
8. Long, 'Down on the Pharm'.
9. A. H. Walters, chapter 14, 'Nitrates in Food', pp. 172–8.
10. Chris Rose, chapter 12, 'Pesticides', pp. 143–64.
11. Waller, 'Britain's Farm Policy, p. 50.
12. David Harris, chapter 9, 'The Mackerel Massacre', p. 117.
13. Erik Millstone, chapter 15, 'Food Additives', p. 184.
14. Alan Irwin and Doogie Russell, chapter 29, 'Fighting Back against Cancer', p. 316.
15. Millstone, 'Food Additives', p. 182.
16. Peter Bunyard, chapter 23, 'The Sellafield Discharges', pp. 252–66; chapter 25, 'Radiation and Health', pp. 273–83.
17. Nick Gallie, chapter 34, 'The Case for Direct Action', p. 356.
18. Peter Bunyard, chapter 26, 'Ignoring the True Cost of Nuclear Power', pp. 284–95.
19. The term 'Britain' is used throughout this book although it is recognized that strictly speaking Britain excludes Northern Ireland.
20. Poll conducted by NOP Market Research Ltd in March 1986.
21. Geoffrey Lean, 'Nuclear Plant Poll Fuels Concern', *The Observer*, 30 March 1986.
22. Peter Bunyard, chapter 27, 'Britain and Plutonium Exports', pp. 296–303.
23. *The Times*, 7 November 1984.
24. John Madeley, chapter 30, 'Britain and the Third World', p. 325.
25. Madeley, 'Britain and the Third World', p. 328.
26. Chris Rose, chapter 11, 'Pesticide Controls', pp. 136–7.
27. Millstone, 'Food Additives', p. 184.
28. Walters, 'Nitrates in Food', *The Ecologist*, vol. 15, no. 4, 1985.
29. Chris Rose, chapter 12, 'Pesticides: An Industry out of Control', p. 160.
30. David Wheeler, 'Britain's Polluted Drinking Water', *The Ecologist*, vol. 16, nos. 2/3, 1986.
31. *The Rayner Scrutiny Committee on Fisheries Research and Development*, Draft Report, 1982.
32. Nigel Dudley, chapter 8, 'Acid Rain and British Pollution Control Policy', p. 96.
33. Millstone, 'Food Additives', p. 184.
34. Angela Singer, chapter 17, 'Asbestos', pp. 198–209.
35. *Report of the Food Additives and Contaminants Committee on Aldrin and Dieldrin Residues in Food*, London; HMSO, 1967.
36. Maurice Frankel, chapter 32, 'Environmental Secrecy', pp. 333–41.
37. Singer, 'Asbestos', p. 202.
38. Price, 'Lead Astray', p. 192.
39. Millstone, 'Food Additives', pp. 185–6.
40. Rose, 'Pesticide Exports', pp. 329–32.
41. Irwin and Russell, 'Fighting Back against Cancer', p. 317.
42. Irwin and Russell, 'Fighting Back against Cancer', p. 318.
43. Irwin and Russell, 'Fighting Back against Cancer', p. 318.
44. Alice Coleman, chapter 2, 'The Loss of Productive Land', p. 37.
45. House of Commons Select Committee on Energy, *Report* (3 vols), London: HMSO, 1981.
46. Chris Kaufman, chapter 13, '2,4,5–T', p. 166.
47. Singer, 'Asbestos', p. 205.

48. Price, 'Lead Astray', p. 190.
49. Dudley, 'Acid Rain', p. 101.
50. Fred Pearce, chapter 20 'Dirty water under the Bridge', p. 231.
51. The Royal Commission on Environmental Pollution, *Tackling Pollution: Experience and Prospects*, London: HMSO, 1984, p. 80.
52. Price, 'Pollution on Tap', p. 239.
53. Pearce, 'Britain's Dirty Beaches', p. 210.
54. Gallie, 'Case for Direct Action', pp. 352–60.
55. Jim Slater, chapter 24, 'Dumping Nuclear Waste at Sea', p. 267–72.
56. Kaufman, '2,4,5–T', pp. 165–6.
57. Singer, 'Asbestos', p. 207.
58. Jonathon Porritt, chapter 33, 'Beyond Environmentalism', p. 346.
59. Porritt, 'Beyond Environmentalism', pp. 345–6.
60. Quoted in Manfred A. Max-Neef, *From the Outside Looking In: Experiences in 'Barefoot Economics'*, Uppsala: Dag Hammarskjold, Foundation, 1982.

1

The Destruction of Britain's Urban Heritage

Ken Powell

Towns and cities are, by their very nature, places of change. Manchester in the 1840s, Chicago in the 1930s, Tokyo in the 1980s – all have been symbols of relentless growth and the destruction of traditional values. Yet urban life has also been seen as the embodiment of the truly civilized existence, in contrast to the 'rustic idiocy' of the countryside. Florence and Milan, Amsterdam and Bruges, Seville and Bordeaux – the great historic cities of Europe are living proof of the continuing vitality of European urban civilization. Our own large cities seem, in comparison, to be in a state of crisis. The optimism which accompanied the post-war redevelopment boom seems to have evaporated. The tower blocks of the City of London, the high-rise housing estates of Glasgow and Salford, and the entire rebuilt city centre of Birmingham are the objects of public opprobrium. As places to live, British towns and cities seem to fail on almost every count – indeed, it seems almost that the British have forgotten the art of living in towns. Are the British a basically anti-urban nation, or have we simply made our towns difficult places to inhabit? Are there grounds for hope that they can be reclaimed?

The destruction which has overtaken so many of our towns and cities has its origin in the Second World War. Long before that, small groups of 'conservationists' (a term then unknown) were fighting for buildings and areas of beauty and historic interest. The destruction of terraces, squares and great mansions in London between the wars was horrifying, and led to the formation of the Georgian Group in 1937. Whilst William Morris's Society for the Protection of Ancient Buldings (formed in 1877) had been chiefly concerned with rural buildings – farms and churches – the Group was urban in emphasis. Vast destruction in London and elsewhere was caused by wartime bombing, but it was to be eclipsed by the 'second blitz' of the developers. After years of depression and war, the nation took eagerly to the idea of rebuilding. This was particularly the case in the urban centres of the midlands and the north, which had suffered during the 1930s in a way unknown in the south east. Manchester, Leeds, Newcastle, Liverpool, Bradford and other commercial cities embarked on an era of massive reconstruction. Housing was a major issue, and everywhere thousands of houses were earmarked for demolition – 90,000 were deemed 'unfit' in Leeds alone. A campaign of clearance which is only now grinding to a halt began throughout Britain, dispersing the inhabitants of the old inner-city areas to sprawling suburban estates or pushing them into tall blocks of flats. City centre development went hand in hand with housing clearance. The 1963 Buchanan Report, *Traffic in Towns*, concluded: 'We shall have to make a gigantic effort to re-plan, re-shape, and re-build our cities . . . There is nothing that need frighten us . . . our cities, most of them, are pretty depressing places.' In the city of London, where over a quarter of the buildings had been destroyed or severely damaged by wartime bombs, a programme of rebuilding, aimed at the removal of the 'intolerable confusion' of the old city, began in the late 1940s. Within a decade the space lost through bombing had been more than replaced by new buildings, though really high buildings had yet to be seen. The city was divided up into 'redevelopment units', more than 50 in all, where the ancient townscape was ruthlessly destroyed. The intimate quality of the old courts, alleys and churchyards were unappreciated by the planners working under Lord Holford. Attention was concentrated on achieving grand settings for the principal monuments, especially St Paul's cathedral. Holford's desperately bleak Paternoster Square, north of the cathedral, is typical of this approach and one must give a cautious welcome to current proposals for its redevelopment. East of St Paul's, buildings which had survived the war, including most of the gutted Wren church of St Augustine, Watling Street, were swept away for the feeble new Choir School. New roads were a central element in the rebuilding process. The Barbican/London Wall area and the drab canyons of Upper and Lower Thames Street are the result, while some of the fine Victorian banking *palazzi* of Bishopsgate were demolished to widen the road by a few feet. The quality of the post-war architecture of the city is a national scandal. Only very rarely – as in the case of the new Lloyds building, currently being constructed to designs by Richard Rogers – has a new building of positive interest replaced what has been lost.

Where the city led, others followed. Offices and shops were the base of urban redevelopment. Some provincial cities had, of course, been badly blitzed, providing a head start for developers. The new centres of Bristol, Plymouth and Exeter, all recognizably 'historic' cities, are a half-way stage between the Garden Cities and the worse horrors to come.

(Plymouth's wide, wind-swept and tweely named Armada Way provided a striking contrast to the cosy, congested streets of the pre-war city.) Other cities, particularly those in the north, remained untouched and full of what planners were beginning to call 'obsolescent fabric' until the 1960s. Leeds, for example, suffered hardly at all from the war and for 15 years afterwards, despite elaborate plans, there was little redevelopment. Between 1961 and 1976, however, over a million square feet of shopping space was built in Leeds, which competed eagerly with other cities to attract new offices. With the decline of traditional industries, cities like Leeds saw the future in 'service' industries. In Newcastle-upon-Tyne, the vast new enclosed shopping centre took the name – and the site – of the historic Eldon Square. Manchester's vast Arndale Centre, extending over many blocks of the city, was dubbed 'the largest public convenience in Europe' on account of its unappealing cladding of ceramic tiles. In Birmingham, it became possible to arrive at the rebuilt (and tomb-like) railway station and to walk across much of the central area through a connecting series of shopping centres without touching ground level or breathing fresh air. The great shopping centres of the 1960s and early 1970s typified a spendthrift approach to resources, not just of existing buildings, but of energy too. Artificially heated, ventilated and lit, dependent on banks of escalators and lifts, and profoundly inflexible, they will in due course be very difficult structures to re-use. Already large sums have been spent on some of them not only to repair premature decay but to stimulate flagging public interest.

Smaller towns were not immune from destructive redevelopment. Gloucester, a town associated, for those who have never been there, with the splendid cathedral and the charming illustrations in Beatrix Potter's *The Tailor of Gloucester*, has suffered terribly. Since 1945, the city has been subjected to systematic ruination, with medieval, Georgian and later buildings casually cleared away. A large, bland shopping development now dominates the central core. Despite the large amounts of capital poured into its rebuilding, Gloucester today has a scruffy, slightly depressed air, with the old buildings that remain isolated as islands in the sea of rebuilding. It was in 1967 that the Civic Amenities Act empowered local authorities to designate Conservation Areas to protect not just individual buildings but entire historic areas. In too many towns and cities, designation came too late, while elsewhere it has not led to proper planning policies for preservation. It is inconceivable that historic cities in France or Italy could be subjected to the injuries inflicted on Gloucester – or on Worcester, Lincoln, Derby or Hull. French legislation, for example, to protect the setting of listed buildings (*monuments classés*) has had a potent effect in protecting the character of towns.

No greater contrast to the cheerful, bustling small towns of provincial France could be found than the once charming market towns of rural Lincolnshire. Spalding and Gainsborough have been badly ravaged. Grantham, a red-brick town dominated by the noble spire of the parish church of St Wulfram, seems to have succumbed to the chain stores, with its main street covered with a rash of brash and mass-produced fascias and much of the hinterland cleared for an architecturally commonplace shopping centre. Lincolnshire is truly a bleak area for the national heritage.

It is fair to say that great and irreparable damage has been done to that heritage over the last 40 years. Since 1950, at least 1,000 Georgian buildings have been destroyed in Bath (some 400 of them listed). Nationally, over 8,000 listed buildings have been destroyed in the last 20 years (though the rate of loss is now declining, and there are 310,000 listed buildings). The destruction of the old environment is linked, of course, to the growth of the new – increasingly a suburban environment necessitating the steady erosion of the countryside. Every five years, an area the size of Buckinghamshire is built over. But the picture is not entirely gloomy. Public disenchantment with modern development has never been more profound. Historic areas are now often the liveliest parts of towns and cities. London's Covent Garden is a prime tourist attraction, and it is hard to believe that the whole district faced clearance so recently. In Liverpool, the refurbishment of the great warehouses of Albert Dock will do more good for that depressed city than any post-war development scheme. Liverpool, like Bristol, may follow the example of US cities in turning to its waterfront for a source of urban revival. The famous Quincy Market development in Boston, Massachussetts, a refurbishment of redundant market buildings, has become the most successful shopping centre in North America and every other city is seeking to emulate it. In Nottingham, the once depressed Lace Market is coming to life, its factories and warehouses converted to offices, workshops and housing. Developers are taking a serious interest in riverside warehouses in the centre of Leeds, proposing their conversion to flats. Halifax, a mill town crippled by the decline of the woollen industry, is finding a new identity as a tourist centre.

Our towns and cities are reviving, it seems, and the trend towards city-centre housing is especially welcome. The new Historic Buildings and Monuments Commission places proper emphasis on urban conservation and more than £60 million is spent annually on conservation and environmental enhancement. Progress, however, depends on new policies at a national level. During 1984, public attention was concentrated on issues of conservation versus redevelopment, of new versus old by the long public inquiry into Peter Palumbo's 'Mansion House Square' scheme. The outcome is well known – the rejection of the plans for a tower block designed by the late Mies van der Rohe and a vast piazza replacing the traditional street plan. The applicant was invited, however, to come forward with a new and different set of proposals for redevelopment. At the time of writing, James Stirling's new scheme is awaited eagerly. It may well involve the demolition of many of the listed buildings on the site, though (unlike the previous scheme) it will be a genuinely up-to-date design by an architect revered more widely abroad than in his own land. Meanwhile, the vast 'Broadgate' development at Broad Street (where the Victorian railway terminus was ruthlessly demolished) is rising rapidly. It is devoted very largely to the great dealing floors now essential in any major City of London development. The plan to erect a colossal (929,000 square metres) office complex at Canary Wharf in the Isle of Dogs has alarmed the city corporation. Fearing the loss of investment to the docklands, where an enterprise zone suspends the normal planning constraints, it is seeking to throw out its own draft local plan as excessively conservationist.

Oddly, the city, with a vast income and a highly paid and highly educated workforce, is one of the poorest of working environments. Overloaded and antiquated public transport facilities contrast with the endless tide of road transport flowing unchecked down every street and alley. Tiny gardens, often former churchyards, are the only open spaces. Light and fresh air are at a premium. It is staggering that, in these circumstances, the city proposes to change the plot ratio requirements to allow up to 17 million square feet of new office space to be built. In part, this may replace dated and uninteresting buildings of the recent past – such as the 1950s slabs along London Wall. But there are threats also to the historic core. A large part of Ludgate Hill, part of the ancient processional route to St Paul's, faces demolition and redevelopment. Nowhere in all this rebuilding is the issue of the *character* of the city apparent – it is a character which often derives from small buildings and simple streets.

Whilst there are positive and humane trends in current architecture they are not yet widely manifest in the new buildings of our city centres. At Hammersmith in west London, local community groups commissioned the Terry Farrell Partnership (a practice associated with Post-Modernism and an imaginative approach to old buildings) to design an alternative to a truly inhuman development scheme. Good as it is, the Farrell scheme seems to have little chance of being built, since the local authority, eager to improve a drab and run down part of the town centre, is prepared to allow the developers free rein. 'Community' is, of course, a term we now hear much about – though it is often ill-defined. 'Community architecture' has been endorsed by HRH The Prince of Wales, who has called for people to be given a wider degree of choice in their surroundings. In the mill-town of Macclesfield, Rod Hackney, the prime practitioner of community architecture, has been at the centre of a campaign of regeneration. The exact relationship between community architecture and conservation is uncertain. Macclesfield's simple terraced houses have been thoroughly modernized with little regard for period details. In Liverpool, community architecture, in the form of a series of housing co-operatives, has saved decaying Georgian houses. However, the city council, which is dedicated to a renewed programme of council house building, views the co-operatives with suspicion. Some have even seen them as sinister forces undermining the democratic structure of local government.

In the end, community architecture must possess strong links with conservation since it generally seems to represent opposition to radical changes in the urban fabric. Community architects – if such a breed exists – can, however, be as merciless in their treatment of old buildings as any other architect. One of the most positive tasks for this movement, if such it is, must be the recycling of the huge stock of decaying council housing, often multi-storey flats, which it is hard to sell off individually even at knock-down prices. In much of the inner-city, 'community', in the old sense personified by the television series *Coronation Street*, does not really exist. Communal efforts at environmental improvement may actually help to create feelings of community. As such, they are conservationist in the broad sense. The community may not, of course, produce good architecture. Indeed, Sir John Summerson (in his book *Georgian London*) defined the handsome architecture of the Georgian age as a blend of enterprise and taste – rare commodities. But any movement which fights against the continuing degradation of our cities must be welcomed wholeheartedly.

Community involvement should surely extend beyond the narrow confines of housing policy. The worrying revival of large-scale development schemes for town centres has produced some deeply unpopular proposals. In towns as diverse as Trowbridge, Abergavenny, Hull and Torquay, schemes affecting the entire character of those places have been pushed through against local opposition. In several cases, public inquiries have been refused, despite the scale of the proposals. (It is, of course, remarkable that the government has refused to hold an inquiry into the Canary Wharf scheme, the largest of its kind in Europe.)

Clearance and rebuilding still seem to hold an attraction in Britain, whereas conservation and conversion are seen as a second-best. As a nation, we discourage the refurbishment of old buildings by imposing VAT on the work – the tax was lifted, after protests, on listed buildings but still applies to those in Conservation Areas. All repair work is subject to VAT, even those on listed buildings, so that rebuilding, rather than repair, is encouraged. The tax incentives used in the USA to encourage conservation work are quite lacking here. Our planning policies, especially the vestiges of the old zoning provisions, discourage the mixture of uses which is vital to the health of towns. Our housing policies encourage the social division of towns and the creation of poor and underprivileged ghettos. Conservation is a peripheral activity for our planners. Most seriously, Britain is a country where public and private enterprise rarely seem to work well together. (The contrast with the USA is again striking.) Public expenditure, often vital as a catalyst to private investment, is seen as an evil by some on the Right. On the other side, the dogmatism of the Left would allow buildings to rot rather than see them sacrificed to the demands of 'profit'. People are the pawns in the middle of the argument, as we have seen in Liverpool and some of the inner London boroughs. Our towns and cities reflect the futility of a political system which is unhealthily obsessed with versions of materialism. In the midst of it, the quality of life seems to be an irrelevant issue.

2

The Loss of Productive Land

Alice Coleman

Many of Britain's environmental problems spring from faulty land use, and it is natural for concerned people to press for more and better land-use control. Unfortunately the experience of the last 40 years (during which time, government intervention has been unprecedented) does not support the assumption that more planning means better land use. On the contrary, it is mainly those areas that have received the most planning attention and quango activity that have deteriorated most rapidly.

The Failure of Official Land-Use Surveys

One reason why control tends to be counter-productive is the practice of adopting policies *in advance* of collecting firm data on how land in Britain is actually used. The problem has been compounded by the failure to test policies in the light of hindsight. Successive governments have received poor unintegrated land-use advice from their civil servants, who have presided over a series of incredibly defective land-use surveys (see table 2). The first government survey required local authorities to produce maps which were a mixture of fact (uses to be retained) and fiction (new uses planned, but not necessarily ever achieved). No checks were made as to the accuracy of the data supplied and less than 10 per cent of the country was surveyed, so all possibility of obtaining national statistics was ruled out. The government's second attempt imposed an elaborate classification of 630 categories, which many counties have not been able to complete a decade later. The third survey, organized centrally by the Department of the Environment (DoE) itself, consisted of a mere five categories, and even those were so perversely defined that accuracy was impossible; errors of up to 30 per cent have been found by independent academics.[1] The

Table 2 A comparison of university and government land-use surveys

Survey agency	Survey title	Date	Data source	Number of Surveyed	categories Published	Coverage of England and Wales	Accuracy level
London university geographers' surveys	I Land Utilisation Survey of Britain	1930s	Field Survey	25	7	100%	Over 95%
	II Second land Utilisation Survey of Britain	1960s Resurvey in 1970s, 1980s	Field Survey	250	70	100%	Over 99%
Department of the Environment systems	I Local Planning Authorities	1940s–50s	Field logging	40	40	Under 10%	Not checked
	II Local Planning Authorities	1970s–80s	Field logging	630	–	Unknown	Not checked
	III DoE Map of Development Areas	1969 data 1967–8	Air photos	5	5	10%	70%–80%
	IV DoE Monitoring Land Use Change	1983–4 start	Field survey	19	–	2.8%	Under 70%

fourth survey accepted a 30 per cent error level from the outset but over-looked additional sources of error which made it even more inaccurate.

It is difficult to avoid the conclusion that all these surveys were designed, consciously or unconsciously, to conceal information rather than reveal it, which may reflect the fact that the Department responsible for advising on land-use policies (the DoE) also designs the surveys that might expose the adverse consequences of its advice. It would have been wiser to have established an independent, single purpose Land Utilisation Survey, staffed by professionals without axes to grind.

Independent Surveys

There have, in fact, been two independent surveys, directed by academics. The first was undertaken by L. Dudley Stamp in the 1930s and the second by myself in the 1960s, with subsequent sample resurveys taking place since. Numerous checks, including a joint exercise with the Ordnance Survey, proved that the error level was tiny – under 1 per cent – and the combination of clearly readable maps and new methods of analysis has produced both national figures and insights into local problems, which can now be studied in greater depth. As we shall see, the results have revealed a massive gap between planning aims and achievements, and also, which is worse, the fact that many 'successfully' implemented plans have done more harm than good.

Planning and the Inner Cities

The primary area of planning attack was the inner city where the urban ecosystem was dislocated on an unprecedented scale. Admittedly, there were aspects of its rich diversity that needed adjustment, but that was already happening spontaneously before the Second World War and, if given a chance, would have accelerated during the affluent post-war period. Instead, homes and workplaces were ruthlessly demolished for comprehensive redevelopment, and extensive empty niches were created in the form of long-term derelict sites. Residential diversity was replaced by the monoculture of utopian blocks of council flats that were completely lacking in any sense of 'human scale' and oblivious of other naturally or culturally evolved traditions of the human habitat. As a result, the plight of the planned inner city is much worse than that of its unplanned predecessor. A land-use survey entitled *Utopia on Trial* has shown that design features recommended and subsidized by the DoE are related to increases in litter, graffiti, vandal damage, excrement in the entrances of buildings, children in care, juvenile arrests, and many categories of crime.[2]

In human ecosystems, nutrient and energy cycles are joined by another important circulation, the flow of money. This has been completely destabilized in the inner city, not only because of the high cost of redevelopment, but also because of the policy of shrinking all forms of land use that yield rates income and employment, while expanding those that need subsidy. Many aspects of inflation have been refuelled by inner-city land-use mistakes; as a result of population dispersal, for example, wages have risen to fund longer journeys to work, and public spending on new out-of-town schools and hospitals has risen to replace inner-city closures.

Loss of Agricultural Land

Land-use changes ripple through the human ecosystem and, as with natural ecosystems, less damage is done where small multiple changes lead to small multiple adjustments than where there is a single source imposing changes of plague proportions. In the land-use system, that single source is officialdom, which not only determines large-scale mutations, but also robs individuals of the power to adjust spontaneously. 'No land-use may change without planning permission', thundered the 1947 Town and Country Planning Act, and if we have land-use problems today, officialdom must be held responsible.

The knock-on effect of decreasing urban densities is the sterilization of farmland by new development. It was here, in defence of the vital soil resources, that strong planning was genuinely needed, but the maps of the Second Land Utilisation Survey revealed that half a million hectares had been lost between 1933 and 1963.

That figure disguises both the rate and extent of land loss. First, the period during which the 500,792 hectares were lost included 12 years of food rationing, with a 10 year period of no net loss of farmland. The real loss thus effectively took place over 20 years – at a rate which, if continued, would mean the total extinction of all our agricultural land within 400 years.

Secondly, the figure of 500,792 hectares represents a *net* loss of agricultural land. A great deal more was in fact taken, but this was largely replaced by 'up-grading' scrubland to farmland. Thus between 1933 and 1972, in the Kyloe Hills of Northumberland, the area classified as 'farmscape' went up from 67.7 to 75.5 per cent, and that classified as 'marginal fringe' from 11.7 to 13.1 per cent, whilst that classified as 'wildscape'

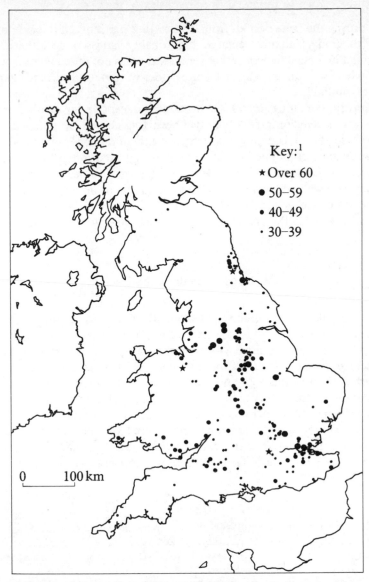

Figure 1 Reported types of vandalism on farms

Note: 1 A questionnaire on urban and recreational pressures on farmland elicited responses from all over the country, but the worst-hit farmers (who reported 30 or more types of vandalism) were conspicuously concentrated in urban fringe areas.

Source: Second Land Utilisation Survey of Britain.

fell, during the same period, from 20.6 to 10.2 per cent. In this way fourth or fifth grade land has replaced high-quality land – a fact that is not reflected in official figures. What is more, there is not much wildscape left, and when we run out the net annual loss of agricultural land will rise correspondingly.

Thirdly, the figure of 500,792 hectares is deceptive for another reason: part of our remaining farmland has been fragmented by planned urban growth so that it is exposed to trespass and to damage by vandals. For instance, boys from one housing estate are reported to have climbed into an adjacent field and cut the tails off all the cows. At least 22 per cent of our agricultural land is now affected by fragmentation. This has led to the reduction of yields and, as reflected in the increase in 'wasteland' in and around towns, to the abandonment of much otherwise good agricultural land.

Resurveys Confirm the Trend

Critics of the Second Land Utilisation Survey tried to argue that the Survey's figures for land loss reflected the squandering of land during the six pre-war, pre-planning years – and that land is now efficiently conserved as a result of planning control. Resurveys during the 1970s, however, showed this to be no more than a pious hope. Table 3 sets out the results of the resurvey and shows the acceleration of farmland loss, with extremely short periods to total extinction if continuing unchecked.

The resurveyed areas are not administered by particularly profligate planning authorities. On the contrary, they are 'conservation' minded. Nevertheless, the rate of farmland loss has accelerated, and often for no

Table 3 Farmland losses, 1960s–1970s, in resurveyed areas

	Thames Estuary	Merseyside	Surrey	Buckinghamshire
Date of resurvey	1972	1976	1977	1978
Area resurveyed (ha)	85,050	65,375	168,050	188,188
Improved farmland 1960s (ha)	36,700	25,413	78,638	146,638
Net loss per annum (ha)	420	384	553	638
Farmland extinction period (years from 1978)	81	51	127	215

useful purpose. In the Thames Estuary, for example, only one-third of the loss was converted to urban uses. The other two-thirds merely became wasteland, where farmers could no longer make agriculture pay because of the impact of urban pressures. As the wasteland remains in farm ownership it does not figure in the official 'land-take' statistics, and consequently the rate of loss is much faster than officially acknowledged. In fact, the 400 year estimate of the 1960s had to be halved to 200 in the 1970s, and although there has been less actual building during the present recession, the rate of advance planning permissions on farmland has continued to increase each year.[3]

Why Conserve Farmland?

Agriculture is essential for human survival, yet as an open air industry it is extremely vulnerable to the vagaries of the weather. A year of low production and high prices can bring great hardship to consumers, whereas a glut year with low prices can bankrupt many farmers and accentuate scarcities in lean years. Keeping on an even keel benefits producers and consumers alike, as was demonstrated by the British system of price supports adjusted annually to avoid over-production of any commodity. Unfortunately, the EEC is less flexible and allows escalating subsidies for excessive production. Taxpayers are therefore challenging the need to conserve farmland, while farmers who have been misled into heavy investment in subsidized enterprises are increasingly cut off by quota restrictions from any hope of repaying loans, and bankruptcies are increasing.

In the broader land-use context, the twin policy of conserving farmland and of increasing our self-sufficiency makes good sense, with great potential benefit to Britain and also to the world at large. World hunger does not need underlining, and Britain is a massive food importer, absorbing world surpluses that are sorely needed elsewhere. The increasing efficiency of British farmers has decreased our food imports and doubtless saved many lives. It has also greatly assisted the balance of payments, offsetting the decline in manufacturing exports. At present, the balance of payments problem has been thrust into the background by oil revenues, but these will not last forever, and when they taper off we shall wish we had conserved the soil resource that is capable of sustained renewable production in perpetuity.

The combination of shrinking farmland and a policy, until now, of greater self-sufficiency could be met only by intensification on the residual area and compensatory expansion of farmscape into wildscape. This has involved grubbing up hedges and copses, filling ponds, draining wetlands and increasing the use of herbicides and pesticides. Further, as average farm incomes have declined, while other incomes have continued to rise, farmers have pruned their costs by measures such as burning stubble rather than ploughing it in. All these things have incurred the displeasure of conservationists, who are fighting back vigorously, but often with the wrong weapons.

Reconciling Conflicting Interests

A better understanding of land use could help conservationists to present their case more efficiently. It is essential to recognize that there are three major groups of uses – settlement, farmland and vegetation – which all have legitimate interests. At present all three are in mutual conflict, and it would be more constructive to reconcile them than to exacerbate them by constant recriminations. For the conservationist there should be two objectives: to halt the unnecessary destruction of rural land by urban sprawl, and to replace the internecine strife among rural interests by a spirit of co-operation and encouragement.

Properly surveyed land-use maps show that urban expansion is unnecessary. There is enough unused urban land to absorb many years of development, especially if we abandon the myth that low density is beneficial – a myth which continues to mesmerize in spite of being repeatedly disproved. Derelict eyesores could, in innumerable cases, be eliminated by building houses with gardens which would increase urban wildlife and allow more people to enjoy gardening as a hobby. If more people had first-hand experience of the vulnerability of growing things, then at least the inadvertent destructiveness of visitors to farms might decline. And it would also become easier to reconcile the needs of agriculture, forestry, wildlife, game and water conservation if their respective protagonists no longer felt they were scrambling for a share in a diminishing rural cake.

Conservationists could promote reconciliation in a variety of ways. They could give favourable publicity to the many farmers who are creating new hedges, copses and ponds, and special praise where multi-species planting produces richer ecosystems than the old hawthorn hedges of the enclosure period. They could also present the conservationist case more lucidly. A more mature approach would admit that farmers incur losses, but would demonstrate the clear factual evidence that there are also greater and more diverse gains to be had from conservation that they have realized.[4]

Reconciliation requires enough information to understand each other's point of view; yet successive governments have never commissioned a proper, comprehensive urban and rural land-use survey. All the evidence to date suggests that ministers have been ill-advised on this subject by their civil servants, who have consistently designed successive surveys to be smokescreens of disinformation. Until this ineptitude is finally eradicated, many of the barriers to a more ecologically based land-use policy will remain.

Notes

1. D. Rhind and R. Hudson, *Land Use*, London: Methuen, 1980.
2. A. Coleman with S. Brown, L. Cottle, P. Marshall, C. Redknap and R. J. Sex, *Utopia on Trial*, London: Hilary Shipman, 1985.
3. R. Grove-White, unpublished MS, 1984.
4. F. Terrasson and G. Tendron, 'The Case for Hedgerows', *The Ecologist*, vol. 11, no. 5, 1981, pp. 210–1.

Select Bibliography

Coleman, A., 'Is Planning Really Necessary', *Geographical Journal*, vol. 142, 1976, pp. 411–37.

Coleman, A., 'Vanishing Farmland in England and Wales', *Association of Agriculture Journal*, no. 29, 1977, pp. 3–11.

Coleman, A., 'Land-Use Planning: Success or Failure', *Architects' Journal*, vol. 165, no. 3, 1977, pp. 91–134.

Coleman, A., 'The Loss of Farmland: The Evidence from Maps', in A. W. Rogers (ed.), *Urban Growth, Farmland Losses and Planning*, Institute of British Geographers, 1978, pp. 16–36.

Coleman, A., 'The Death of the Inner City – Causes and Cure', *The London Journal*, vol. 6, no. 1, 1980, pp. 3–22.

Coleman, A., 'Dead Space in the Dying Inner City', *International Journal of Environmental Studies*, vol. 19, no. 2, 1982, pp. 103–7.

Coleman, A., 'Population Pressures vs. Agricultural Land', in M. R. Brett-Crowther, *Food and Climate Review 1982–1983: Prospects for Self-Sufficiency*, 1983, pp. 16–22.

Coleman, A. and Feaver, I. J., 'Farm Vandals – Who Carries the Can?', *Farmers' Weekly*, 4 July 1980, pp. 114–20.

Coleman A. with Brown S., Cottle, L., Marshall, P., Redknap, C., and Sex, R. J., *Utopia on Trial*, London: Hilary Shipman, 1985.

Department of the Environment, *Strategy for the South-East: 1976 Review*,

1976. (Edited by P. H. Walls, M. I. Waldron and P. F. Thomas. See chapter 4 by A. Coleman, pp. 45–60.)

Department of the Environment, with Local Authorities' Management Services and Computer Committee, Scottish Development Department, *National Land-Use Classification*, London: HMSO, 1975.

Hooper, M. D. and Holdgate, M. W., (eds), *Hedges and Hedgerow Trees*, Monkswood Experimental Station, 1970.

Ministry of Housing and Local Government, Housing Cost Yardstick, Design Bulletin, no. 7, 1963.

Rhind, D and Hudson, R., *Land Use*, London: Methuen, 1980.

Stamp, L. D., *The Land of Britain: Its Use and Misuse*, London: Longmans, 1962 (third edn).

3

Britain's Farm Policy

A History of Wanton Oblivion

Robert Waller

It is impossible to understand the tortuous and unpredictable changes of farm policy in Britain since the war unless those changes are seen in their historical context.

In 1875 British agriculture suffered a dramatic collapse, as Disraeli had predicted 30 years earlier when the Corn Laws were repealed.[1] His policy had then been to prevent Britain from dividing into a rich urban society and a poor rural one – one aspect of his 'One Nation' ideal. He believed that the Whig policy of *laissez-faire* would lead to the 'counting house' mentality dominating all our political decisions and shaping our way of life. It was not understood by the townspeople that many of the problems of the towns originate in the countryside; a bankrupt countryside over-populates the towns and creates ghetto areas. This problem was not so intense in nineteenth-century Britain as it is today in developing countries, since many dispossessed rural people were able to emigrate to the colonies and dominions. Nevertheless it unbalanced the relation between town and country. The drift to the town of agricultural workers still continues, but today it is a consequence of 'technological unemployment'.

Disraeli was prime minister when the agricultural rural collapse he had predicted took place (as the result of floods of imports of cheap food from the countries where the British rural refugees were now farming). He did nothing to impose his former policy, however, either because he recognized that to reintroduce protection in a nation besotted with an absolute *laissez-faire* ideology would be to jeopardize his political authority, or possibly because he had been converted to the new ideology. He was, in any case, absorbed in his oriental dream of making the Queen Empress of India. If he had not succumbed to the forces of the industrial revolution (which aimed to make Britain the workshop of the world and the Empire its farm) but had maintained the strength of character to stand by his original insight into the effects of undisciplined industrial development, Britain might now be a more balanced and environmentally stable and attractive country. There were two Disraelis: the young idealist who saw the truth and the pragmatic, elder statesman. Today's Tories pay lip service to the first and follow in the footsteps of the second.

The subordination of farming and the countryside to the free-trade industrial interest continued (except for a brief respite when we needed to grow more food during the First World War), until the Second World War when food once again became essential for victory.

George Stapledon: A Man of Vision

The fact that the cities grew richer so far as money in the bank was concerned (through hardly richer aesthetically and as human habitats), while the countryside became poorer (though aesthetically and spiritually more attractive) disturbed many countrymen such as George Stapledon, who sought for a means of correcting the imbalance. He felt that, failing any political moves to restore justice to the countryside, the best way would be through improved agricultural technology. While a lecturer at the Royal College of Agriculture at Cirencester in 1910, he observed that the leys which the farmers sowed in the Cotswolds died out before the wild grasses growing in the hedgerows. On investigating why this should be, he discovered that the seeds were imported and were not suited to the length of our growing season. They had been purchased as part of a commercial deal to sell our manufactured goods abroad. So he started a campaign to make it illegal to sell seeds without stating their place of origin; this led to the Seeds Act of 1920. Meanwhile he had dug up the indigenous grasses and started to breed from them. They formed some of the first breeding stock of the Welsh Plant Breeding Station which he inspired and of which he was the first director when it was founded in 1919. These famous Aberystwyth grasses put millions in the pockets of farmers, which is probably why he was knighted in 1939. It is unlikely that he would have been honoured for his ecological contribution to farming.

When the war broke out Stapledon crusaded for his ley-farming plough-up campaign. Sir Reginald Dorman-Smith, Secretary of State for Agriculture in 1937, wrote to *The Times* on Stapledon's death in 1960, stating that without his agricultural victories, 'We would most certainly have been starved of food and there would have been no military victories.' The vision of one man was pitted against the mistaken political philosophy of a nation. His experiences after the war expose more clearly than anything else what has gone wrong with our agricultural policy.

The Agriculture Act of 1947

At the end of the Second World War, both political parties agreed that we must change our farming policy so that we were never again in danger of being starved out, and that it was a mistake to let the countryside be a poor relation of the town. The Agriculture Act of 1947 was intended to secure prosperity for the farmer. It was assumed that this would also lead to a prosperous rural community and an attractive landscape. Why did this not happen – as we all, including myself, believed it would?

In 1947, food was still rationed and was bought and distributed by the government; it was therefore easy to give farmers guaranteed prices for staple foods. What was not anticipated was that guaranteed prices lead to surpluses and impose an impossible burden on the taxpayer, or, as is the case today now that we are in the EEC, on the consumer as well. When rationing was abandoned in 1952, the government seized the opportunity to open up our markets to food imports. To guarantee the farmers' incomes, minimum food prices for the farmer were negotiated annually with the National Farmers Union (NFU). If world food prices sank below the minimum price, the farmer was reimbursed the difference. The absurd situation was then established whereby the national finances benefited from high world prices, since the Treasury did not have to fund massive deficiency payments. This did not please the taxpayers who were paying twice for their food – once in the market and again in taxes. As a consequence, the government kept the minimum price levels every year slightly below the rising production costs. This, no doubt, kept farmers on their toes and made them aim at greater efficiency in order to stay in business. But they complained that they could no longer finance greater production out of retained profits. The government was not concerned; on the contrary, since the problem was surpluses, it told the farmer it wanted less food at lower cost. The catchword was to be 'efficiency'. 'Farming is a business now, not a way of life' was the slogan nailed to the masthead of the ministry. The era of production at any cost was over; arguments in favour of that were no longer convincing.

Reasonable as all this seems in economic terms, it turned out to be the cause of today's disastrous farming methods and of the subsequent rape of our countryside. Yet only Stapledon among his contemporary agricultural scientists seems to have understood that efficiency would drive out good husbandry, as well as contribute to the destruction of rural society.

Agribusiness and Environmental Destruction

Astute urban business men realized that the government's policy could only be successful if mass production methods were introduced into agriculture. New technology – much of it originating from research begun in the period when Stapledon saw improved technology as the only hope – was brought on to the farms and funded by the large-scale agro-industrial enterprises. Many small farmers were driven out and a new surge of farm workers left the land for the cities. Though this was socially regrettable, it was generally agreed that only large-scale farming business could revolutionize farming and make it profitable. Research was concentrated on the most cost effective ways of using the new machines and fertilizers, pesticides and herbicides. Rotations were banished from many 'progressive' farms, the land was laid out in the new prairie pattern of huge hedgeless fields; stock was kept intensively indoors and separated from the grasslands where it had previously been a natural contributor to fertility.

Worse still, it was not thought necessary to research the effects of these changes before they were undertaken. Environmental research was virtually ignored and has remained neglected to this day. In 1984, not one representative of a non-governmental environmental body is to be found on any of the committees that decide how research projects shall be chosen and funded, or at which research stations they shall be carried out.[2] The biggest customers shape the policies of government research stations.

The government and the agricultural lobby cannot, however, say they were not warned about the dangers. In 1956, Stapledon told the International Grassland Conference, of which he was president: 'The politicians, administrators and businessmen are deft pastmasters in the business and art of wanton obliviousness. The decisions of expedience are taken under the shadow of this wanton obliviousness and they come home to roost.'[3] He then described how as a young man in the conditions of 'those remote times' he had believed in technology but that:

Today technology has begun to run riot and amazingly enough nowhere more so than on the most progressive farms. The red lights, if as yet, only on the sub-thresholds, are there for those who can discern them . . . Man in putting all his money on a narrow specialisation and on the newly dawned age of technology has backed a wild horse which, if given its head, is bound to get out of control.

It would have been better, he said, if we had paid more attention to learning and scholarship, then we would have realized our ignorance and come to terms with the innate interdependence between the two poles of knowledge and ignorance. As it was, we acted according to a law which he called 'The Law of Operative Ignorance'; as a result, we would have to strive to hold in check all the unpleasant and destructive effects of applying new knowledge without understanding its relationships to nature as a whole. He concluded: 'Science will never brave up to this uncomfortable situation until it has vastly broadened the base of its outlook and totally revised the basis of a great number of its techniques.'

These prophetic words fell, as can be imagined, on deaf ears. Scientists did not like to be cricized in this way. Besides, how could technology do any harm? Was it not massively increasing productivity and profit?

Growth of the Agricultural Lobby

By this time, the agricultural lobby was becoming all powerful; it was a conglomerate of large farmers who dominated the NFU, the Country Landowners' Association (CLA), the multinational suppliers of agricultural ancillaries and the Ministry of Agriculture. The lobby was a deft pastmaster at treating ecological doctrines with contempt – as so much 'muck and mystery', as it dubbed it. In this way, its members frightened off scientists, farmers, educators and students who dared to show a sympathetic interest; they were afraid to be labelled cranks yearning for a long lost rural arcadia. The lobby's financial power – it could give or withhold grants to universities and research stations, invisibly fund conferences on fertilizers, new machines and modern 'husbandry' – and its 'chummy' liason with the government (which made no decisions without its advice and consent) entirely dominated farm policy.

In this era of 'wanton oblivion', we have – according to the Nature Conservancy Council – lost: 25 per cent of our hedgerows; 50–60 per cent of our heathlands; 80 per cent of our chalk grasslands; 95 per cent of our hay meadows; 80 per cent of our lowland bogs; 80 per cent of our ancient woodlands; 30 per cent of our upland grass and moors; and 90 per cent of our ponds since 1945. As for the amount of top soil we have sacrificed, we shall never know because it has never been monitored and the government is determined that it never will be, cutting the grant to the Soil Survey by half. In assessing this appalling loss of 'heritage' we must remember that the loss involves the destruction of an ecological habitat for specific flora and fauna which are often threatened with extinction because they cannot establish themselves anywhere else. Since this wildlife contains genetic reserves of value for plant and animal breeding and for medicine, Sites of Special Scientific Interest (SSSIs) have been devised for protecting some of these species. But farmers are not yet legally obliged to accept them. Conservation is seen as a restraint on production, so the farmers who pay attention to safeguarding the SSSIs or the landscape place themselves at an economic disadvantage compared to their less ethical rivals. Unless protection has a legal bite, it cannot stop the so-called 'cowboys' who often farm on an agribusiness scale. In addition to this wanton destructiveness, access to the countryside has become more and more restrained – most noticeably in the uplands where ramblers once roamed at will. Farmers now fence off areas of former permanent pastures, which they plough up and reseed with specialized grasses or even use to grow cereals. Neither of these enterprises would be profitable were it not for high guaranteed prices and development grants. Indeed the profits of upland farmers are equal to their grants; without these financial aids, the hill lands would revert to wilderness. Unfortunately, the grants are not for farming so as to conserve the landscape, *but to increase production by spoiling it*. Headage payments for livestock encourage overgrazing.

The Separation of Husbandry and Environment

The industrial dogma of maximum productivity has led to financial assistance to farmers being divided into two parts: one part to encourage productivity and another part for conservation. The disparity between the two sets of grants is huge, so that the farmer has little incentive to give priority to the environment. The administrators and bureaucrats who apply the grants are so conditioned to thinking of farming as only a business that they misread the wording of agricultural acts, always assuming that grants and subsidies must apply only to productivity and not the environment, unless the wording of the act specifically says so.[4] Surely 'the business of farming' includes the conservation of the environment? How can it be a healthy 'business' if it erodes the soil, contaminates our waterways and our groundwater and even the food it produces?

It might well be asked how scientists and farmers square their consciences with what they are doing. One reason is that it has long been assumed that the use of fertilizers and chemical aids cannot damage the fertility of the land on which they are applied. Yet if fertilizers are not applied, the declining nutrient status of the soil is soon exposed. In fact, when nitrogen fertilizer leaches out of the soil (as 40–50 per cent of it does), it carries with it by chemical action (electrolysis) some of the other plant nutrients in the soil. Owing to the crude theory that plants only need nitrogen, phosphate and potash, these effects were not considered significant (even if they were known at all).[5] In other words, crop yields have been inflated like money and should the cause of the inflation be removed they would drop drastically.

The Rise of Public Indignation

The cumulative effects of the degrading and dangerous practices which flow from the doctrines of efficiency and productivity, together with an increasing awareness that agriculture is not as efficient as it has been made out to be (if it is measured in terms of energy and resources used, instead of short-term profit) have led to growing public indignation. At last, despite the comforting advice of the agribusiness lobby that all is well, and that any alternative policy could only lead to dearer food, the government has begun to lose its nerve and suspect that it must have a 'green policy' before its political rivals cash in on the green votes. So the Wildlife and Countryside Bill was presented to parliament after unimaginable rows in committees and debates in both Houses.

To those critics who understood what was involved, the bill appeared to be a simple publicity exercise, intended to delude the public into believing that the countryside would in future be protected from the adverse effects of industrial farming. The critics fought hard to get its provisions radically tightened up, but though they won a few battles, they failed to impose significant changes. Once the bill became an act, however, it was soon clear that its critics had been right – so much so that it now looks as if they might finally prevail and ensure that the act is only an interim measure. It cannot be a final solution for it only applies to a very small part of the farmed landscape. Nor do we yet know what financial provisions will be made for protecting environmentally vulnerable sites, or whether the existing system of grants and subsidies will be revised so that development and the environment are treated as one, rather than separately as at present. Agriculture reformers are agreed that, ultimately, the only solution is a return to a more traditional, low input/low output system. As the output would be of a higher nutritional value, however, would it really be lower? This is far from how the agricultural lobby sees the situation. Its members think the answer is to intensify still further present farming methods in the lowland and leave the rest of the countryside to the environmentalists: thus, just as we now have industrial towns, so we would have industrial farm areas. These would mass produce the raw ingredients that the food-processing industry would then transform into convenience foods. No doubt these industrial farms would become in due course as barren and repulsive as today's inner-city areas.

As it stands, the act is a Highwayman's Charter that says in effect 'Your money or your landscape'. It was drawn up by a political party that believes property rights are absolute and have priority over communal or public rights. The landlord, it is assumed, has the natural right to do what he likes on his own property (no mention of 'stewardship') and if he makes any concessions, such as agreeing to abide by the conditions of an SSSI, he must be compensated annually for the profits he is 'deemed' to have foregone. With the present high levels of guaranteed prices maintained by levies on imported food, and with development grants available, these assumed profits are out of all proportion to the real value of the crops alleged to have been forfeited. The conditions of the management agreement are likely to be fiercely contested and only changed with great reluctance. Are the agreements going to remain voluntary? It is likely that, in due course, management agreements will have to be abolished so far as compensation is concerned in order to bring agricultural planning into line with town planning, where, if a planning committee vetoes a development plan, no compensation is paid to the developer for 'deemed' profit lost. If the lobby refuses to accept changes, the government has a powerful weapon at its disposal – it may threaten it with free trade!

Casting the mind back on the wanton development of farming policy, we recognize that it can only be satisfactorily reformed if a new attitude of mind is brought to bear on it. In the first place, the industrial dogma of productivity is not rational, since its single vision fails to take into account factors essential to the sustainability of both farming and the environment on which it depends. Secondly it denies the rights of the community – including the right to a broader spread of the ownership of farm land. It is doubtful if a prosperous agricultural system has ever before coexisted with a decayed rural community such as we have today. Converting the way those in authority 'see' the reality of the relationship between both farming and environment and country and town will mean educating them out of their nineteenth-century outlook; they might begin by reading Stapledon's *Disraeli and the New Age*.[6]

Fiddling with the Wild Life and Countryside Act, important as that is as a first step, will not solve the fundamental problems caused by agricultural industrialization.[7] Only the most fundamental changes in the priorities of our contemporary politicians can achieve that. I believe that events will eventually force these changes upon them.

Notes

1. The Corn Laws were an unsatisfactory means of protecting agriculture, since they gave unjustified privileges to the cereal growing aristrocracy at the expense of the poor, who had to pay high prices for bread, as did the livestock farmers for their cattle feed – which is true today. Much subtler methods than that could have been adopted.
2. The absence of environmental representation is pointed out in the *House of Lords Select Committee on Science and Technology Fourth Report on Agricultural and Environmental Research*, London: HMSO, 5 July 1984.
3. The text of this address is repeated in my biography of Sir George Stapledon, *Prophet of the New Age*, pp. 257–8. When this book was published in 1962, most agriculturists thought the new age had something to do with grassland and agricultural prosperity. In fact the new age was the Age of Ecology. I remember Sir George saying to me 'I hate grass', because his identification with grass distracted attention from his ecological philosophy.
4. Professor O'Riordon examined this in detail in the 'Centre for Agricultural Studies' Paper 5, *Investing in Rural Harmony*, Reading University, May 1984. He questions the verbal interpretation of some agricultural legislation and EEC Directives.
5. The effect of fertilizers on soil nutrients is analysed by Michael Blake, *Concentrated Incomplete Fertilizers – Their use and abuse in Price Review Agriculture*, London: Crosby and Lockwood, 1967.
6. *Disraeli and the New Age* was published by Faber and Faber in 1943. It is a study of what Disraeli said ought to be done, not what he did. Disraeli is reported as telling a friend 'I was never so powerless as when I was Prime Minister.'
7. The traditional village as we imagine it does not exist any more. In the average village the remaining rural inhabitants live in slums; their children have left because they cannot afford the high price of houses – driven up, alas, by environmental legislation making building land scarce. The prosperous part of the village is inhabited by ex-urbanites, retired or commuting or owning second houses. This is analysed in detail by Howard Newby's *Green and Pleasant Land?* – an outstanding study of the rural scene, republished with an epilogue by Wildwood House, 1985.

4

Conservation and the Conservatives

Chris Rose

In a 1985 pamphlet headed *Conservation: No Left Wing Monopoly*, the Conservative Party's Central Office wrote: 'The present government is firmly committed to bringing about a sensible balance between the interests of farmers and conservationists.' As evidence of this commitment, the pamphlet cited:

* The Wildlife and Countryside Act 1981; *'a much needed statutory framework within which conservation could be actively promoted'*.

* An EEC Agricultural Structures Directive which is a *'carefully constructed package of incentives'*.

* The reformed structure of capital grants, with £18 million being spent in areas like Halvergate Marshes *'to discourage farming practices that might be particularly harmful'*.

* The introduction of new and stronger model bylaws *'in order to control the burning of straw and stubble each summer'*.

* Changes to planning policy, now being studied by the Department of the Environment, which *'in no way threatens green belts'*.[1]

Such a list of concerns includes both the politically expedient, and those which pressure groups have managed to force on to the political agenda and which cannot therefore be avoided. It is hardly a comprehensive description of the things which really matter in determining the future of wildlife, countryside and our landscape. But it is still worth considering the government's record on each of its own claimed successes.

The Wildlife and Countryside Act

Mrs Thatcher's government can hardly claim credit for bringing the Wildlife and Countryside Act into being, since it was introduced purely to meet Britain's obligations under the EEC's 1979 Birds Directive; namely, to protect the habitat of some rare and vulnerable bird species. In fact it originated as the short-lived Countryside and Wildlife Bill of the Callaghan government. This would have been a rather ineffective measure, designed to prevent the disappearance of wild habitats in only a few places (inspired by the Porchester Report in 1977 which revealed that moorland was fast being converted to grassland on Exmoor) as well as meeting the EEC's demand for new legislation. But when the Thatcher government pushed its own act through, first under Michael Heseltine then under Tom King at the Department of the Environment, an ineffective measure had become a disastrous one. It rested on the extaordinary principle, reminiscent of Danegeld, that statutory protection for key wildlife sites (designated SSSIs) could only be guaranteed by 'compensating' owners and occupiers for the profit they might have made if a threatened development was carried out. So, for example, a million pounds had to be promised to one north Kent farmer, in return for his undertaking not to drain and improve saltmarshes as cereal land in order to keep them as traditional low-intensity (and £2–300 less profit a hectare) sheep and cattle grazing for ten years. In effect, the farmer or forester is guaranteed the maximum cash flow available under the European Common Agricultural Policy or Forestry Commission grant aid and tax schemes, even though the extra crops would be in surplus and it is officially decreed to be in the national interest to conserve the site under threat. When Part II of the act (the section dealing with habitat protection) came into force in 1982, it immediately came under fire from conservationists, who attacked it as unworkable, ineffective and expensive. And so it has proved.

Originally, the act was only intended to protect a few dozen super-sites but under pressure in parliament, the government created an open-ended commitment to pay to protect any SSSI. Finding itself under pressure from conservation organizations, the Nature Conservancy Council (NCC) made it clear that all SSSIs would be treated equally. In brief, the act requires landowners and occupiers to notify the NCC of any intention to carry out any one or more of a set of Potentially Damaging Operations (PDOs), a list of which is issued to the owner or occupier by the NCC.[2] The process of listing the PDOs, which can be anything from a change of grazing regime, to main drainage or grubbing trees, is accompanied by the drawing up of

POND ACTION

THE HAMLET PARTNERSHIP

OXFORD POLYTECHNIC

Supported by WWF

In Association with BBONT and ffPS

A UK2000 Project - MSC Agents
NORTEC TRAINING AGENCY LTD.

POND ACTION

Throughout Britain ponds provide a refuge for water plants and animals and a place where we can begin to appreciate the diversity, intricacy and order of natural ecosystems. Yet between one quarter and half of the ponds that existed in lowland Britain at the beginning of the century have either dried up or been drained and filled in and many of those remaining are threatened by pollution and neglect.

Ponds are more important than ever before for the conservation of water plants and animals in our intensively managed landscape. We urgently require the information needed to assess the effect of the loss and neglect of ponds on aquatic communities. Pond Action will provide this information and will use it to promote the conservation of ponds.

WHY CONSERVE PONDS?

Ponds, particularly those that are old and unpolluted, are refuges for many water plants and animals, including rare species requiring particular protection.

Ponds add beauty to the landscape in both intensively farmed and urban areas.

Ponds are an attractive educational resource which can be used at all stages of environmental education.

Ponds are easily created and are particularly valuable to landowners and farmers wishing to set aside areas for wildlife.

Pond management can provide interesting jobs in rural areas.

Water plants and animals are a potentially useful genetic resource; they can most easily be maintained in ponds supporting diverse semi-natural communities.

POND ACTION

At least 3000 ponds remain in Oxfordshire and Pond Action is beginning its work in the county with a survey of a large sample of these ponds. Pond Action will gather the information needed to improve our understanding of the ecology of these ponds in order that they may be adequately managed and protected.

Pond Action is coordinated by environmental consultants HAMLET PARTNERSHIP in collaboration with OXFORD POLYTECHNIC, NORTEC Training Agency (Banbury), the Berkshire, Buckinghamshire and Oxfordshire Naturalists' Trust (BBONT) and the Fauna and Flora Preservation Society (ffPS). Pond Action, which is supported by the World Wildlife Fund UK, is a UK2000 project funded by the Manpower Services Commission.

Pond Action will:

1. Carry out detailed studies of the ecology of a large sample of Oxfordshire ponds.

2. Provide the technical information required to develop new conservation and management strategies for ponds.

3. Promote the conservation of ponds by encouraging voluntary groups to adopt and manage ponds.

4. Promote the use of ponds at all levels in education.

5. Provide valuable training for unemployed people.

If you are interested in taking one of the 20 places on this project, or require further information about Pond Action, please contact:

Jane Cotton
External Relations Centre
Oxford Polytechnic
Gipsy Lane
Headington
OXFORD OX3 0BP

Tel: (0865) 819071

a detailed habitat map for the SSSI. This is termed 'renotification' for sites already identified as SSSIs under the previous 1949 act, and 'notification' for the new sites. It has proved extremely cumbersome, and deadlines for renotification have been repeatedly put back. But without renotification, there is no protection for a site, and many which meet the NCC's site selection criteria have been lost before renotification, whilst they are in limbo as PSSSIs (Proposed SSSIs). Groups such as the Royal Society for the Protection of Birds have calculated that there are at least as many sites meriting SSSI status as there are sites currently designated as such.

The NCC's slowness in renotifying sites has been due to the byzantine complexities of the bureaucratic procedures laid down in the act and its financial guidelines, to inordinate and unjustified optimism on the part of the government and the NCC itself, and to a shortage of staff. In 1983, the co-ordinating voluntary sector conservation group Wildlife Link reported:

> It is clear that the Government did not fully understand the tremendous strain being placed upon the NCC, in starting to implement the Wildlife and Countryside Act. Ministerial statements as to the speed of the completion of the vital renotification process have been consistently over-optimistic, despite the growing evidence of the problems which the NCC face . . .

> Within six months the estimated completion date for the renotification of SSSIs has been March '83, end of 1984 and now early 1985. Only 11 per cent of the existing SSSIs have yet been renotified and we have doubts as to whether the process will be complete by even the end of 1985. This timetable is unacceptable to conservationists, as sites continue to be damaged and destroyed . . .[3]

In fact 'renotification' is now not expected to be complete until 1987 or beyond.

The financial mechanisms of the act have attracted more criticism than anything else. One writer described the act as 'a hot air balloon kept aloft by burning public money'. Others have pointed out that it established 'property development rights' for those who owned or occupied a SSSI which had long been rejected (as betterment) in conventional town planning. In other words, a landowner or occupier who threatens to destroy a SSSI by development can only be stopped by being bought off with the sum that the development would have earned (a luxury unobtainable if you are refused planning permission for a back garden hypermarket). The owner who cares for a site voluntarily, however, will probably receive nothing.

The NCC initially estimated the cost of the act at £1.2 million, yet the government provided merely £600,000 in 1982. Almost immediately it became obvious that this was utterly inadequate. Far from voluntarily forgoing development as the government had claimed owners would do, they began queuing up to ask for management agreements (compensation), and the NCC's budget had to be expanded to £18 million by 1983–4. A detailed study of the act undertaken by Bill Adams of Cambridge university geography department calculated that the act would eventually cost £42.8 million a year to operate through compensation agreements – about a thousand times the amount spent in 1982–3.[4] As renotification and finally notification proceeds, so the number of SSSIs with agreements increases steadily. Many of these are in the form of annual payments and run for 21 years or longer. The financial implications are staggering, and many people are alarmed that such huge sums should be diverted from other programmes to pay to protect something which under a system of planning controls could be saved at no (or very little) cost. A London university study, by M. Halling, suggests that a planning control system over agriculture and forestry would cost £15 million a year, 75 per cent of which could be recouped in handling charges.[5]

The nature of the NCC's payments to landowners and occupiers has also aroused considerable public anger. At Blair nam Faoileag, a peat bog and flow in Caithness, Lord Thurso was awarded £287,000 for a 99 year lease on an area formally declared a National Nature Reserve. The sum was calculated from the earnings he could have gleaned from a forestry scheme where much of the cash would have derived from tax concessions and grants designed not as an income subsidy but to encourage forestry. At Boulsbury Wood in Hampshire, Lord Cranborne received a 65 year pay-off of £20,000 a year, index linked, not to destroy the ecological value of one of the finest woods in southern England. At a Norfolk site where the owner suggested converting grassland to cereal production, the NCC is now having to fund a £22,141 annual payment of £542 per hectare (the aggregated value of EEC subsidies and MAFF grants) to be reviewed each year. Just one such payment would be enough to pay the salaries of the average County Naturalists Trust's conservation staff for a year.

Inequitable, cumbersome and bureaucratic as it is, the act might be more acceptable if it worked; but it hasn't. A bill had to be introduced by the opposition environment spokesman, Dr David Clark MP, in 1984 to amend the act and close the '3-month loophole' under which sites could be destroyed with impunity in a period of 'consultation' with owners. Even where it does apply, the penalties are too small to deter some determined developers; the act is overridden by planning permission (as at Duich Moss on Islay, where an internationally important wildfowl site is to be destroyed by peat extraction for the Distillers Company). Elsewhere

Table 4 Damage to Sites of Special Scientific Interest and proposed Sites of Special Interest: April 1983 to March 1984

1	2	3	4	5	6	7	8	9
		Nature of damage						
Site status		*Agricultural activities*	*Tree felling*	*Motor cycling, fire, dumping, quarrying, construction works etc*	*Serious*[1]	*Minor*[1]	*Long-term*[1]	*Short-term*[1]
SSSIs awaiting renotification	81 (51.9%)	38 (46.9%)	12 (14.8%)	34 (42.0%)	19 (23.4%)	28 (34.6%)	24 (29.6%)	28 (34.6%)
Proposed SSSIs (including extensions to existing SSSIs)	34 (21.8%)	22 (64.7%)	3 (8.8%)	10 (29.4%)	10 (29.4%)	6 (17.6%)	16 (47.1%) (3 sites lost)	6 (17.6%)
SSSIs damaged after notification	5 (3.2%)							
SSSIs damaged after renotification	18 (11.5%)	8 (34.8%)	4 (17.4%)	14 (60.9%)	3 (13.0%)	6 (26.1%)	7 (30.4%)	10 (43.5%)
Status of site when damage occurred unclear	18 (11.5%)	7 (38.9%)	1 (5.5%)	10 (55.5%)	2 (11.1%)	2 (11.1%)	5 (27.8%)	1 (5.5%)
Totals	156	75 (48.1%)	20 (12.8%)	68 (43.6%)	32 (20.5%) (43.2%)	42 (26.9%) (56.7%)	52 (33.3%) (53.6%)	45 (28.8%) (46.4%)
					(as a percentage of 74 sites)		(as a percentage of 97 sites)	

Note: 1 Figures indicative only: figures in columns 6 and 7 are available for 74 sites only (47.4% of sites damaged).
figures in columns 8 and 9 are available for 97 sites only (62.2% of sites damaged).

Source: NCC Evidence to House of Commons Environment Committee, 1985.

the NCC is deliberately holding back from designating new sites in order to try and meet 'renotification targets'. (By May 1985 it had identified 1,848 PSSSIs and 4,051 existing sites of which, in England, only 28 per cent are renotified.) In other words, the act does not protect the countryside.

In July 1984, Charles Secrett of Friends of the Earth listed 133 examples of SSSI destruction that had taken place since the act became law in 1982.[6] As he remarked, even this was a considerable underestimate. Later that year in its own evidence to the House of Commons Environment Committee the NCC listed 156 SSSIs known to have been damaged or destroyed from April 1983 to March 1984. As can be seen from table 4 and figure 2, the largest single cause was drainage (30 sites), the largest overall source of damage agricultural development (48 per cent) followed by tree felling (13 per cent).

Agriculture Structures and Grants

Capital grant aid only comprises a small proportion (around 5 per cent) of the public subsidies to agriculture, but it has a very direct impact on the countryside. Drainage schemes, grubbing woodlands, building roads – these changes are all examples of what comes under structures policy and capital farm grants. It is true that in the recent revision of the 1972 Structures Directive, Britain's Ministry of Agriculture, Fisheries and Food argued that some of the £3.15 billion due to be spent on structures up to 1990 ought to be available for conservation measures on farms. To anyone familiar with the attitudes of MAFF (which up to 1984 was still funding hedgerow removal) it was an extraordinary decision. Here was the ministry wanting to fund farmers to retain ancient meadows unimproved, unresown and unfertilized; yet, MAFF was still financing the opposing policy, through grants for drainage and other schemes which frequently caused habitat loss.

In truth, the ministry's about-turn was entirely due to public pressure channelled through voluntary conservation groups, such as the Royal Society for the Protection of Birds and the Council for the Protection of Rural England. They had proved far more effective lobbyists than the civil service. At one meeting of ministers in France, the British were astonished to turn up and hear the French putting forward the idea that environment policy should be integrated into farm plans: the conservation groups had got there first. Nevertheless, in the end, the so-called 'British initiative' was lost.[7] After protracted negotiation, the agriculture ministers decided that while any country was welcome to spend its own farm funds on

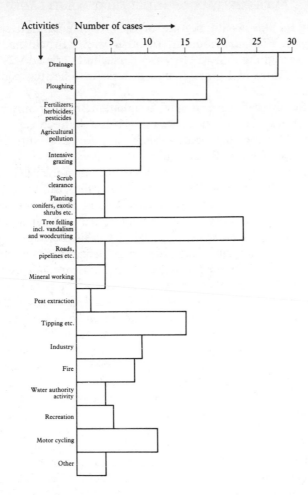

Figure 2 Causes of loss and damage to Sites of Special Scientific Interest: April 1983 to March 1984 on a case basis

Note: A small number of the 156 sites known to have been damaged during this period have been damaged by more than one activity. The 156 sites include: 81 SSSIs awaiting renotification; 34 proposed SSSIs – including extensions to existing SSSIs; 5 SSSIs damaged after notification; 18 SSSIs damaged after renotification; 18 sites where information is not available on the status of the site when the damage occurred.

Source: NCC vidence to the House of Commons Select Committee on the Environment, *Inquiry into the Operation and Effectiveness of the Wildlife and Countryside Act, 1981.*

conservation, this would be one of only two recognized schemes not attracting an EEC contribution. In other words Britain had to pay up herself – something that MAFF had persistently argued was impossible under the Treaty of Rome. It was a decision reflecting badly on both MAFF and the EEC.

Now Britain will have a new designation, that of Environmentally Sensitive Areas. A new agriculture bill is being drafted to effect it.[8] At the end of 1985, some four areas – including parts of Northern Ireland, Scotland, the Norfolk Broads and the Somerset Levels – were being considered for inclusion on a list of ESAs to be put to the EEC. A final set of just six or seven areas will be chosen. In reality, ESAs will be merely another form of farm grant and there will be no planning powers to stop destruction of landscape features or wildlife habitats. This palliative will neither stem the haemorrhage of Britain's wildlife heritage, nor create an economic and ecologically acceptable form of farming which could promote harmony not only between farmers and the landscape and its wildlife, but also between farmers and conservation groups, as the government claims it wants to do. Without some powers of intervention – whether local or national – tinkering with the fringes of grants will do nothing much to save the countryside. Minor cutbacks of the annual £200–300 million that have been spent on capital works in the past years have not stopped farmers carrying out 'improvements' made extremely profitable by the high level of price support for crops. And much of the solution is in the government's own hands: some 75 per cent of all capital grant aid used on British farms is directly provided by MAFF and could be easily redirected. Drainage, for example, could be profitably abandoned along river valleys to create fisheries, wildfowl refuges, and marshes to help denitrify fertilizer-laden water. Planning controls could be introduced to protect habitat features identified on a map (see, for example, Friends of the Earth's 1983 proposals for a Natural Heritage Bill).[9]

Straw-burning

Launched with a fanfare in early 1984 (after the particularly infamous events of summer 1983), the Home Office model bylaws on straw-burning have since been firmly rejected by local authorities, which are increasingly convinced of the need for a legal ban on the practice. Of 160 county and district councils surveyed by the National Society for Clean Air in 1985, 144 wanted straw-burning prohibited.[10] The bylaws simply tried to hide straw-burning from the public by asking that it should not be done at weekends, on bank holidays and so on, while attempting to make it more acceptable by stipulating wide firebreaks. Yet where detailed surveys have been carried out (for example, in Worcestershire where the Naturalists Trust found trees and hedgerows damaged at 31 sites in 1983 and 89 in 1984),[11] it is clear that damage has not been stopped. Straw-burning causes enormous problems. In 1983, it killed two people in a car accident and in 1984 a farmer was himself consumed in a fire he had started.[12] It swamps the ability of the fire brigade to respond to genuine emergencies and has a direct impact on wildlife. Hares are reported to have been burned to death, badgers were seen fleeing a wood in an uncontrollable fire in Cornwall. After burning, the number of insects above ground drops by 85 per cent and the species diversity by 61 per cent. The direct cost of structural, crop and other material damage has been estimated at £2–5 million each year.[13] In 1983, a Forestry Commission plantation caught fire in Lincolnshire; elsewhere, damage was recorded on 55 per cent of the trees surveyed by Friends of the Earth in the Vale of Evesham. In Essex, Fingrihoe Wick Nature Reserve has been set alight.

But it is not these effects or even the ash and other air pollution (straw-burning is a major source of aldedydes, oxides of nitrogen and carbon dioxide) which most angers people. It is the waste and the wanton disregard for simple principles of decent husbandry. The public opposition is based on a more general, less often stated demand for a genuine alternative to 'modern farming'. Straw-burning is, in fact, merely the most irresponsible face of agricultural industrialization. The five to six million tonnes of straw burnt each year in Britain are a waste product which never used to exist. Yields were lower (inputs and breeding have pushed them up at 2–3 per cent a year and cereal tonnages per unit area have trebled since the late 1940s) and there was less cereal land (arable farmland is now around 40 per cent of Britain, while only 17 per cent is naturally suitable for cereals). But most importantly the straw was *used* – in animal bedding for manure and in other ways. With cereal monoculture in the east of Britain and dairying intensified in the west, such joint economies are altogether lost. And it is the rush to sow in the autumn (chasing higher yields with winter instead of spring cereals) on huge farms, where hedgerows have been removed (along with farmworkers) in order to use machinery to grow continuous cereal monocultures,[14] that 'justifies' the burn to most farmers. Burning is the quickest and easiest method of destroying the straw. It creates short- and probably long-term ecological problems and externalizes the cost of pollution – but this is conveniently ignored by the government, which has so far failed to implement the recommendation of the Royal Commission on Environmental Pollution, which in its 1984 report urged that a ban be announced.

Green Belts

Green belts do not protect wildlife habitats and never have done. Conservation of the countryside was not mentioned in the 1955 circular which set them up. They are sorry and confused planning notions with little real statutory backing: sorry because they are designed to keep farming and urban people, and town and country, at arm's length; confused because it is not clear why or how this should be done; and without backing because large and damaging developments tend to get through, while there is no effective promotion of acceptable land-use management. After bowing to pressure from city and property developers to relax planning controls to allow the creation of new commuter towns to exploit the M25 and other new roads, the present Conservative government reacted painfully to attacks from within its own ranks over green belts. Far from developing a transport policy which would encourage people to live and work within a short journey's distance, or even to develop more efficient commuting by train, the government seeks to please both the road lobby and housebuilders at once.

Four Decades of Destruction

Britain's principal political parties are struggling to apply a bit of green whitewash to their nether policy regions. None of them is too keen to tackle the root causes of countryside destruction, which lie partly in the failure to administer and set policy at the farm-ecosystem level. The truth is that Britain's countryside has been a dying patient for at least four decades. The causes are not separable – straw-burning here, hedgerow loss there, pesticides over there and nitrates to be dealt with tomorrow or perhaps not all – but they are interlinked. Ask some farmworkers. They will have pulled up the hedge to drill the seed and apply the spray on the crop that was ordered to pay off the loan made to purchase the nitrogen to promote the growth of the barley for the beef, which is in surplus. Farm policy and countryside protection have grown up separately and failed side by side.[15] The present government still talks of balancing opposing interests – those of farmers and conservationists. Yet farming should be able to provide food and sustain wildlife and landscape at the same time, for, as we used to be told so often by the National Farmers' Union, farming (but of another type) created the countryside we value.

Notes

1. *Conservation: No Left Wing Monopoly*, London: Conservative Central Office, 1985.
2. W. M. Adams, *The Effectiveness of the Wildlife and Countryside Act*, Oxford: British Association of Nature Conservationists, 1984.
3. Wildlife Link, *Habitat Report*, no. 2, 1983.
4. Adams, *Effectiveness of the Wildlife and Countryside Act*.
5. M. Halling, *Agriculture and Planning Controls, Countryside Act*. Unpublished Ph.D. thesis, University of London.
6. C. Secrett, *Sites of Special Scientific Interest 1984: the Failure of the Wildlife and Countryside Act*, London: Friends of the Earth, 1984.
7. D. Baldock, 'Farm structures in Europe: the British initiative', *Ecos*, vol. 6, no. 3, 1985, pp. 2–6.
8. It is ironic that when late in 1985 the MAFF decided to use the bill to assuage long-standing conservationist criticisms by taking on a duty to further conservation, it was prevented from doing so by the Treasury who were, in the words of one civil servant, 'strapped for cash' at the time.
9. C. Secrett, *Proposals for a Natural Heritage Bill*, London: Friends of the Earth, 1983.
10. NSCA, *Last Year's Harvest Burn: The NSCA Survey of Straw/Stubble Burning in 1984*, National Society for Clean Air, Brighton: 1985.
11. Personal communication. Letter sent to the Worcestershire NFU in 1984 by Dr Graham Martin of the Worcestershire Trust for Nature Conservation, giving grid references to damaged trees and hedgerows.
12. Huw Watt, *Farmers Weekly*, 1985.
13. *ENDS Report* (Environment Data Services), no. 104, September 1983. See also, C. Rose and M. Dunwell, *Strawburning: You'd think Farmers Had Money to Burn*, FOE, 1983.
14. 120,000 miles of hedgerows were lost in England and Wales, mainly to agricultural development between 1945 and 1980.
15. Agriculture acts dating from 1947 embody farm policy. These were never connected with the Town and Country Planning Acts (starting 1947) and neither was integrated with the 1949 National Parks and Access to the Countryside Act. The fact that farm policy was in conflict with environmental objectives was first officially acknowledged in a major government statement in the 1979 Strutt Report and the first agricultural act to include nature conservation clauses will probably be the Agriculture Bill due to be introduced to parliament in 1985/6. The Bill will allow for funding ESAs.

5

The Destruction of the Countryside

Chris Rose

Since the Second World War, agriculture has turned from a craft, an art, a science and a way of life into an industrial process. Traditional woodmanship underwent the same transformation rather earlier: it was abandoned in favour of plantation forestry, which produces one uniform raw material rather than a myriad of products, from fruits and grazing to timber and fuel. Characteristically, woodmanship and low-intensity traditional farming systems were accessible to a wide range and large number of people and many rights were common rights. With the industrialization of forestry and agriculture, a process now being repeated all over the 'developing' world, wildlife habitats have been destroyed and impoverished. That destruction is a direct consequence of a process which replaces human skills with chemicals, and human labour and the muscle power of working animals with machine power.

By the time, probably early in the 1950s, when British farming as a whole actually began to consume more energy than it produced in food,[1] it had become a major industrial system. Materials previously recycled and used as inputs for growing subsequent crops have now become pollutants. Together with the residue of the new agrochemicals, they have built up in soil, air and water.

The failure of government policy has been the failure to cope with and control the process of agricultural industrialization, in almost every respect. Farm policy has simply concentrated on encouraging the conveyor belt of food production without any regard for the total costs. Environment policy has neglected to investigate (and then failed to control) agricultural pollution, and for decades has paid no regard to wholesale losses of wildlife habitat.

The Nature Conservancy Council

The Nature Conservancy Council (NCC) is at the sharp end of holding up Britain's environmental record on the conservation of wildlife and countryside. With 600 administrative, field and scientific staff, the NCC's budget was for many years the equivalent of just one Mars bar per head of the population: now, swollen by funds destined for farmers and foresters under the 1981 Wildlife and Countryside Act, the budget has reached the price of a cup of coffee per head. But poor funding has not been the only shortcoming of official attempts at nature conservation. From the time when it was changed from a research quango (as the Nature Conservancy) into a grant-in-aid quango funded by the DoE, the NCC has been bedevilled by an inability to stand up to pressures from the Ministry of Agriculture, the Forestry Commission and other government agencies. In addition, the list of names on its ruling council reads at times more like a page from a *Who's Who* in development than a list of 'appropriate guardians of the natural heritage'.[2] For example, as a quid pro quo for dropping pressure to wind up the NCC and merge it with another government department, the Country Landowners' Association and the National Farmers' Union accepted the appointment of Sir Ralph Verney as NCC council chairman in 1979. As a past president of the CLA, a Forestry Commissioner and farmer, there was no doubt about Sir Ralph's rural interests. Joining him as vice chairman was Viscount Arbuthnott, a chairman of the Scottish Landowners' Federation and chairman of the NCC Scotland committee. As a past land agent for the NCC in Scotland as well as a landowner and forester, Arbuthnott has close links with other NCC committee and council members, such as Lord Dulverton, Dr Jean Balfour, D. G. Badham and others, who all held positions in forestry organizations or forestry companies or were active foresters and landowners at the time of their appointment. Consequently the appearance of a NCC policy paper critical of forestry was delayed from 1978 (or earlier) until the present day.

A Watchdog Without Teeth

While emasculated financially and politically, the NCC has perhaps failed most of all because it lacks teeth. In 1949, when it was created as a 'biological scientific service' under the National Parks and Access to the Countryside Act, hardly anyone conceived of the enormous changes that forestry and agriculture would wreak on the countryside. Despite the fact that the 1947 Agriculture Act set out to support farm production, and the

Ministry of Agriculture established a substantial advisory service aimed at consolidating the wartime plough-up programme which had brought areas like the downs into more intensive, arable production, the conservation agency was given no defence against agricultural development. Indeed it had little say in town and country planning: the planning authorities established under the 1947 Town and Country Planning Act were required to consult the old Nature Conservancy about any development on a Site of Special Scientific Interest but no more than that.

The SSSI itself was the final problem; an arcane device with an inappropriate name that hardly anyone understood. Indeed, until the 1981 Wildlife and Countryside Act made them hot property for landowners and occupiers, hardly anyone had heard of SSSIs. The designation was introduced at a time when the NC felt that allusions to the scientific utility of nature conservation sites (for teaching ecology, for example) would be most persuasive with local and national governments. As years passed and popular support for nature conservation and countryside protection grew (reaching a total of three million people by 1981, with most organizations established in the 1960s and most recruitment occurring in the 1970s), the SSSI became less and less relevant and more and more restrictive. Under the SSSI system, the NCC set out not to protect the countryside and its wildlife but to designate a representative sample of the habitats of flora and fauna. It became clear by the 1970s, however, that this might be the *only* countryside to survive – islands of habitat that were once representative but are now simply relics and remnants. Only about 30 per cent of all ancient woodlands will be designated as SSSIs,[3] for example, although the NCC itself has stated that the remaining 300,000 hectares should all be conserved. Converting it all to intensive industrial plantations would only meet around 2 per cent of Britain's timber needs.[4] Of the ancient semi-natural woodland lost since the last war 70 per cent has fallen victim to forestry schemes while most of the remainder, especially the smaller woods, has been grubbed for farmland.

Loss of Habitats

Although it is easy to be wise in retrospect, it can be seen now that the NC and NCC repeatedly failed to anticipate, to confront or to deal with threats to nature and the countryside by other land users throughout the 1950s, 1960s and 1970s. For example, despite warnings from a few scientists as long ago as the 1940s, hedgerow removal was allowed to progress throughout East Anglia and the east midlands in the early 1960s, into the

County	Habitat or area	Dates	Loss (%)
Avon	Meadows	1970–80	50
Bedfordshire	Wetlands	25 years	70
Cambridgeshire	Woodland	35 years	17
Cheshire	SSSIs	In 1980	7 damaged
Cumbria	Limestone pavement		
Devon	Woodlands	1952–72	20
	Grass and heaths (outside Nat. Parks)	This century	67
Dorset	Heathland	50 years	75
	SSIs	In 1980	32 damaged
Essex	SSSIs	In 1980	9 damaged
Hampshire	Chalk grassland	Since 1966	20
Hertfordshire	Ancient woodland	Since 1850	56
Huntingdonshire	Hedges	30 years	88
Isle of Wight	Chalk grassland	14 years	18
Kent	SSSIs	Since 1951	40 damaged or lost
Lancashire	Lowland bogs	This century	99
Lincolnshire	Ancient grassland	30 years	50
Northumberland	SSSIs	15 years	33 damaged or lost
Nottinghamshire	Ponds	25 years	90
Oxfordshire	Floodplain meadows	3 years 1978–81	20
Powys	Moorland	6 years	7
Shropshire	Prime sites	18 months	3
Staffordshire	SSSIs	10 years	25 damaged
Sussex	SSSIs	15 years	25 damaged or lost
Worcestershire	Ponds	55 years	35
	SSSIs	25 years	17 damaged

Source: NCC Survey, 1981.

Figure 3 Site loss and damage to 1981 in selected counties

Vale of York, south England and west midlands in the late 1960s and 1970s, and into southern Scotland, mid-Wales and the west of England in the 1980s. Woodland loss and conversion to plantation followed a similar pattern. Today it is perhaps most obvious in mid-Wales, where farm roads established with EEC funds in 'Less Favoured Area' are allowing farmers access to hillsides always thought to be too remote to suffer in this way: consequently ancient Welsh oak woods are being ripped up for conversion to nitrate-fed rye grass leys or rape.

An increased rate of habitat destruction followed Britain's entry into the EEC in 1973/4, as food subsidies encouraged farmers to grow more at any cost (as the market was guaranteed) and the government aided this process by deliberately setting out to make Britain a cereal-surplus nation. Losses of moorland, old meadows and rough grassland were rapid throughout the late 1960s and 1970s and although they have since slowed down in many areas this is because most sites have already been destroyed (see figure 3). In the uplands such as mid-Wales, large grassland conversions are being conducted, even at over 2,000 feet.

In 1974, water authorities in England and Wales were instructed to prepare five-yearly advance plans for improving land drainage. The principal effect of this was the introduction of major drainage schemes in river valleys: this allowed agricultural drainage to follow and resulted in about 100,000 hectares per annum being drained in the 1970s (as opposed to around 10,000 hectares in the early 1940s).[5] The lowland flood meadow, characteristic of much of England and parts of Wales, effectively disappeared within ten years.

The story was the same along coastlands. Unfettered by any controls and despite cereal surpluses, drainage for conversion to arable land consumed huge areas of wildlife habitat and transformed the landscape. In the north Kent marshes, for instance, 14,750 hectares of saltmarsh and freshwater grazing marsh were lost between 1935 and 1983, all but 13 per cent to agriculture.[6] With cereal prices effectively subsidized at a level of 166 per cent (through price support, export subsidy and tariffs against imports),[7] there was no economic reason for farmers to resist the temptation and deliberately forgo the opportunity to drain, especially when lured on by drainage grants of 50 per cent or more from the Ministry of Agriculture.

Roads and Concrete

While these changes swept across the landscape unchecked, bricks, mortar and reinforced concrete were also taking their toll of wildlife and countryside. Perhaps the most obvious examples are large roads (see figure 4), frequently held up as an example of 'bad planning' but actually quite outside the control of local authorities and built at the whim of the Department of Transport.

Despite the widespread impression (fostered by the road industry, oil companies and the Department of Transport) that new roadsides are 'valuable wildlife refuges', major roads frequently destroy irreplaceable

SSSI	Locality/County
Aston Rowant	Oxford/Buckinghamshire border
Brettenham Heath	Norfolk
Barrow Green Block	Nr Oxted, Surrey
Bucken Park	Sussex
Bullick Alder Carr	Northants
Chilington Park	Wolverhampton, Staffordshire
Charing Heath	Kent
Coles Hill Pool and Bog	Warwickshire
Cooper's Hill	Ampthill, Bedfordshire
Darenth Wood	Northdown, west of Dartford, Kent
Epsom/Ashstead	Leatherhead, Surrey
Epping Forest	Waltham Abbey, north London
Ferndown	Wareham, Dorset
Gailey Pools	Staffordshire, south of Stafford
Hare's Down and No Stone Moors	Devon
New Forest	Hampshire
Oxleas Wood	London
Roughdown Common	Hertfordshire
Ruxley Gravel Pits	Bromley, London
Snells Moor	Newbury, Berkshire
Southfield Farm Marsh	Northants
St Catherine's Hill	Nr Winchester, Hampshire
Stoke Ferry Fen	Norfolk
Thetford Heath Breck Common	Norfolk
Thetford Golf Course and Marsh	Norfolk
Titsey Wood	Nr Oxted, Surrey
Thatcham Reed Beds	Newbury, Berkshire
Wreakin-Ercall	Telford New Town

Source: Preliminary results of a Friends of the Earth Countryside Campaign Report, 1984.

Figure 4 Road schemes: recent, planned and probable damage to SSSIs (trunk roads and motorways)

ancient habitats and merely create rough grassland and secondary woodlands where only the most adaptable species which are under no threat from man's activities survive. A good example of a highly threatened woodland is Oxleas Wood in south east London. Lying on the clay and gravel summit of Shooters Hill in the London borough of Greenwich, Oxleas is an ancient semi-natural woodland (that is, it has always been wooded and has vegetation derived directly from the original natural

forest). Oxleas is especially rich in fungi: there are over 210 species
recorded from the wood of which 16 are 'notable' and six locally rare. Of
the 200 different species of fly found there, over 40 are 'notable', and six
locally rare. There are over 70 species of plant bug and leaf hopper
recorded from within the wood. A secondary wood – that is, a planted
wood or self-sown wood on wasteland – will only hold around a third of
the commoner species from most of these groups. But Oxleas is most
interesting for its trees, especially the wild service, a rare maple family
native which is found there in unusual abundance. Its fruits are known as
'mulders' and were formerly used to produce a potent brew; (their
chequered appearance has given the name 'The Chequers' to many
Wealden pubs) and the tree has a beautiful orange-red foliage in autumn.
Reflecting the rich diversity of conditions on the forest floor and the
mosaic of vegetation, 210 species of fungi were located on just one short
visit in 1982 and further study would doubtless increase this number.

Putting a motorway through Oxleas, as the Department of Transport
intends to do, would eliminate a number of these species. Within ancient
woods, some plants and animals are invariably restricted to just a few
spots while others are dependent on the whole area of the wood and will
be lost if it is reduced in size. Indirect effects on drainage and from
pollution (80 per cent of the pollution will fall out over around a 30
metre band on either side of the road-cutting) will also have detrimental
consequences. The road will eliminate the only colony of the orange tip
butterfly in the wood and will directly consume at least 10 acres (and 500
standard trees) of land. The 20 acres block left isolated in one corner is
likely to be too small to provide a viable habitat for many species. At the
southern end of the wood a fine hornbeam stand will be destroyed.
Ironically, the wood is currently being designated a Site of Special
Scientific Interest but this affords no protection from the plans of the
Department of Transport. And is the road necessary? Numerous
objectors argue that it is not. It will funnel traffic from the M11 down
into central south London creating new traffic jams. Local suburbs will
be less pleasant. People will be likely to move out to Kent or Essex and
commute to the area along the new motorways. Their car exhausts will
add to Britains's growing acid rain damage to trees.

The advice issued to local planning authorities on nature conservation
matters amounts to just one DoE circular dating from 1977, which is both
vague and ineffectual. As a result, the spread of retirement homes and
commuter suburbs, paricularly in southern England, in the 1960s and
1970s, has proved excessively damaging to habitats such as heathland. In
1980, one-third of all heathland SSSIs in Dorset were damaged or
destroyed: since then destruction of heathland by speculative building in

the Southampton–Bournemouth–Poole area has been allowed to continue, with particularly serious consequences for species such as the adder, nightjar, smooth snake, sand lizard and several rare dragonflies. The incentive for local authorities to allow such development is stong, not least because of the increased rateable value that accrues from retirement villages or hypermarkets on 'green field' sites.

Thirty-six Years of Destruction

As a result of these inadequacies, the NCC has presided over decades of habitat loss. In 1983, it published *The Objectives and Strategy for Nature Conservation*, which summarized 36 years of failure in these terms:

* *Lowland neutral grasslands including herb-rich (i.e., flower-filled) hay meadows*: '95 per cent now lacking significant wildlife interest and only 3 per cent left undamaged by agricultural intensification'.

* *Lowland grasslands of sheep walks on chalk and Jurassic limestone*: '80 per cent loss on significant damage, largely by conversion to arable or improved grassland (mainly since 1940)'. By 'improvement' the NCC refers to the fertilization, draining and reseeding, which converts semi-natural species-rich grassland into an artificial ley or pasture that can be highly productive (responsive to nitrogen imputs) but supports very few species.

* *Lowland heaths on acidic soils*: '40 per cent loss, largely by conversion to arable or improved grassland, afforestation and building . . .'

* *Limestone pavement in northern England (flower-rich areas of bare rock with humid fissures in its surface)*: '45 per cent damaged or destroyed, largely by removal of weathered surfaces for sale as rockery stone, and only 3 per cent left undamaged'.

* *Ancient lowland woods of native broadleaved trees*: '30–50 per cent loss by conversion to conifer planation or grubbing out to provide more farmland'.

* *Lowland fens, valley and basin mires*: '50 per cent loss or significant damage through drainage operations, reclamation for agriculture and chemical enrichment of drainage water'.

* *Lowland 'raised mires' (mosses)*: '60 per cent loss or significant damage through afforestation, peat winning, reclamation for agriculture or repeated burning'.

* *Upland grasslands, heaths and blanket bogs*: '30 per cent loss or significant damage through coniferous afforestation, hill land reclamation and improvement . . .'[8]

Loss of Species

As each of these habitats has declined, so the characteristic species which relied upon them have dwindled, and sometimes vanished. Once common species, such as lady's smock which was so abundant and widespread that it merited over 50 local names, are now becoming rarities. Species which were always very localized because they could only tolerate unusual combinations of physical conditions (many invertebrates, for example), or needed large areas of continuous habitat (such as many predators), or which depended on specific biological interactions with other species, have been the first to go. The large blue butterfly, dependent on a certain type of chalk grassland, symbiosis with one species of ant and the right microclimate, became extinct in 1979. The merlin, a small bird of prey of the uplands, is in steady decline as its moorland habitat is continually eroded by forestry and farming development.

Other species which have drastically declined were, a generation or two ago, simply 'normal' parts of a rural scene. The corncrake is now extinct as a breeding species in Britain in almost every county except the Western Isles. Even here, on the northwesterly seaboard of Europe, its hay meadow habitat is threatened with agricultural development through the EEC-funded Integrated Development Plan. The cirl bunting breeds in fewer and fewer parishes of southern England each year. The stone curlew was once common across England's chalk belt, breeding from Dorset to Norfolk on stony arable fields. Today it is a rarity, confined to a small area of the East Anglian Brecks, and even here it is threatened by changing practices, and probably by pesticides such as aldicarb which destroy its earthworm food. As a breeding bird, even the familiar lapwing has slumped severely. The British Trust for Ornithology has found that the number of successfully hatched eggs fell from 75 per cent to just under 10 per cent in areas where there had been a switch from predominantly grassland to arable farming between 1960 and 1980. In counties experiencing an overall halving of pasture since 1960 (in favour of arable) there had been a five-fold reduction in the number of breeding lapwings.

Such a change is tragic. The lapwing (green plover or peewit) is a beautiful and useful bird, feeding much of the time on insect larvae which are agricultural pests. But it is in decline like most of the rural flora and fauna because it cannot survive the machine and chemical age of farming.

Other groups of species which have suffered badly include butterflies. Although the use of pesticides has led to a decline in hedgerows and other habitats where they still survive, the outright loss of semi-natural vegetation has had a drastic effect on many species. Of those using grasslands, the adonis blue and silver spotted skipper are now considered 'vulnerable', the chalk hill blue, dingy skipper and grizzled skipper and small blue are declining, and the large blue is extinct. The large tortoiseshell is considered endangered, probably through loss of hedgerow trees, although its ecology is not (and now may never be) properly understood. The silver studded blue is classed as vulnerable following a decline due to heathland loss and the abandonment of traditional grazing on chalk grassland sites. Woodland butterflies are also suffering from the end of traditional coppicing and other management practices, the conversion of native broadleaved woods to conifer plantations, and outright grubbing. The chequered skipper, brown hairstreak, dark green fritillary, duke of burgundy fritillary, heath fritillary, high brown fritillary, pear bordered fritillary, purple emperor, silver washed fritillary and wood white, are all primarily or wholly woodland species which are declining or vulnerable. The grayling (a species which lives on heaths, sand dune scrub and chalk grassland) is disappearing as conifers are planted on its breeding sites, or they are put to the plough. The white letter hairstreak, reliant on elm, is also regarded as 'vulnerable'. In all, of Britain's already small butterfly fauna of around 60 species, at least 20 are in decline or have already been reduced to scattered populations on 'habitat islands'. For example, the black hairstreak depends on areas of blackthorn along wood edges or on very large hedges, which are at least 200 years old. The black hairstreak is increasingly confined to isolated sites, and in such a situation populations are unable to 'recolonize' one another's areas; as a result of otherwise unimportant natural, local disasters, species become first locally, then regionally, then nationally extinct.

Turning to other groups, the story is much the same. Birds are relatively mobile, adaptable species, and, in Britain, are quite well protected by law. Here if anywhere, government conservation policies should have worked. But, in reality, although some rare species have done well at a few closely guarded sites, once common species have, in the 'normal' countryside, become rarer and rarer.

The lapwing is not the only species to have been hit by agricultural industrialization. The NCC believe from field studies that the ploughing of old grassland and the loss of rough scrub and heath has seriously reduced populations of the cirl bunting, tree pipit, common snipe, yellow wagtail, redshank, and rook, while the overall loss of habitat variety is blamed for a decline in the little and barn owl, along with the grey partridge. In all, some 18 species show a serious decline as breeding birds since 1950: six on heathlands (including specialists such as the nightjar), three on open downland and short heath, including the chough. This red billed crow once lived along the coast of south-west England but is now restricted to parts of Wales and the Isle of Man as its clifftop feeding grounds are ploughed up or overgrown with bracken as sheep grazing is abandoned in favour of arable. In woods, the nightingale is in decline; in marshes, lakes and rivers, the kingfisher, marsh and montagu's harriers, water rail and bittern. On coastlands, the little tern is still losing its breeding habitat of undisturbed, shingly beach hinterlands as it is forced onto beaches to nest by coastal developments, and on the uplands, at least six species (including the golden plover) have shown marked decline since 1950, mostly as a result of afforestation.

Dragonflies are one family which is particularly susceptible to pollution as well as habitat loss. Of some 43 species found in Britain in the 1950s, three or four are now extinct, six are classed as vulnerable or endangered and five have declined dramatically.

A Divided House . . .

For a country which likes to regard itself as acutely conscious of the beauties of landscape and wildlife, Britain's environmental record in these respects is plainly lamentable. The major failure – to arrest the wholesale change that has affected the countryside – owes something to the illogical separation of wildlife and landscape protection and agricultural policy. The split stems from the time when the Nature Conservancy and the National Parks Commission (NPC) were set up in 1949, one firmly set in the science research camp, and the other in the planning profession. The 1968 Countryside Act which established the Countryside Commission and Parks Authorities (in place of the NPC), and the 1973 Nature Conservancy Council Act, only served to solidify the divide. As Ann and Malcolm MacEwen wrote in 1981:

> Britain seems to be the only country in the world to have two separate agencies for landscape and for nature conservation . . . In theory at least the Commission's job is to conserve both nature and beauty, although in practice it tends to lose interest in nature unless it believes it to be beautiful.

What then is the job of the NCC? It is to conserve nature, but not its beauty! Nobody is coordinating the two.[9]

The plight of the Yare Marshes of the Norfolk Broads – and particularly the Halvergate part of these marshes – exemplifies the failure of the divided policy. It does not take the ordinary person very long to see that this is a remarkable area. In summer, the lush grasslands grazed by cattle are alive with snipe, yellow wagtails and redshank, the ditches are full of different waterweeds (*Potomageton* species) and, where pollution has not reached, unusual water plants such as the water soldier. They are also home, in places, to the rare Norfolk aeshna dragonfly. The summer landscape can be torpid, tranquil, soporific; a few months later, the marshes are chilling, bleak and intimidating as the winter east winds blow in from the centre of Europe and continental Russia. The winter brings flocks of wildfowl such as bean geese, which frequent the flat fields, with only a few passing trains, the boats along the distant Yare and skeins of bewick swans from the arctic for company. But with EEC money (and MAFF grants), this unique landscape could be profitably converted to cereal fields, drained of both water and wildlife, character and landscape. Yet when it came to protecting the marshes, the authorities were powerless. The whole area could not be made a SSSI because of the restricted criteria used by the NCC. The Countryside Commission was effectively impotent because planning gave no protection against agricultural improvement. In June 1984, in the face of government assurances that Halvergate was 'safe', a local farmer ploughed a huge V-sign through the pasture preparatory to drainage works. As Dave Baldock commented, the V-sign seemed to symbolize 'the shortcomings of the Wildlife and Countryside Act in one deft gesture'. It was also symbolic of the basic failure of government policy, which simply could not and would not protect the countryside. The 'Environmentally Sensitive Area' to be set up at Halvergate will provide a stay of execution if some farmers take the extra grants in return for inaction: locally it may even be a success.

But there are other Halvergates in the wings, and there can only be a limited number of botched-up compromises and one-off solutions. The whole basis of agriculture and land use must be overhauled. The present system cannot be sustained. But waiting for the wasteful, polluting, resource hungry, ecologically unbalanced British agricultural system to bleed itself to death is like waiting for a bull to do the same in a china shop. In its death throes and despite its lacerations, it will do vast and irreparable damage. First, therefore, we need immediate controls. Then we need fundamental reforms.

Notes

1. P. Chapman, *Fuels Paradise*, Penguin, London, 1975. See the discussion of energy yields, from crops under different input systems.
2. National council and country committee members are the Secretary of State's appointees.
3. K. Kirby and J. Spencer, NCC Chief Scientists Team, personal communication, 1985.
4. K. Kirby and G. F. Peterken, *Ancient Woodlands Inventory*, Peterborough: NCC, 1984.
5. Royal Society for the Protection of Birds, Conservation Planning Section, *Land Drainage in England and Wales, An Interim Report*, RSPB, Sandy, Bedfordshire, 1983.
6. RSPB, *Land Drainage*, RSPB, Sandy, Bedfordshire.
7. R. Body, *Agriculture: The Triumph and the Shame*, London: Maurice Temple-Smith, 1982.
8. NCC, *Objectives and Strategy for Nature Conservation*, Peterborough, 1983.
9. A. MacEwen and M. MacEwen, 'Nature and Landscape; Why the great divide?' *Ecos*, vol. 2, no. 2, 1981, pp. 24–9.

6

Soil Erosion in Britain

The Loss of a Resource

R. P. C. Morgan

A sudden storm on a light sandy or sandy loam soil can have dramatic effects. On slopes of only two or three degrees, fields can be scarred by a dense network of well-defined channels or rills, 20 to 25 centimetres deep and 30 to 50 centimetres wide. The material washed out of these channels fills localized depressions in the landscape and covers the gentler sloping land on the valley floors, frequently burying crops. Where the material discharges into rivers, it is a pollution hazard, both from the sediment itself and from the chemicals adsorbed to it. Such a scene is all too common in the midlands, and in the southern and eastern counties of England on land devoted to the continuous production of cereals, sugar beet and vegetables.

The widely held assumption that soil erosion is not a problem in Britain because of the low intensity of rainfall and the high proportion of the land under grass and trees is invalid. The effects of erosion can be seen in the spring by the lighter coloured patches on hillsides where ploughing has turned the subsoil on to the surface. Deposits of fine material, often a metre or more thick, in the valley bottoms are further evidence. As more and more subsoil is exposed on the slopes above, however, these deposits become coarser and the lower land is covered by infertile sandy material.

Since erosion is a natural process, the problem is not that it occurs but the rate at which it takes place. The question of what constitutes an acceptable rate of erosion is much debated. It depends upon the rate at which new soil forms, on the amount of organic materials and fertilizers being added to the soil, and on the depth of the top soil. American scientists in the early 1960s adopted an annual rate of 11 tonnes per hectare as the maximum acceptable soil loss, but they recognized that this figure was probably too high for areas of thin, highly vulnerable soils. Top soil depths on the sandy soils in Britain are often only a few centimetres

"No Zir, nearest water'ole be at Bury-St-Edmunds zum twenny moil away"

and soil scientists suggest that the annual rate of new soil formation probably averages less than 0.1 tonne per hectare. Thus, even allowing for the mixing of the top soil with the subsoil by cultivation and the addition of chemicals through fertilizer application, adopting a figure of one tonne per hectare may prove too high for British conditions. For a typical sandy loam soil, this is equivalent to a loss of about seven millimetres of top soil in a hundred years. Since this will have hardly any effect on crop yield, however, it seems a reasonable figure to adopt.

Erosion in Britain

Until the mid-1970s, data on erosion rates in Britain were scarce. Most measurements had been made by geomorphologists in areas with only limited interference by man. Not surprisingly, the rates of erosion were low – generally less than 0.1 tonne per hectare as an annual average. Virtually no data existed for arable land. Measurements made between 1973 and 1979 for a variety of soils and land use in the Silsoe area of Bedfordshire thus gave cause for concern (See table 5).

Table 5 Average annual soil losses in the Silsoe area, Bedfordshire

Type of soil and land use	Tonnes per hectare[1]
Bare sandy loam soil	10 – 45
Cereals on sandy loam soil	0.6 – 24
Cereals on chalky soil	0.6 – 21
Cereals on clay soil	0.3 – 0.7
Grass on sandy loam soil	0.1 – 3
Woodland on sandy loam soil	0.01

Note: Values are for slopes of 7° to 11°, except for the woodland where the slope was 20°, in undulating terrain.

Not only, as expected, was the erosion rate excessive on sandy loam soils kept bare of crops, but sandy loam and chalk soils under winter wheat and spring barley were experiencing erosion at more than 20 times the acceptable rate. By implication, this was also true for land under market gardening, which generally provides less protection to the soil than cereals especially where two crops per year are obtained and there is a period in mid-summer when the land is bare at the time of heavy thunderstorms.

Although the Silsoe study provided evidence of serious soil erosion, the rates obtained are rather conservative. The highest rate for a single storm was 19 tonnes per hectare, whereas storm soil loss rates of 156 tonnes per hectare have been measured by Dr A. H. Reed in east Shropshire and 195 tonnes per hectare by Dr R. Evans and Dr S. Nortcliffe in north Norfolk. Observations in the Silsoe area show, however, that erosion is a frequent event. Storms with soil loss greater than two tonnes per hectare from bare ground occurred in the Silsoe area in four out of the seven years of study, including the drought year of 1976.

Comparable data are not available for wind erosion but studies by Dr S. J. Wilson in the Vale of York indicate that the rates are similar to those for water erosion.

By combining information on soil properties, land capability, rainfall and wind velocity it is possible to estimate the area where the risk of soil erosion is above acceptable rates. This is about two million hectares or about 37 per cent of the arable area of England and Wales. It should be stressed that this figure is an estimate of the problem on land classified as suitable for arable farming. It is not therefore a reflection of the misuse of the land but rather of the failure to manage that land in an appropriate manner. Misuse of the land (for example, by bringing land of lower class into arable production), will increase the area vulnerable to erosion still further.

Current management practices on arable lands are unlikely to sustain land quality. An erosion rate of 20 tonnes per hectare each year will result in a loss of 150 millimetres of top soil in 55 years. Crude estimates of cereal yields in relation to soil depth suggest that with an initial top soil depth of 15 centimetres, the result will be a decline in yield from three to two tonnes per hectare. Organic content is likely to diminish from 1.5 to 0.75 per cent over the same period, with consequent deleterious effects on structural stability, increases in surface compaction and sealing, decreases in infiltration rate, increases in run-off, reductions in the amount of water available for plant growth and further increases in erosion rates. Under these conditions, the effects of compaction by tractor wheels are compounded. The successive passage of tractors over the same piece of land can be devastating with tractor wheelings developing into deep gullies. By contrast, the effects of hedgerow removal are less important. Only where the hedges are at right angles to the dominant wind or on the contour are they an effective barrier to erosion.

Controlling Soil Erosion

Soil erosion can be controlled through agronomic measures, soil management and the use of mechanical protection works such as terracing. The latter are generally inappropriate for this country. The erosion rates are not severe enough to warrant terraces; it should be possible to reduce erosion to acceptable levels through crop and soil management alone. Also, the size and shape of many of the fields would make the installation of terrace systems, or even contour farming, uneconomic by producing a large number of rows with very small working lengths.

Recent research has brought into question the widely accepted principle that establishing a crop cover will always help to protect the soil from erosion. Erosion is a two-phase process consisting of detachment of soil particles from the soil mass and the transport of those particles downslope, downstream or downwind. One way of controlling erosion is to limit the rate of detachment; wihout the supply of loose detached material on the soil surface, there is no sediment for water and wind to transport. Different crops give different degrees of protection from detachment. The deachment rate on sandy and sandy loam soils under cereals decreases with increasing canopy cover, so that it is about 30 per cent of that on bare soil with a 10 per cent cover and about 3 per cent of that on soils with a 75 per cent cover. With large-leaved vegetables, such as Brussels sprouts, the detachment rate initially declines as the crop grows reaching 70 per cent of that for bare soil with 15 to 20 per cent canopy cover. With further crop growth, the effect of increasing interception of rainfall by the plant canopy is offset by the increasing role of leaf drip. Studies at Silsoe College have shown that leaf drips, 4.5 to 6.0mm in diameter, are capable of detaching up to 30 grams of a sandy loam soil per Joule of drop energy whereas a typical raindrop of 2.1mm diameter detaches only 13.1 grams per Joule. The greater detaching power of *leaf drips* and their increasing importance as the crop grows means that, by the time 40 to 50 per cent cover is attained, the detachment rate equals that recorded without a crop. Since increased detachment is usually associated with soil aggregate breakdown, surface crusting and increased runoff, this may explain why erosion is frequently observed under vegetable crops even after they have reached maturity.

Since rates of erosion are very low when the soil is covered by grass, the simplest solution to the problem is to reintroduce rotations with grass leys. These will not only reduce erosion to acceptable levels but are also the only effective way of preventing the decline in organic content on the sandy soils. If government policy continues to favour continuous arable, the alternatives are to accept the consequences of not being able to sustain cereal production on the lighter soils far into the next century because of the effects of erosion on yield, or to make more money available for research to develop an acceptable soil conservation technology. Unfortunately, very little research is conducted on soil conservation in this country. Initiatives are left to innovative farmers and to trials on experimental husbandry farms. Although frequently successful, the measures are pragmatic and local. Without the fundamental studies to understand how and why they work, it is impossible to predict the likely success of transferring them to another area.

If mechanical methods and agronomic measures are not the bases for solutions, the potential role of soil management through tillage must be investigated. Minimum tillage has its advocates for erosion control but its effectiveness cannot be demonstrated in this country because few measurements have been made of the runoff and soil loss associated with its use. Alternative approaches might be just as effective and more acceptable to farmers. One midlands farmer has adapted conventional tillage to produce a surface which is resistant to wind erosion using a system of ploughing, rolling and pressing of the soil. This technique is now used by many farmers on the sandy soils in the east midlands. Research is now in its early stages to see whether a similar approach could be developed for water erosion control.

Agricultural soil erosion is severe enough to warrant attention. Complacency will only lead to a deterioration in the quality of the country's soil resources. The problem can be solved if its importance is recognized and a policy developed for its control. Such a policy requires funding so that the present limited efforts of individuals can be properly supported by a co-operative programme involving research scientists in the universities and government, the advisory services and farmers. The development of acceptable and effective conservation measures is a matter of priority.

7

The Environment in Forestry Policy

Colin Price, Christine Cahalan and Don Harding

Forestry is a long-term matter. Foresters have traditionally been educated to this, but modern forestry (and Britain has been pre-eminent in establishing this philosophy) is becoming a business, and as such it has imbibed the business ethic of fast results. This transformation has come about through a century in which there has been a continual shift in objectives and, perhaps more insidiously, in the way those objectives are interpreted and implemented.

A Century of Changing Objectives

At the close of the nineteenth century, the small remaining area of forest in Britain was largely in the hands of wealthy landowners and farmers. The farmers valued the woodland for the shelter and forest produce it provided, often on land which was uncultivable with existing agricultural technology. The landowners called upon a long tradition of land management, in which the perpetuation and improvement of their estates were major objectives. They prized their trees as an amenity, a rich variety of species and long rotations being the norm. The conservation value of woodlands was maintained partly by non-intervention and partly by traditional forms of management, such as coppice-with-standards or long rotation shelterwood systems which, though intensive, created varying light conditions at the forest floor and were conducive to species diversity, while maintaining an essentially woodland environment.[1]

The strategic crisis of the First World War, and the social changes of its aftermath, wrought a revolution in forestry policy. The creation of the Forestry Commission brought not only extensive state ownership, but increasing oversight of private woodlands. The new preoccupation with building up a strategic reserve of timber was long term in its ends, but speed was of the essence, and it was to fast-growing exotic species that British foresters turned as a means of afforesting upland acquisitions, and often in restocking devastated broadleaved woodlands.

Inevitably, a policy aimed at maximum physical production brought conflict with environmental interests. The scale of opposition to forestry began to increase as early as the 1930s, well before the introduction of site protection legislation. The adverse effects of afforestation on the landscape of the Lake District led to the creation, by agreement, of the first 'no-go' area for forestry in Britain, covering the core of what was to become, some 15 years later, the Lake District National Park.[2] The problems of introducing conifer plantations into the uplands are great, and opposition to forestry on landscape grounds has continued unabated since the early insensitive plantings of monocultures within the geometric shapes of the acquisition boundary lines. Meanwhile, opposition to forestry on nature conservation grounds has grown up in the lowlands, largely because of the practices of planting up small areas of environmental significance and underplanting broadleaved woodlands with conifers.

The Zuckerman report of 1957 foreshadowed the relacement of strategic by economic and social objectives.[3] There were two prongs to this more business-like attitude to forestry: first, pressure to cut costs by implementing new techniques, and, second, the adoption of discounted cash flow (DCF) methods for appraisal of silvicultural investments. The new approach was epitomized by the use of chemical poisons to kill young oak plantations so that faster-growing conifers could be grown underneath – perhaps the nadir of forestry's environmental consciousness.

During the 1960s, explicit environmental objectives were declared in forestry policy, backed by the appointment of a landscape consultant to the Forestry Commision and by increasing research and dissemination of literature on conservation. Yet the DCF mentality pervaded even environmental assessments: the Treasury's 1972 cost-benefit study of forestry in Britain discounted the recreational and hydrological value of forests;[4] a 1978 Forestry Commission internal paper on landscaping was cast in DCF terms;[5] more recently, compensation for management agreements made under the Wildlife and Countryside Act has been calculated as forgone discounted profit.

In the late 1970s, two studies pressed the case for a massive further expansion of afforestation, largely on the grounds of anticipated world timber shortages with associated price rises.[6] The environmental implications of the suggested doubling of forest area over 50 years would clearly be serious if implemented, particularly in the uplands where the planting would be concentrated. It is difficult, for example, to see how 35 per cent of the remaining area of rough grazing in the national parks could be afforested without detriment to the landscape.[7] Most national park authorities now have policies to control afforestation and, although these arrangements have on occasion been criticized,[8] there is no doubt that they represent a significant restraint on forest expansion in the parks. There have been signs of a hardening of foresters' attitudes in the face of environmental constraints on afforestation, however, and it is worth considering that *the national park landscape arrangements have no statutory force*: a private owner without environmental scruples is free to plant any land he wishes, regardless of the judgements of forestry panels. The most contentious conservation issue is likely to be the effect on bird populations of further moorland recession in areas such as upland Wales.[9] Much resentment still exists in both the Nature Conservancy Council and the Royal Society for the Protection of Birds over the loss to forestry of Llanbrynmair Moors, a very important area on ornithological and botanical grounds. Major expansion can only avoid repetition of such conflicts within agreed land-use strategies, but attempts to devise strategies are likely to receive little support from owners and occupiers of land. It has been suggested that changes in attitudes to management agreements are occurring, but this change is predominantly confined to farmers.[10]

Forestry and the Physical Environment

What kind of forestry has emerged as standard British practice as a result of all these influences – strategic, economic and environmental? How compatible is it with the maintenance of ecosystems? The requirement for a quick return on investment, and for large areas of a uniform product for the processing industries, leads inevitably to the establishment of plantations of those species which grow most rapidly on sites available for extensive afforestation, regardless of other biological considerations. As a result, the two major commercial species, Sitka spruce (*Picea sitchensis*) and lodgepole pine (*Pinus contorta*), grown in monoculture over large areas of upland Britain, are often planted on sites to which they are ecologically unsuited, and where their susceptibility to damage, pests and disease is increased.

Wind damage to coniferous plantations growing on exposed and waterlogged ground is a major problem in British forestry. The danger of windthrow dictates the use of a 'no thin' regime in approximately 60 per cent of Forestry Commision forests, yielding no revenue from thinnings and producing fewer trees of sawlog size at the time of final harvest than conventionally thinned crops. Short rotation silviculture or the use of ground preparation techniques other than ploughing may solve the problem on some sites, but on others, where exposure and high water tables continue to determine both the choice of species and the silvicultural regime, it makes little ecological sense to plant trees at all.[11]

Outbreaks of previously untroublesome insect diseases have occurred on average once every five years during the period of post-war afforestation in Britain. [12] In the late 1970s, populations of pine beauty moth (*Panolis flammea*) reached epidemic proportions in plantations of lodgepole pine, and aerial spraying of 4,800 hectares of forest with the insecticide fenitrothion was necessary to prevent a potentially devastating attack.[13] Of more recent concern is the introduction to Britain of the great spruce bark beetle (*Dendroctonus micans*), and, although biological control by the natural predator *Rhizophagus grandis* is likely to prove successful, this will not eliminate the beetle completely, and annual losses of a proportion of the growing stock will have to be accepted while Sitka spruce continues to be grown on any scale.

The long-term effects of forest monocultures on soil properties are not fully understood. Peat soils planted with logepole pine undergo irreversible drying, which increases aeration and rooting depth but may cause losses in soil strength.[14] On lowland soils conifers appear to cause 'podzolization', a process which results in the acidification of soils.[15] Recovery from the soil compaction which occurs during harvesting operations may take many years, and the initial growth of trees planted in compacted or disturbed soil is reduced.[16] It is not yet clear whether these changes are accompanied by decreases in soil fertility, but the nitrogen deficiencies which develop late in the rotation on some nutrient-poor soils have led to concern about long-term nitrogen supply on these low-quality sites.[17]

Forests may have important long-term influences on water yields and quality.[18] Experiments at the Institute of Hydrology's Plynlimon research station in mid-Wales have shown that water losses from a forested catchment are much greater than those from a grassland catchment,[19] although further research is needed – and is being undertaken in Scotland – to compare forests with other common upland vegetation types such as heather moor. Sediment yields from the forested Severn catchment area have been significantly higher than from the grassland Wye area, a surprising result attributed to the fact that drainage ditches dug at the time of ground preparation some 35–40 years ago are still eroding under conditions of intense run-off.[20] It is also likely that felling at the end of the first forest rotation will increase sedimentation significantly, although the effects may not be as dramatic as at other research sites in temperate areas.[21] The Institute of Hydrology will be monitoring this as the harvesting phase is entered at Plynlimon.

There has been a considerable increase of research into acidification of water-courses in the last five years. Harriman and Morrison, and Stoner, Gee and Wade have described the spatial differences of pH values between streams draining areas of conifer forest and moorland.[22] Ormerod and Edwards report a trend towards decreasing pH of soft water streams draining from afforested catchments which is consistent with the results from these spatial surveys.[23] Recognition of this increased acidity has now led the Forestry Commission to end all planting of conifers close to water-courses, though the effectiveness of this in reducing acidification has yet to be clarified.[24]

The Political Economy of Forestry

Despite what is obviously a great deal of uncertainty and dissension, the prevailing pattern of forestry has received remarkably consistent support from various governments. Although landowning interests are heavily represented in the forestry lobby, backing has not been confined to conservative groups and politicians. And, somewhat curiously, the level of fiscal subvention for private forestry has been greater under socialist than Conservative governments. Apart from threats to its funding in its early days, the expansion of the state forest enterprise was maintained over the first 60 years of its existence, by governments of both persuasions.

Yet there is an inherent contradiction underlying this consensus on forestry policy. Forestry is a long-term matter, and bases its claim for governmental support on the need for an extended time horizon. But the political directive to adopt discounting procedures, with their explicit devaluation of the long term, can be seen as the clearest possible expression of short-term political thinking, whether as embodied in the 10 per cent discount rate used in the 1972 interdepartmental study under Conservative rule, or in the 5 per cent rate set by the 1978 Labour administration. These rates reflect both the return expected from investment in the growth economy and putative reduction in the value ascribed to the increasingly trivial products of that economy:[25] they have tenuous relevance if the prospects of long-term economic growth are uncertain, and no relevance at all to environmental benefits.[26] Their validity in the appraisal of natural resources is being increasingly questioned.[27]

Paradoxically, however, the effects of DCF appraisal are not uniformly hostile to the environment. On the one hand, it favours short rotations, makes the growing of broadleaves financially suicidal and draws a veil across the very long-term effects of land use upon soil and ecosystem. On the other, it decreases the optimal level of investment in intensive silviculture and reduces the compensation payable for management agreements. At high discount rates, the conversion of scrub and coppice to conifer forest cannot be justified – the poisoning of young oaks was a financial as much as an environmental error. Nor can the afforestation of low-grade upland sites, often of high conservation value, show a profit.

A further twist is that recent governments, while imposing high discount rates for silvicultural decision-making, have made the dispensation that rates of return as low as 1 per cent are acceptable for upland afforestation, on the grounds that it maintains employment and hence the social fabric of remote rural areas.[28] This social justification is mildly ironic in view of the fact that over the last 30 years, the adoption of highly mechanized technology has led the Forestry Commission to reduce its field workforce to a third of its pre-Zuckerman level, on a forest area which has since doubled in size. The environment, it seems, must suffer the worst of both worlds: a lenient test of profitability for deciding *whether* commercial forestry should be undertaken; and a discount rate which only takes account of short-term factors when determining *how* it should be practised.

These considerations show that the abandonment of DCF would not of itself afford the environment maximum protection. Specific recognition of a wide range of long-term effects in the ecosystem, and political directives on their appraisal, which go further than the mere declaration of a benevolent policy, are also required.

Back to Capability Brown?

The extent to which environmental objectives are incorporated into forestry policy is, of course, a political judgement and, arguably, successive governments have failed to act early enough to ensure that account is taken of environmental issues: changes in policy have tended to follow on public concern rather than to pre-empt problems before they arise. Recently, the Forestry Commission has increasingly involved itself in landscaping and conservation measures, but no government has tackled in depth the question of the performance of private forestry.

It must be a matter of concern, therefore, that a major switch in emphasis towards the private sector has been occurring since 1979. In the current political climate, there is even support for the abolition of the Forestry Commission and all state involvement in forestry. According to one proponent of this viewpoint the interests of future generations can be left to 'a chain of entrepreneurs' acting in response to expected price signals, while recreation (and other environmental values?) can be sustained as appropriate by charging for access to woodland.[29] It must be said, however, that in other sectors of the economy there is little evidence that an untrammelled private enterprise has been effective in these ways.

A number of specific developments are of major concern to environmental interests: forestry grants now emphasize planting rather than management of woodlands; the stabilizing influence of the Dedication Scheme has been abandoned:[30] and measures are in train for reduced surveillance and control of private forestry. The decision to sell off 10 per cent of state holdings to the private sector has been heavily critized as being short-sighted and damaging, and the future of broadleaved woodland has been a matter of considerable concern. Restrictions on public access in privatized woodland are likely to be greater and many woodlands of aesthetic or ecological importance could fall into the hands of unsympathetic owners. In some cases the Forestry Commission has been criticized for failing to give adequate notice of intending sales of such areas,[31] with the result that conservation agencies and organizations have found it difficult to raise funds and to liaise effectively over purchase 'packages'. As for the income tax arrangements which have been inherited barely modified from the days of the strategic reserve, they make no reference to the long-term state of the environment. Indeed, they perversely reward owners of short-rotation, even-aged blanket forests and penalize those who make concessions to the environment by, for example, planting broadleaves or leaving areas unplanted.[32] The consultative arrangements remain as a precondition for grant aid, but the fact that recent private afforestation has been undertaken without grants shows that the arrangements can be circumvented when a planting proposal is expected to draw environmental opposition: the tax advantages of forestry remain unaffected.

Conclusions

It is quite clear that trends since 1979 are extremely worrying from an environmental viewpoint and that continual political action to strengthen the private sector will increase the concern felt for the future of environmental conservation. The possible restoration by the end of this century of a predominantly private pattern of woodland ownership seems unlikely to return forestry to the physical condition prevailing a hundred years ago. In the lowlands, landowners have turned to business methods and are preoccupied with DCF and energy-intensive technologies, while in the uplands the struggle for survival often precludes long-term considerations. So, in the lowlands, the remaining woodlands and hedgerows would continue to be grubbed up, in the uplands grazing pressure would take an increasing toll on regeneration and the wood pastures would decay still futher. The current fiscal incentives for forestry are of little interest to these long-established landowners. But they are attractive to high income earners and financial institutions with no tradition of land management. The extent to which environmentally sympathetic land use, or even sound production silviculture, would be practised by these new owners hangs on the balance of influence between forester and accountant, while the cost of environmental concessions depends capriciously on the marginal tax rate dictated by the policy of the government of the day. Much uncertainty surrounds the future of support grants for upland agriculture, and there is a widespread view that current levels of support in Less Favoured Areas cannot be sustained.[33] The prospect of a substantial expansion of forestry by a private sector facing weak competition from agriculture, and largely uninhibited by environmental constraints, is not an attractive one, and would severely damage upland landscapes and habitats.

In virtually every country in the world, there is state involvement in forestry, largely on the grounds of its long-term nature and its impact upon the environment. It would be a fatal error to assume that the pursuit of private profit and pleasure will accommodate these values and reproduce the idealized land-use patterns and landscapes of the eighteenth century, when so much has changed, technologically, economically and politically.

As for technical aspects, there is a strong case for reversion to a more ecological approach to silviculture in Britain, one which would take into account local variations in environmental conditions and design silvicultural regimes accordingly.[34] Only by adopting such an approach will it be possible to maintain the productivity and health of Britain's forests in the long term.[35] Foresters are beginning to show signs of boredom with the standard fare of upland afforestation, and of concern at the extent to which British forestry has allowed itself to be dominated by short-term economic and ideological considerations. The revival of interest in diversified, long-term, ecologically sensitive silviculture will produce few results, however, unless it is backed by money and controls which specifically favour the environment. High-sounding policies are not enough.

Notes

1. G. F. Peterken, *Woodland Conservation and Management*, London: Chapman Hall, 1981.
2. H. H. Symonds, *Afforestation in the Lake District*, London: Dent, 1936.

3. S. Zuckerman, *Forestry, Agriculture and Marginal Land*, London: HMSO, 1957.
4. Treasury, *Forestry in Great Britain: an Interdepartmental Cost/Benefit Study*, London: HMSO, 1972.
5. Forestry Commmission, *Landscaping, Forestry Commission Planning and Economics Paper 16*, Edinburgh: Forestry Commission, 1978.
6. Forestry Commission, *The Wood Production Outlook in Britain*, Edinburgh: Forestry Commission, 1977; Centre for Agricultural Strategy, *Strategy for the UK Forest Industry*, Reading: Centre for Agricultural Strategy, 1980.
7. C. Price, *Right Use of Land in National Parks*, Bangor: Department of Forestry and Wood Science, University College of North Wales, 1975.
8. Snowdonia Park Society, *Annual Report*, Gwynedd: Dyffryn Mymbyr, 1980.
9. H. Williams and D. M. Harding, 'Towards a Land Use Strategy for the Uplands of Wales', *Quarterly Journal of Forestry*, vol. 76, 1982, pp. 7–23; Royal Society for the Protection of Birds, *Hill Farming and Birds: A Survival Plan*, Sandy, Beds: RSPB, 1984.
10. T. H. Thomas, D. M. Harding and R. Hattey, 'Land Use in Upland Wales; the Prospects for Integration', paper read at conference on Upland Farming, Co. Galway, Ireland, October 1984. (Proceedings in press.)
11. E. D. Ford, 'Can we design a short rotation silviculture for windthrow prone areas?' in D. C. Malcolm (ed.), *Research Strategies for Silviculture*, Edinburgh: Institute of Foresters of Great Britain, 1980, pp. 25–34; D. Seal and K. Miller, 'What's happening about Windthrow?' *Forestry and British Timber* vol. 13, no. 10, 1984, pp. 24–7.
12. D. Bevan, 'Coping with Infestations', *Quarterly Journal of Forestry*, vol. 78, 1984, pp. 36–40.
13. J. T. Stoakley, 'The pine beauty moth, Panolis flammea', *Forestry Commission Report on Forest Research*, London: HMSO, 1979, pp. 36–7.
14. D. G. Pyatt and M. M. Craven, 'Soil changes under even-aged plantations' in E. D. Ford, D. C. Malcolm and J. Atterson (eds), *The Ecology of Even-Aged Forest Plantations*, Merlewood: Institute of Terrestrial Ecology, 1979, pp. 369–86.
15. I. C. Grieve 'Some effects of the plantation of conifers on a freely drained lowland soil, Forest of Dean, UK', *Forestry*, vol. 51, 1978, pp. 21–8.
16. D. G. Pyatt, 'Classification and Improvement of Upland Soils', *Forestry Commission Report on Forest Research*, London: HMSO, 1983, pp. 23–4.
17. H. Miller, 'Forest fertilization: some guiding concepts', *Forestry*, vol. 54, 1981, pp. 157–67; W. O. Binns, 'Nutrition' in D. C. Malcolm (ed.), *Research Strategy for Silviculture*, Edinburgh Institute of Foresters of Great Britain, 1980, p. 50–7.
18. D. M. Harding, 'The Hidden Input: Water and Forestry' in Centre for Agricultural Strategy, *The Future of Upland Britain*, Reading University, 1978.
19. Institute of Hydrology, 'Water Balance of the Headwater Catchments of the Severn and Wye', *Institute of Hydrology Report*, no. 33, Wallingford, Oxon, 1976.
20. M. D. Newson and J. G. Harrison, 'Channel Studies in the Plynlimon experimental catchments', *Institute of Hydrology Report*, no. 47, Wallingford, Oxon, 1978.
21. M. Newson, personal communication, 1985.
22. R. Harriman and B. R. S. Morrison, 'Ecology of streams draining forested and non-forested catchments in an area of central Scotland subject to acid precipitation', *Hydrobiologica*, 88, 1982, pp. 251–63. J. H. Stoner, A. S. Gee and K. R. Wade, 'The effects of acidification on the ecology of streams in the upper Tywi catchment in West Wales', *Environmental Pollution*, A, vol. 35, 1984, pp. 125–57.
23. S. J. Ormerod and R. W. Edwards, 'Stream acidity in some areas of Wales in relation to historical trends in afforestation and the usage of agricultural limestone', *Journal of Environmental Management*, vol. 20, pp. 189–97.
24. W. O. Binns, 'Acid Rain and Forestry', *Forestry Commission Research and Development Paper*, no. 134, 1984.
25. Forestry Commission, 'Comparison of rates of return in forestry and industry', *Forestry*

Commission Planning and Economics Paper, no. 34, 1979.

26. C. Price, *Landscape Economics*, London: Macmillan, 1978.

27. C. A. Nash, 'Future Generations and the Social rate of discount', *Environment and Planning, A*, vol. 5, 1973, pp. 611–17; T. Page, 'Equitable use of the Resource Base', *Environment and Planning, A*, vol. 9, 1977, pp. 15–22; C. Price, 'Project appraisal and planning for over-developed countries: (I) The costing of nonrenewable resources', *Environmental Management*, vol. 8, pp. 221–32, 1984.

28. Ministry of Agriculture, Fisheries and Food (MAFF), *Forestry Policy*, London: HMSO, 1972.

29. R. Miller, *State Forestry for the Axe?*, London: Institute of Economic Affairs, 1981.

30. Under the Woodland Dedication Scheme, the owner of a specified woodland area undertook to retain tree cover in perpetuity, and to manage it in accordance with a plan of management agreed by the state forest authority. In return, he or she received grant aid, advice on management and freedom from repeated application for felling licences. From 1974, the Dedication Agreement could include, where appropriate, the granting by the owner of recreational access, and the requirement to fulfil other environmental obligations.

31. A. Jones, personal communication.

32. C. Price, *Right Use of Land in National Parks*, Bangor: Department of Forestry and Wood Science, University College of North Wales, 1975.

33. Thomas, Harding and Hattey, 'Land use in upland Wales'.

34. J. D. Mathews, 'The Case for Regional Silviculture', *Forestry*, vol. 51, 1978, pp. 3–4; D. C. Malcolm, 'The Future development of even-aged plantations: silviculture implications' in E. D. Ford, D. C. Malcolm and J. Atterson (eds), *The Ecology of Even-Aged Forest Plantations*, Merlewood: Institute of Terrestrial Ecology, 1979, pp. 481–504.

35. House of Lords Select Committee on Science and Technology, *Scientific Aspects of Forestry*, vol. 1: Report, London: HMSO, 1980.

8

Acid Rain and British Pollution Control Policy

Nigel Dudley

Well I must say that in 1976 to 1977 we had . . . we were working together, in fact they stayed with us here in Tovdal. We did our experiments together, we had the same explanations, we saw the same thing. Then suddenly in – I think it was around 1978 – something happened that – well, they found other reasons for the results they got themselves and we got . . . In my opinion, it was more obvious political reasons for it than scientific reasons.

Norwegian researcher talking about the Central Electricity Research Laboratory, Leatherhead, on a *Horizon* BBC television programme, 1982

In August 1983, the Central Electricity Generating Board announced that before it took any action to reduce sulphur dioxide (SO_2) emissions, the Royal Society would carry out a five year research programme into acidification in Scandinavia, funded jointly by the National Coal Board and the CEGB itself.[1] The Scandinavians responded with surprise and anger. After decades of patient investigation, they were finally receiving international acceptance for their claims that European air pollution was acidifying their lakes and killing fish. Britain's intentions, which amounted essentially to checking the Scandinavian research, were regarded as a slap in the face; one Norwegian scientist described it as 'dirty money',[2] and the Swedish conservation groups accused the CEGB of buying time.[3]

The angry exchanges were further signs of an increasing breakdown in relations between British and European researchers, and of the poor reputation of the British government with respect to environmental issues. Sharp divergences of opinion have also emerged within Britain, between the CEGB on the one hand and many scientists in universities and the civil service on the other.

The seriousness of pollution from SO_2 and nitrogen oxides (NO_x), including acid precipitation, the dry deposition of gases and the formation of photochemical ozone, is no longer seriously open to doubt. The widespread acidification of freshwaters in the north, the catastrophic damage to central European forests and the rapid deterioration of many ancient buildings and monuments have made the issue an international priority. Until quite recently, however, it was thought that Britain's chief role was as a polluter, exporting SO_2 and NO_x to Europe via the release of gases from high chimneys, rather than a victim of acidification. Moves to reduce air pollution would in that case have been mainly for the advantage of countries abroad and were tacitly regarded as being of little vote-pulling potential at home.

Arguments about Science

Many scientists believe that this attitude has had a strong influence on the scope and direction of British research and especially on the CEGB which, as Britain's largest SO_2 emitter, has considerable vested interests in acidification. The Board's avowed aim has been as a Devil's Advocate, testing the various theories put forward elsewhere, and it has certainly helped refine these by showing up inconsistencies in hurried research work. Taken as a whole, however, the CEGB research has angered many other scientists who regard it as nit-picking; worrying away at the edges of the subject rather than taking an objective view. Writer Steve Elsworth summed up this attitude in his book about acid rain: 'The CEGB's scientific research is framed so that it does not ask the question "what causes acid rain" . . . but rather "what apart from sulphur oxide emission could cause acid damage to the environment?" '[4]

Britain's Own Pollution Problems

The political situation altered dramatically in 1983, when the possibility that Britain might also be suffering from acid rain was first widely publicized. In retrospect, the fact that few people had apparently considered British acidification problems is itself remarkable, especially as the country has a very long history of SO_2 damage. John Evelyn wrote a tract against air pollution called *Fumifugium* as early as 1661,[5] while another Englishman, Matthew Smith, invented the phrase 'acid rain' in 1865, in a book examining the various air pollution effects which had resulted from the industrial revolution.[6] Conifer plantations were abandoned in parts of the Pennines during the 1930s because of high pollution levels which killed trees;[7] sulphur dioxide was identified as the culprit here and its role in the decline of British foliar lichens has been carefully documented.[8] Papers suggesting a link between freshwater acidification and air pollution were published over a decade ago, but largely ignored thereafter.[9]

In the event, it was left to a visiting Norwegian team to publish the first report overtly looking at the effects of acid precipitation on British freshwaters. Members of the Oslo-based Sur Nedførs Virkning På Skog Og Fisk (SNSF) project, accompanied by scientists from the Freshwater Fisheries Laboratory in Pitlochry, visited Scottish lochs in habitats similar to those where acidified lakes occur in Scandinavia, and found that acidification had apparently already occurred in parts of Galloway in south-west Scotland.[10] This news was again largely ignored by the media and politicians until it featured in a 1983 *Observer* article,[11] which was itself inspired by a sponsored press tour of an acidified area of Norway organized for British journalists by the 'Stop Acid Rain Campaign' of Norway. Meanwhile, analysis of diatom species in the sediment of Galloway's lochs suggested that accelerated acidification had been occurring there throughout Britain's industrial period.[12] Since 1983, unnaturally rapid acidification has also been discovered in Wales,[13] the Lake District,[14] and parts of southern England,[15] while researchers believe similar effects may occur in the Pennines, Charnwood Forest and areas of Scottish and English heath.

Information on the acidification of British freshwaters is still very incomplete. About 40 lakes in Galloway have been identified as being without fish and in some of these fish populations have disappeared within living memory.[16] This correlates with diatom studies which suggest that acidification in Galloway progressed gradually for a century or so and then suddenly became more acute in the last two decades.[17] Studies in Wales suggest the interaction of acidification from conifer plantations and air pollution in the disappearance of fish from streams and rivers such as the Teifi.[18]

A more serious problem for fish is often caused by an 'acid pulse', when a sudden peak of acidity is caused by melting snow or heavy rains following drought; in both cases large amounts of acid can be washed into freshwaters, releasing aluminium into solution which later settles out again once pH has returned to normal. If the aluminium is subsequently deposited onto fishes' gills it can cause death, resulting in the possibility of a large kill of fish in rivers and lakes where acidity is usually not a problem.[19] Fish kills of this sort have been seen in the rivers Duddon and Esk in the Lake District,[20] and acid snowmelt has been blamed for salmon deaths in a Scottish hatchery.[21]

Attitudes to forest death are also changing very rapidly. In March 1983, a Forestry Commission report on dieback in German forests was sceptical about the role of air pollution, pointing instead to climatic factors such as drought. It stated: 'There is undoubtedly an element of neurosis involved in the readiness of many foresters and some workers to attribute any decline or dieback in forest tree species to the combined effects of atmospheric pollution and acid rain without adequate critical examination.'[22] Whilst there may be some justification for this remark, the rapid deterioration of the Black Forest led to a fairly quick re-appraisal published early in 1984,[23] and by September of that year one of the authors, William Binns, stated at a conference that damage of a type previously unknown in Britain had been detected in March 1984.[24]

Earlier, a German forester had visited a number of British sites and claimed to have seen dieback similar to that observed in the Black Forest. The Forestry Commission replied that it was impossible to assign observed effects to any particular cause without many years of observation in any one site and knowledge of other environmental variables, but the lead helped persuade Friends of the Earth to seek and obtain funding from the World Wildlife Fund for a more detailed study of British tree damage. In the summer of 1985, FOE volunteers carried out a survey of beech and yew trees, using assessment methods devised by Chris Rose and Mark Neville,[25] based on information from German researchers. Yew and beech are believed to be good indicator species for Britain, analagous to spruce and fir in continental Europe, with yew showing a 'tinsel syndrome' of yellowing leaves, while beech develops clusters of twigs due to reduced crown growth, chlorosis (yellowing) of leaves and premature leaf-fall of green leaves. FOE's survey was launched by a visit from Bengt Nihlgard, a forest ecologist from Lund university in Sweden and an expert in beech damage, who said that effects in woods he visited were worse than in Sweden and similar to those in central Europe.

The results of the FOE survey, which involved 500 volunteers, including professional ecologists and forest scientists, have sparked further controversy. FOE claim that more than half Britain's beech and yew trees are showing signs of acid rain damage and that symptoms are present in at least ten other species.[26]

The Forestry Commission are still equivocal about the cause of this damage, however, after their own preliminary survey of some 20 beech woods, and tend to attribute most of the observed damage to the effects of the 1976 and subsequent droughts. There are, nonetheless, some notable shifts in the official position, with John Gibbs, the Commission's chief pathologist, conceding that air pollution damage is not ruled out as a causal factor. Forestry Commission scientists admit to being baffled by the 'unusual and quite widespread' yellowing of beech leaves which occurred in August 1985, and which does not fit into the expected pattern following the preceding wet and cool summer.[27]

The dispute has occasionally become heated. Binns dismissed Nihlgard's opinions about forest damage: 'His report bears no resemblance to the country we see around us. I find it quite extraordinary . . . we don't need to adduce air pollution.'[28] Chris Rose, then FOE's acid rain campaigner, in turn accused the Commission of choosing to survey trees whose age and location left them less vulnerable to pollution[29] (both admit that older trees, which were omitted from the Commission's survey, show greater signs of damage), and said that 'we do not know whether the Commission is genuinely ignorant of the damage or simply refuses, or is not allowed, to admit it.'[30]

A continuing source of disagreement is the role that pollution might play in weakening trees so that other factors, such as drought or disease, can have greater effects than they would on a healthy tree and be the final cause of death. In West Germany, Otto Kandler, a botanist at Munich university, ascribes the damage to a biological pathogen rather than pollution,[31] although most other scientists argue that the damage results from a more complex interaction of many factors including various air pollutants, disease, geography, climate and the ways trees have been managed. The eventual consequences for the forests of Europe are still not apparent. It is significant, however, that the main impetus for research is coming from the voluntary sector rather than the establishment bodies.

Similar changes in attitude seem likely to develop for air pollution damage to crops. It has long been accepted that high SO_2 levels can retard crops, but SO_2 pollution in Britain was not considered severe enough to make a significant impact on agricultural productivity. Recognition, however, of important synergistic effects (that is, the accelerated damage which results from the combined action of several pollutants), and the discovery of high ozone levels,[32] themselves formed largely by NO_x, have opened the possibility that pollution is already making a measurable reduction to crop yields in some areas.[33]

Acid rain's rapid rise to notoriety has brought other problems into the limelight. Deterioration of sandstone and limestone has occurred for decades in areas such as London, other large cities and the Trent valley, but recent evidence of large-scale building damage in Europe has highlighted the issue. The CEGB's failure to examine corrosion damage to cathedrals was strongly criticized in a recent House of Commons Select Committee report,[34] and corrosion to well-known landmarks such as St Paul's cathedral has already been well documented.[35] The arguments are by no means decided, however, with some researchers believing that the effects seen today are the result of historical damage rather than fresh pollution, and the issue further confused by the effects of poorly applied and unprofessional restoration, which can occasionally cause more damage to the buildings than the pollution itself. Roy Butlin of the Building Research Station is now conducting a detailed survey of three buildings – Lincoln cathedral, Wells minster and Bolsover castle – and helping co-ordinate a national monitoring system and a listing of important buildings which may be at risk.[36] Although frequently underplayed in discussions which centre on ecological damage, the implications of any pollution damage to buildings go beyond the risk to heritage. The US Council for Environmental Quality estimated that structural damage causes $2 billion worth of damage a year in the USA.[37]

Similarly, effects on wildlife are also more serious than was previously thought. Swedish research has linked acidification with the decline of fish-eating birds like the osprey,[38] and with eggshell thinning in passerine species (that is, birds of the order *Passeriformes*, including most perching and song birds of about sparrow size).[39] In Britain, the dipper is declining in areas where acidification of streams has reduced aquatic invertebrates.[40] The flora of part of the Pennines is known to have undergone major changes as a result of SO_2 pollution.[41] The Warren Spring Laboratory released a report on acidity of rainfall in 1984,[42] showing that precipitation acidity in Britain was artificially high, and the Nature Conservancy Council published a report on the wildlife effects of acid rain which are no longer seriously open to doubt.[43]

The Reasons for Britain's Response

The direction of British research is important in two ways. First, it has caused widespread doubts, both inside and outside Britain, about the impartiality of nationalized industry and the government in setting the tone of the scientific debate. It is hardly a new phenomenon to find arguments about bias in nationalized or privately-owned industries with respect to environmental issues, but these criticisms have seldom been expressed as vehemently as over the acid rain debate in the 1980s. Secondly, it is sobering to see how a major problem can develop undetected, even in a country with relatively high levels of research funding, if there is no incentive to investigate it. (Of course, the reverse is also true, and 'problems' can be found where none exist if enough people believe in them, but the evidence for acidification effects within Britain is now compelling.) It is significant that even given the amount of publicity afforded to acid rain in the recent past, it has taken visiting European scientists, and British pressure groups, to initiate several lines of research into freshwater, tree and material damage.

Some of the reasons for this are historical. Britain's introduction to the problems of air pollution came early, as did the first steps to control it, and the Clean Air Act which followed the great smog disaster of 1952 has been used as a model of legislation by many other countries. Cutting out the visible signs of pollution and building tall chimney stacks to disperse the rest undoubtedly improved urban air quality, leading to improved human health and limited recolonization of inner cities by pollution sensitive plant and animal species.[44] Having been convinced that air quality was improving, people have found it very difficult to accept that the dispersed air pollutants are having more subtle, but arguably more important, effects elsewhere.

Political Response

The rapid escalation of damage has produced a flurry of legislation in individual countries and moves towards international pollution control agreements and laws within the EEC and the United Nations. An important step was the formation of the '30 per cent Club', comprising 20 nations with a commitment to a 30 per cent reduction on 1980 SO_2 pollution by 1993. The EEC is proposing several new directives, including a 60 per cent reduction in SO_2 and a 40 per cent reduction in NO_x from large industrial and electricity generating plant by 1995 and a proposal to limit NO_x from car exhausts to Japanese levels within ten years.

During the various conferences and negotiations, Britain's attitude has become increasingly isolated. Britain is the largest SO_2 emitter in western Europe and its high stack policy has long angered the Scandinavians who argue that British SO_2 affects their own freshwaters. For many years the CEGB claimed that no British pollution reached mainland Europe. While this has now been well and truly disproved, the amounts crossing the North Sea are still disputed and the CEGB says high stacks only add 15 per cent to long-range pollution.[45] Britain's major role as a polluter is not open to doubt, however, and has produced strong European pressure for a reduction in emissions. This Britain has steadfastly resisted, blocking resolutions within the UN and European parliament wherever possible.

Initially Britain had an influential ally in West Germany. Catastrophic forest dieback resulted in a spectacular West German turnaround in 1984, however, and the commitment of $5 billion to halve the emission of SO_2 within ten years. The Germans are also reducing NO_x by compulsorily fitting catalytic converters to car exhausts and proposing similar EEC legislation. Now Britain is alone among major polluters in objecting to EEC power station legislation; the other countries to seek exemption –

Ireland, Greece and Luxembourg – do so on the reasonable basis of their small level of industrialization. While France and Italy also oppose car exhaust legislation the unilateral West German move seems to be forcing car manufacturers to fit catalytic convertors in any case.

Britain's Arguments for Delaying Pollution Legislation

Given that Britain is now effectively isolated within Europe, it is worth evaluating the arguments that the government uses to support this stance:

1   *The link between air pollution and ecological damage is unproven*

There will inevitably be disagreement with any theory, and the tiny minority of researchers who dispute any connection between pollution and acidification have been used to justify funding research rather than pollution control. A further £30 million was allotted to research in October 1984. The frustration of this attitude is summed up by a Canadian minister in 1982: 'It's a bit like saying it looks like a skunk, it walks like a skunk and it's stinking the house out like a skunk, but we're not prepared to commit ourselves that it is a skunk without four more years research.'[46] It is believed in Europe that sufficient information is now available to justify any risk in funding pollution control measures, when compared with the risk of hoping the problems will disappear, or have some other, unknown cause; Britain is still officially advocating holding off action while more information comes in, but this is looking, increasingly, more like a holding operation than realistic research.

2   *Sulphur dioxide is less destructive than nitrogen oxides or ozone*

The discovery of forest dieback at a time of steady SO_2 emissions and rising NO_x levels has prompted long arguments about their relative importance and, in turn, the decision as to which one should be reduced first. In practice, both are apparently involved in most effects and those countries taking pollution control seriously are reducing sulphur and nitrogen oxides simultaneously. This is a necessary trade-off for our incomplete knowledge of pollution pathways and effects. Given the lack of data on the separate effects of both pollutants, and the even greater dearth of data on their additive and synergistic effects, the scale of damage occurring dictates that both pollutants be cut back to minimize the effects of dry and wet deposition.

3   *Pollution free nuclear power will soon be replacing coal stations*

The nuclear power issue is frequently raised at conferences and may well be an underlying factor in the CEGB's strategy. The most wildly optimistic nuclear scenario, however, relies on an enormous coal input for many decades to come and no one in government can seriously expect a distant and uncertain switch to nuclear power to be acceptable politically as an excuse for ignoring coal pollution at the moment. Even if a new nuclear power plant were commissioned immediately, it would not be on stream early enough to affect SO_2 emissions within the timescale of the proposed EEC legislation.

4   *The relationship between sulphur dioxide and rainfall acidity is not linear*

It has been argued that the emission of hydrocarbons is the most important factor in determining the levels of atmospheric acidity, so that reducing SO_2 may not initially make very much difference. Recent research suggests that this is untrue. The reduction in Britain's sulphur emissions over the last few years has led to a measurable decrease in average Scottish rainfall acidity.[47] These experimental results back up expected figures calculated from long-range transport model.[48] It appears that reducing sulphur dioxides does indeed reduce acidity.

5   *Britain has already reduced sulphur dioxide levels by about 30 per cent since 1980*

It has, but largely because of the recession. If the promised recovery takes place, SO_2 levels can be expected to rise again. Increased use of natural gas and changing industrial fuel mix have also made an impact which is likely to last longer but the availability of low SO_2 gas is uncertain, again raising the possibility of future SO_2 increases.

There is one other crucial political factor in Britain's ability to resist legislation; the US administration has also completely ignored pleas from Canada, and from within the USA, to reduce sulphur pollution. At present Margaret Thatcher and Ronald Reagan are supporting each other and observers believe that a policy change in one country would isolate the other so much that they would be forced to introduce greater pollution control measures.

It is still by no means certain which will be the first to break. The CEGB's chairman, Lord Marshall, responded to the House of Commons Select Committee report advocating greater pollution control by saying that he had 'never seen a select committee report where the written statement so clearly contradicted the evidence they received',[49] and the government duly rejected the report's recommendations in December 1984.[50] A cabinet committee did approve the new EEC vehicle exhaust standards in summer 1985, albeit reluctantly, after William Waldegrave, Minister of State for the Environment, had gone beyond his brief by agreeing to them in Brussels.[51] This agreement does not appear to be legally binding, however, and Britain shows signs of ignoring it for the time being at least.

There was a notable set-back for hopes of an early British commitment to pollution reduction when Britain joined the USA and Poland in refusing to sign an emission abatement protocol under the Convention on Long Range Transport of Air Pollutants, held under the auspices of the United Nations Economic Commission for Europe, and agreed by 18 other countries from western and eastern Europe and Canada. The protocol binds signatories to reduce emissions or 'transboundary fluxes' of SO_2 by 30 per cent of 1980 levels by 1993. The decision seemed unnecessarily contentious, as the government claims that we have already reduced emissions by 25 per cent from the 1980 baseline. Refusal appears to be a backlash against the car exhaust decision, which struggled through cabinet because Sir Geoffrey Howe, who had promised to support the Department of Environment in proposing that Britain sign, failed to attend the meeting.[52]

Meanwhile, the scientific battle of words parallels that taking place in parliament. In a lecture to the Royal Institution, Peter Chester of the Central Electricity Research Laboratory appeared to suggest that fish only survived in Scandinavian lakes in the first place due to the neutralizing effects of calcium dust in air pollutants, and that the removal of this had paved the way to acidification. 'It is quite possible that fifty years ago, our production and export of neutralising mineral dust, including calcium, equalled or even exceeded our production and export of sulphate . . . it is interesting to speculate whether the low levels of calcium in (southern Norway) lakes could in any way be related to . . . environmental improvements'.[53] This seems an extraordinary statement to make in the light of the CEGB's claims that very little pollution reaches Norway (quite apart from the other dubious scientific principles contained within it), and will inevitably enrage the Scandinavians even more.

Also in late 1985 the CEGB released, and vigorously promoted, a film called *Acid Rain*, which it claimed 'examines the complexities and uncertainties surrounding acid rain. It demonstrates the need to investigate every possible route for an effective solution.'[54] The Norwegian government is considering making a formal complaint to the British government about the film and Professor Hans Seip, a leading expert on acid rain, says 'the film gives a very unbalanced and biased view of the effects of acid precipitation.'[55] Our own Department of the Environment has also apparently distanced iself from the film.

Meeting the EEC requirements on large plant pollution would add about 4 per cent on to electricity costs,[56] a figure which the CEGB has not denied. There is also scope for reduction by way of a more comprehensive energy conservation policy. Despite the outward appearance of a hard line, there are sharp divisions within the Conservative Party about this and both Labour and the Alliance are committed to reducing SO_2 pollution. The issue has also been significant in the fissures it has opened up within the scientific community. Environmental campaigners, used to being out on a limb, have had the rewarding experience of hearing academics and civil servants castigating some of the CEGB scientists, albeit off the record. The speed with which West Germany changed its policy on SO_2 and NO_x pollution may be an indication of how fast Britain will move when pressure builds up enough, and it is widely thought that the current attitude disguises a delaying tactic to avoid fitting scrubbers on to old plants with little working life left. Future developments will depend on how far the various pressure groups are prepared to push. As one government official said: 'The evidence for taking action on lead was not nearly as good as that for acting on acid rain.'[57]

Notes

1. National Coal Board Press Release, '£5 million "acid rain" research', 5 September 1983.
2. News conference for British journalists, September 1983.
3. Anon., 'Britain is Buying Time', *Acid News*, October 1983.
4. Steve Elsworth, *Acid Rain*, London: Pluto Press, 1984.
5. John Evelyn, *Fumifugium*, 1661, reprinted by the National Society for Clean Air, 1961.
6. R. A. Smith, *Air and Rain, Towards a Chemical Climatology*, London: Longman, Green and Co. 1872.
7. Roger Lines, *Acid rain and forestry in Scotland*, TGS/NRS meeting, 12 November 1982.
8. D. L. Hawksworth and Francis Rose, *Lichens as Pollution Monitors*, London: Edward Arnold 1976.
9. D. W. Sutcliffe and T. R. Carrick, 'Studies on mountain streams in the English Lake District 3, Aspects of water chemistry in Brownrigg Well, Whelpside Ghyll', *Freshwater Biology*, no. 3, 1973, pp. 561–8.
10. Richard E. Wright and Arne Henriksen, *Regional survey of lakes and streams in southwestern Scotland, April 1979*, SNSF Intern rapport IR 72/80, Oslo; Sur Nedførs Virkning På Skog Og Fisk, 1980.
11. Geoffrey Lean, 'British lakes poisoned by acid rain', *The Observer*, 18 September 1983.
12. Richard W. Batterbee, 'Diatom analysis and the acidification of lakes', *Philosophical Transactions of the Royal Society*, series B 1985. Diatoms are unicellular algae with a silaceous cell wall which is preserved in the sediment of lakes. Diatoms change in type and proportion of species present in water of differing acidity; therefore an analysis of the diatom cases present in a core of lake sediment can allow a picture of changes in acidity over time to be built up.
13. J. H. Stoner, A. S. Lee and K. R. Wade, 'The effects of acid precipitation and land use on water quality and ecology in the upper Tywi catchment in West Wales', Welsh Water Authority, 1984.

14. R. F. Prigg, 'Juvenile salmonid populations and biological quality of upland streams in Cumbria with particular reerence to low pH', North West Water (Rivers Diversion, Scientists Department, Biology North), BN 77–2–83, 1983.

15. G. L. A. Fry and A. S. Cooke, *Acid deposition and its implications for nature conservation in Great Britain*, Shrewsbury: Nature Conservancy Council, 1984.

16. Fry and Cooke, *Acid deposition*.

17. Batterbee, 'Diatom analysis'.

18. Stoner et al., 'The effects of acid precipitation'.

19. H. Leivestad and I. P. Muniz, 'Fish kill at low pH in a Norwegian river', *Nature*, no. 259, 1976; pp. 391–2.

20. Prigg, 'Juvenile salmonid populations'.

21. Matt Mundell, 'Acid snow hits fish farm', *The Galloway News*, 28 October 1982.

22. W. O. Binns and D. B. Redfern, 'Acid rain and forestry decline in West Germany', *Forestry Commission Research and Development Paper*, no. 131, 1983.

23. W. O. Binns, 'Acid rain and forestry', *Forestry Commission Research and Develoment Paper*, no. 134, 1984.

24. Comments made at Scottish Wildlife Trust Conference, September 1984. Quoted from Friends of the Earth acid rain briefing paper no. 1 by Claire Holman and Chris Rose, December 1984.

25. Chris Rose and Mark Neville, 'Tree Dieback Survey – an action guide', London: Friends of the Earth, 1985.

26. Chris Rose and Mark Neville, 'Final Report: Tree Dieback Survey', London: Friends of the Earth, 1985.

27. Anon., 'Air pollution and trees', *ENDS Report*, no. 128, September 1985.

28. Roger Milne, 'Fighting over the corpses of trees', *New Scientist*, 29 August 1985.

29. Milne, 'Fighting over the corpses of trees'.

30. John Ardill, 'Tree survey warns of acid rain disaster', *Guardian*, November 1985.

31. Anon., 'German tree deaths blamed on virus', *New Scientist*, 15 August 1985.

32. Anon., 'Crop damage linked to ozone interaction with acid rain', *ENDS Report*, no. 101, 1983, pp. 6–7.

33. Environmental Resources Ltd, *Acid Rain*, London: Graham and Trotman, 1983.

34. House of Commons, *Fourth Report from the Environment Committee session 1983–84, Acid Rain, Volume 1*, Report together with the Proceedings of the Committee relating to the Report, 30 July 1984.

35. John E. Yokum and James B. Upham, 'Effects on economic materials and structures' in Arthur J. Stern (ed.), *Air Pollution*, London: Academic Press, 1977.

36. Fred Pearce, 'Acid eats into Britain's stone heritage', *New Scientist*, 26 September 1985.

37. John Fox, 'Acid rain – acid tears', *Canadian Consumer*, no. 13 pp. 7–13

38. Mats O. Eriksson, Lennart Henrikse and Hans R. Oscarson, unpublished paper from University of Gotteborg, 1983.

39. N. Erik Nyholm, 'Evidence for the involvement of aluminium in causation of defective formation of eggshells and of impaired breeding in wild passerine birds', *Environmental Research no. 26*, pp. 363–71.

40. Roscoe Howells, 'The effects of acid precipitation and land use quality and ecology in Wales and the implications to the Authority', Welsh Water Authority Memo, 1983.

41. Patricia Ferguson and John Lee, 'Past and present sulphur pollution in the Southern Pennines', *Atmospheric Environment, 17*, pp. 1131–7.

42. Warren Spring laboratory, *Acidity of Rainfall in the United Kingdom – final report*, London: HMSO, 1984.

43. Fry and Cooke, *Acid deposition*.

44. C. I. Rose and D. L. Hawksworth, 'Lichen recolonisation in London's cleaner air', *Nature*, no. 289, pp. 289–92.

45. P. F. Chester 'Perspectives on acid rain study', *Royal Society of the Arts Journal*, 81 pp. 587–603.
46. David Fishlock, 'CEGB plans £30m acid rain study', *Financial Times*, 10 October 1984.
47. Ron Harriman and David Wells, 'Causes and effects of surface water acidification in Scotland', paper given to the Institute of Water Pollution Control Conference, reported in *ENDS Report*, no. 116, September 1984.
48. B. Fisher and P. Clark, 'Testing a statistical long range transport model on European and North American observations', 14th International Technical Meeting on Air Pollution Modelling and Its Implication, Copenhagen, September 1983.
49. Andrew Gowers, ' "Acid rain": now the debate moves to centre stage', *Financial Times*, 7 November 1984.
50. Department of the Environment, *Acid rain: the Government's reply to the Fourth Report from the Environment Committee*, December 1984.
51. Anon, 'Acid rain Meeting', *ENDS Report*, no. 126, July 1985.
52. 'Acid rain Meeting'.
53. Fred Pearce, 'When pollution helps Norway's lakes', *New Scientist*, 31 October 1985.
54. Text of an advertisement used by the CEGB.
55. Anon., 'Norway is angry about film on acid rain', *New Scientist*, 31 October 1985.
56. Nigel Dudley, Mark Barrett and David Baldock, *The Acid Rain Controversy*, London: Earth Resources Research, 1985.
57. Gowers, 'Acid rain'.

9

The Mackerel Massacre

David Harris MP

Alfred John – 'A. J.' – Pengelly is the father figure of Cornish fishing. For 65 years, he has been involved in the industry and has seen boom times and bad in those fishing communities, like his own port of Looe, which are part of the county's heritage. He is a wise, well-travelled (from the days when he was a crewman on Sir Thomas Lipton's *Shamrock* during the America Cup races), quiet man, who shakes his head with sadness as he contemplates the sorry tale of the mackerel off the south west.

It is a saga which epitomizes the inability of successive governments – and now the EEC – to act with speed or effectiveness when fish stocks are threatened. Although this particular story concerns one species and is dramatic because of the sheer quantity of fish caught in a short space of a few years, it illustrates how difficult it is to get a policy based on conservation. Too often, it is too late.

This is how Mr Pengelly describes what has happened:

The facts are that, in 1966, we renewed our handline fishing for mackerel using a few boats from Mevagissey and Looe. Over the following 10 years, a considerable fleet built up in Cornwall with so much catching power that overseas markets had to be found; the prospects looked good to us.

In the winter of 1977/78, midwater trawlers from the South West ports began to appear. At first, the pilchard shoals were attacked. Not content with that, these trawlers began to fish mackerel with success. This encouraged the large trawlers from Hull and the bigger craft from Scotland. It turned from a reasonable fishery to a massacre of fish stocks, with hundreds of tons going to grub meal. In selecting the fish for size – in order to land the best fish – the sea bed was strewn at times with dead, undersized fish. Far from not harming the Cornish handline fishermen, this over-fishing has driven most of us out of the mackerel fishing.

In these last five years, almost all our fishing fleet at Looe has been replaced with craft fitted for bottom trawling so we can secure a living. In turn, this has put a strain on the stock of bottom fish, with too many craft at the same kind of fishing.

Fishing authorities and related interests have to learn to work with Nature and not against it. Modern methods of fishing are so scientific and organised that no stock of fish can survive except through strict conservation.[1]

No reasonable person can doubt that Mr Pengelly is right when he talks about the 'massacre of fish stocks' off Cornwall which has taken place in recent years. That has been obvious even to the layman. In recent winters, one had only to stand on Pendennis Point at Falmouth and look out at the incredible scene in Carrick Roads to the left and the bay to the right to realize that fishing on an enormous scale was taking place. It was not unusual to have a fleet of some 30 or so factory ships – Russians, Romanians, Bulgarians, East Germans, Polish and Egyptians – working flat out to handle the mackerel being caught by over 60 Scottish and East Coast trawlers and purse seiners. The frantic activity went on at night with the processors – the 'klondykers' – all ablaze with lights. For a few weeks each year, Falmouth took on the appearance of a frontier town, as off duty sailors from the factory ships came ashore (sometimes under the watchful eyes of their commissars) to sample our consumer society. Men in fur hats returned from Truro or the Falmouth shops with bulging plastic bags, while luxury coaches and chartered aircraft took Scottish fishermen up north for the New Year or a home weekend break.

It was a fascinating sight, and good television for the regional news programmes. But the local fishermen looked on with anger which turned to desperation. At first, no one seemed to be listening to their warnings. Then, the authorities themselves – particularly the ministry scientists – became alarmed. But by that time, conservation and commercial interests were in open conflict. The result was an unsatisfactory compromise.

The Extent of the Massacre

The Ministry's own statistics give an idea of the rise and fall of the south-west mackerel fishery. In 1970, only 100 tonnes of mackerel were landed at Falmouth. By 1975, the figure was just over 7,000 tonnes. In 1979, landings and transhipments to the factory ships totalled 153,000 tonnes. The figure was down to 34,000 tonnes in 1983, and, after that, it was minimal.[2]

But the official figures only tell part of the story. Despite attempts by MAFF inspectors to monitor the transhipments, no one really knows how much mackerel was caught off Cornwall in those 'boom' years. First, as

Mr Pengelly says, the amount of mackerel landed or sold to the klondykers was only part of the catch. Untold amounts of small fish were dumped on the sea bed in recent winters, thus reducing future prospects. Moreover, in the early years of the 'bonanza' when there were completely inadequate checks on transhipments, everyone knew that sales to the factory ships were far in excess of the amounts recorded. True, there was a belated tightening up on inspections (a major step forward was the licensing of the klondykers and a requirement for them to provide details of loads purchased), but, even now, a combination of quotas and tax liability might not result in completely accurate returns by the fishermen in the Scottish waters where transhipments continue. A letter I received from a radio operator who worked in the Scottish fishery said that his first instruction was never, never to give tonnages handled.

Nor, of course, was the fishery confined to Falmouth and its surrounding waters. Other south-west ports also played a part in the mackerel bonanza, particularly Plymouth, while there was the similar, although earlier in the year, Scottish fishery centred on Ullapool. Less obvious, but more serious in that it has been subjected to little or no control, has been the bulk catching of mackerel by EEC boats in western waters outside Britain's territorial limits.

By the early 1980s, the alarm bells were really ringing. One of the clearest warnings came from Dr Stephen Lockwood, the scientist who specialized in mackerel at MAFF's fisheries research centre at Lowestoft. In 1981, in his assessment of the western mackerel stock to the south west and west of Britain, Dr Lockwood wrote:

In 1960, the landings from the area west and south of Britain were 50,000 tonnes, and even 10 years later they were only 70,000 tonnes, but in the intervening years two changes had occurred. One was the arrival of the Eastern Bloc fleet and the other was the change in the mackerel's overwintering habits. By 1975, the Eastern Bloc was taking more than 300,000 tonnes of mackerel. Prior to 1966, the only winter mackerel fishing was by drift nets either in mid-Channel south of Plymouth or well to the west of Land's End. During 1966, the summer handline fishery continued into the autumn and throughout the winter of 1966–67 as dense shoals remained in west Eddystone Bay. The real causes of this change are not known but it marked the beginning of the English winter mackerel fishery which reached its peak in the late 1970s when over 300,000 tonnes were landed – thereby equalling the Eastern Bloc effort which had left the area following the decision of European Community member states that they would extend their fishery limits to 200 miles.

Other nations' catches have also increased during recent years; notably those of the Netherlands, Ireland, Denmark and the Federal Republic of

Germany. This overall increase in effort has pushed the international landings to over 600,000 tonnes in the past two years, almost double the recommended level! As a result, the stock has declined from about 3 million tonnes to less than 2 million tonnes in the four years 1977–80 and inevitably the recommended Total Allowable Catch has become less.

It is vitally important for the continued existence of the Western stock and its dependent fishery that the international catches are brought into line with the scientific advice. Failure to do so can only result in the continued succession of decreasing TACs until the inevitable advice is given for a ban on mackerel fishing west of Britain.[3]

Dr. Lockwood then presented some startling figures on the western mackerel stock, reported catches, and the probable catch as estimated by a discarding (see table 6).

Table 6   Western mackerel stock, reported stock and probable catches

	1978 (weights in thousands of tonnes)	1979	1980	1981
Spawning stock	2,562	2,258	1,786	1,500
Total allowable catch (TAC)	450	435	330	353
Reported catch	504	601	604	589[1]
Probable catch	550	660	660	640[1]
% excess on TAC	22	52	100	81

Note: Working group of scientists' estimate of reported catch 'in the continued absence of enforcement'.

The Official Response

The government and the EEC tried to control the fishery by stepping up their efforts to monitor transhipments and landings, by shortening the season for bulk catching and by introducing, in 1980, a mackerel 'box' off and around part of Cornwall and stretching out into the western approaches off Land's End. This was an attempt to protect the young mackerel and it represented a conservation area which was open for a comparatively short period only at the height of the winter season.

But, by 1981, Dr Lockwood and his fellow scientists on ICES, the International Council for the Exploration of the Sea, were saying that stocks would continue to decline unless new protection was given to

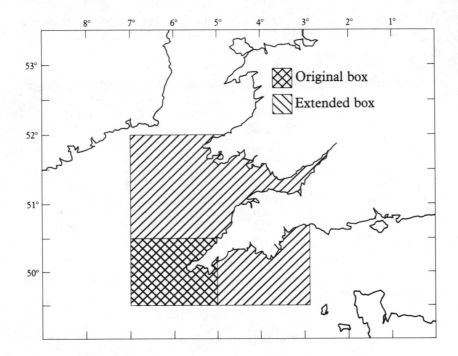

Figure 5 The extended south-west mackerel box

immature stocks in the 'nursery area' off Cornwall. They said that measures introduced under the 'international management regime' had not been adhered to in practice. A disturbing new factor was that the shoals of big, mature mackerel – the 'jumbo' fish – which had begun to overwinter off Cornwall in the mid-1960s, and which gave rise to the phenomenal upsurge in fishing, had moved away. In 1980, they were only caught in large quantities west of Ireland, mainly by the Dutch and the Irish. The fish caught off Cornwall were largely young mackerel.

This led Dr Lockwood to write: 'Cyclical changes in pelagic fish distributions are something that occur world-wide and there is nothing that can be done to alter them, but the bonus of abundant immature fish (1978 and 1979 had been good spawning years) can be used to safeguard the future if they are protected.' He said that this protection must be given by further reducing the fishing effort off Cornwall or by introducing other measures.

The outcome of the concern in ICES was the enlarged mackerel box introduced in November 1983 (see figure 5). Instead of starting near Falmouth, the eastern boundary of the new box was pushed out to a point near Lyme Regis, and the protected zone now covers 32,500 square miles, instead of the original 6,500 square miles. The huge area almost extends to the Channel Islands, goes right down to the western approaches west of the Isles of Scilly, across to Ireland and back to Pembroke. Our small, local fishermen saw the move as the salvation of the mackerel stocks as purse seiners – the real killers – were banned completely during the closed seasons and trawling for mackerel was also to be prohibited, with one exception.

It is that one exception – bottom trawling – which provided a later twist in the story. A few large trawlers began to catch mackerel inside the new box in some quantity, and claimed that they were 'bottom trawling'. Others quickly joined in, making an absolute nonsense of this conservation measure. Pseudo bottom trawling was a blatant dodge.

This particular episode sums up for me the difficulty in getting the authorities to move in time. To my simple mind, there seemed to be a clear-cut case for making, as a matter of urgency, an amendment to the EEC regulation on the box to close what had become the bottom trawling loophole. Surely, if the box was necessary to protect the juvenile stock then there should be an immediate removal of a derogation which undermined the effectiveness of the measure?

Conflicting Interests

As the then Member for Cornwall and Plymouth of the European Parliament, I took the case for this direct to the Fisheries Directorate

General in Brussels. Two Cornish handline fishermen, Robin Hadlow and Robbie Curtis, came over with me to the Commission headquarters (their colleagues from all the ports around Cornwall had raised the money for their fares) to argue the need for an amendment. With other MPs from Cornwall, I also put the case in the Commons and was opposed vigorously by Scottish Members. Of course, there was, as always, a conflict of interests. The Scottish big boats wanted access to what had become a useful winter fishery for them, once the autumn and early winter fishery of the Minches was over.

The precious few processors of any size we have in the south west also wanted, quite naturally, the benefit of a cheap bulk supply of mackerel on their doorsteps. But the hand-liners, who had been given an authorization to continue fishing in the box as it was considered that their traditional method posed no serious threat to stocks, pressed strongly for an end to the bottom-trawling derogation, pointing out that mackerel are not a fish normally caught on the sea bed and claiming that the trawlers were really 'mid-watering' with their trawls well above the bottom.

Lobbying was intense, both in Whitehall and Brussels. The main recipient of the conflicting 'advice' and representations was John MacGregor, the then fisheries minister. In the Spring of 1984, the matter was on the agenda of a Council of Fisheries Ministers in Brussels but no decision was taken. It came up again in September when the Commission proposed an amendment which would deal with the loophole and which it wanted to come into force in October. Mr MacGregor argued instead for a 'limited fishery' inside the box for British boats only. But, quite understandably, our EEC partners were not having that. He then successfully pressed for the Commission's amendment not to be implemented until 1 January 1985, and for the ban on bulk catching in the box to last until the beginning of 1987, and then be reviewed.

That postponement until New Year's day, 1985, pleased no one. The processors and trawler owners were screaming that they faced disaster once the box was completely closed to them, while the hand-liners reckoned that further damage would be done in the meantime if licences were granted for bulk catching inside the area up to the New Year deadline. The 1 January decision was a classic political compromise, but it can hardly be said to have been based on principles of good conservation.

And that has been the trouble with so many governmental decisions – European or national – concerning fishing. Of course, a minister or a Commission official has a number of factors to take into account. Scientific advice is not always conclusive, and, in any case, has tended to alter with the passage of time. But, on top of that, there are economic, social, legal, and, yes, political issues to be weighed in the balance. There are some

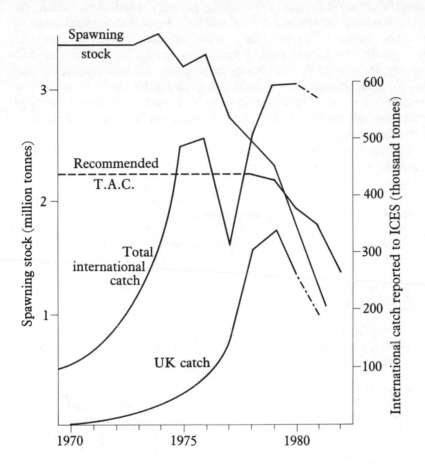

Figure 6 Changes in the western mackerel spawning stock, weight, reported catch and recommended total allowable catch (TAC)

Source: Fishing Prospects, MAFF, 1982.

powerful pressure groups in the fishing industry (particularly among the Scottish pelagic fishermen), and, in addition, there is the natural desire of member states to protect their own national interests. This was particularly true of the long, long period during which the Common Fishery Policy (CFP) was being negotiated. All too frequently, the difficulty of accommodating conflicting claims has been 'resolved' by creating 'paper fish' through the granting of unrealistic quotas and inflated total allowable catches (TACs). That is no answer at all. On the contrary, it is a dangerous folly.

In January 1983, the Commission's expert advisers, the members of the Scientific and Technical Committee for Fisheries, said:

> The attention of Commission services is again drawn to the following points with respect to the western mackerel stock.
>
> (a) All informed scientific opinion is in agreement that the spawning stock biomass has shown a continued decline over recent years.
>
> (b) Recommendations in recent years have been made with the intention of *halting not reversing* this decline.
>
> (c) Despite this, recommendations for TAC have been consistently and vastly exceeded in practice and therefore the decline in spawning stock size has continued unabated.
>
> In the context of the latter statement the Committee wishes to express its concern that last year, despite the fact that the Commission was explicitly informed that on scientific grounds the western mackerel TAC should on no account exceed 272,000 tonnes, the actual recommendation to emerge was for a TAC of 375,000 tonnes. The latter value was then greatly exceeded in practice.

They can say that again.

Even though the box has brought some temporary relief for mackerel stocks off the south west of England, the species has continued to be persecuted at an alarming rate in other waters. By 1985, the ICES mackerel group was recommending a cut in the EEC total allowable catch from 375,000 tonnes to 250,000 tonnes. But it was also extremely worried that around 80,000 tonnes of western mackerel were being taken in the Norwegian Sea and south of the Faroes – that is, outside EEC waters where no catch limits applied. They gave a warning that to take a catch of 500,000 tonnes in 1985 would increase the fishing mortality rate on the western stock to about twice the level advised by ICES and would drastically reduce the spawning stock biomass. And so it goes on.

Will We Ever Learn?

Perhaps it is unfair to blame individual ministers for what is largely an inherited situation. For one thing – and this cannot be good – there has been a rapid turnover in British fishery ministers. Within the space of less than two years, Alick Buchanan-Smith had handed over the portfolio to John MacGregor who, in turn, passed on the responsibility to John Selwyn Gummer in the 1985 government changes. All three were recognized to be able ministers – MacGregor was promoted to the cabinet as Chief Secretary to the Treasury from the fisheries post – but the chopping and changing cannot be good. What is more, the fishery minister is the senior Minister of State in the Ministry of Agriculture, Fisheries and Food and, as such, has a wide range of responsibilities in the farming field, including doing most of the day-to-day departmental work on the Common Agricutural Policy. To make matters worse, there is also quite a turnover in the ranks of the senior civil servants working on fisheries policy.

This is in contrast to the position in some European countries. What has to be recognized and appreciated is that the main political and administrative decisions affecting the management of fish stocks are now made in the EEC. Persistence and patience are the qualities needed to make sure the right decisions are taken in that forum, and that means working over a long period of time. No new minister – however brilliant – can expect to go to the European Fisheries Council and bring in the necessary changes overnight. But what the ministers on the Fisheries Council must do – and I would hope Britain would take the lead – is to have a deep rethink on policy and practice. There is a dreadful tendency to think that having established the Common Fisheries Policy, there is no need to tackle the root cause of the problem, the built-in overcatching capacity of the deep water fleet. We are not only talking about the build-up of the Scottish purse seiners of a few years back or the dramatic increase in the number of 'beamers' in the south west of England, but of the development, against all sensible thinking, of the nomadic fleets of other European countries. There will be even more pressure for 'fishing opportunities' with Spain, with its huge catching capability, as a member of the EEC.

We saw what happened to the herring. Over-fishing brought it to the verge of commercial extinction, and a complete halt had to be called to enable stocks to recover. The boats then turned on the mackerel, and we have had a similar story. The fact is that nearly all traditional species within easy reach are under intense pressure.

Conservation must be paramount in these conflicts. Yet some of the policies actually mitigate against conservation. Thank goodness, it is a long time since government grants were given towards the building of large purse seiners – but there is still a bias towards the larger boats so far as European grants are concerned. The EEC's scheme for 'withdrawal prices' encourages fishermen to catch fish which the market does not want and which end up as fish meal, so reducing a finite resource; while the appointment of a mere 13 EEC inspectors to attempt to 'police' the enforcement of the Common Fisheries Policy is ludicrously inadequate.

The dream I have is of a balanced fleet with a good spread of processors on shore in fishing regions where jobs are badly needed, adding to the value of the catch. Indeed, we need to build a strong regional dimension into the outline structure of the Common Fisheries Policy. A system of 'regional preference' under which local boats were given greater access to restricted stocks than those given to big, go-anywhere, bulk catchers from other parts of the country would, in my opinion, make sense. After all, local fishermen tend to look after their stocks as their future livelihoods depend on them. Regional preference would help the local processors to overcome the undoubted difficulties posed by such measures as the mackerel box, measures which are really acts of desperation on the part of the authorities who have failed to take sensible steps at the right time to control over-fishing.

The basic problem is easy to define: too many boats – or, rather, too much catching capacity – chasing too few fish. (At the time of writing the latest addition to the Scottish fleet is the biggest purse seiner in Britain, a 190 feet long vessel with a hold capable of taking 1,100 tonnes of fish. Her new owners intend to use her to fish for mackerel and blue whiting.) Modern technology – not least in the echo-sounding field – makes the big boats deadly hunters. Some say they are 'efficient', others that they are 'greedy'. What cannot be argued about is the effect they have had on stocks. Yes, we do need some bulk catchers but their activities have to be controlled in a sensible way and in good time. The fleet must be tailored to the stocks because the stocks cannot be tailored to the fleet.

It is a tragedy that the opportunities presented by the south-west mackerel fishery were missed through over-exploitation, and through the failure of the authorities to check this overfishing in time. It has been a textbook lesson in how not to run a fishery, with crazy intense bulk catching seasons allowing so many fish to be caught in such a short space of time that the only way to handle the catch is to call in the factory ships. Like others before it, that particular bonanza is over. A few fortunes have been made, and stocks have undoubtedly been damaged. No wonder Alfred John Pengelly, and many others like him, shake their heads as they wonder whatever happened to the mackerel.

Notes

1. Letter, *Western Morning News*, 22 September, 1984.
2. Figures given to author by MAFF.
3. Ministry of Agriculture, Fisheries and Food, *Fishing Prospects 1982*, Lowestoft: MAFF, Directorate of Fisheries Research.

10

Down on the Pharm

Alan Long

Nowadays, longevity is an unlikely prospect for a farm animal. Harsh economics imprison the farmer almost as much as his animals in a highly-capitalized, intensive system that dispenses with the tender, loving care of the kindly stockman and depends instead upon torrents of 'farmerceuticals' (plied by our new archetype, Pharmacist Giles) in order to extract maximum yields. That reliance on drugs is aided and abetted by a minority of unscrupulous vets and encouraged by reps from the drugs industry, who masquerade in the dissembling guise of purveyors of 'animal health products'.

Production Diseases

Britain's animal production industry effectively consists of enterprises devoted to 'finishing' fatstock, with stockmen doing their utmost to fatten the more lucrative cuts of meat from the hindquarters, and dairy farmers manipulating reproductive processes to obtain maximum yields in gallonages of milk and fecundity in calves. Pig producers exploit the sow in the same way. Such 'brinkmanship' farming methods, in what is verily a bum-and-tit industry, push the animals' physiological resilience to the limits – or beyond. Any good stockman knows that animals need space if they are to escape the ravages of infection. He would not be surprised that bizarre feeds and breeding tricks, zero grazing, harsh conditions of housing and standings, a lack of straw and feet paddling in evil slurries would invite retribution: the modern dairy cow does not last into her fourth lactation – 'production diseases' such as mastitis (an infection of the lining of the udder, which introduces clots and blood in the milk from a catarrh-like discharge), reproductive failure, lameness, and fatty liver take her to an early cull. This relentless industry 'burgers' cows prematurely; likewise, the 'pig-improvers' have consigned sows to five hectic farrowings in two years, after which they are carted off for sausage-meat.

In traditional farming the calf would run with the dam until, between six and nine months of age, the pair would be separated, at a natural time for weaning. The youngsters would be fattened on grass for a couple of years more before they had the finish that the butcher would expect of typically beefy breeds – Aberdeen Angus, Hereford, Galloway, Highlanders and Longhorns. The male calves were castrated before they were weaned, for several reasons: to ensure the farmer had them for meat rather than the reverse; to induce a eunuchoid growth, in which 'marbling' fat laid down in the muscles made for succulent cuts and joints; and to avoid the 'sex taints' – actually the pheromones that would arouse cows in natural mating and intercourse – associated with mature males.

Such systems survive, especially in highlands and other disadvantaged areas, with plenty of room for free range. They depend on heavy subsidies however. The milkmaids are no longer prepared to go into the fields to take the milk left by the calves, so the farmer might buy calves at the market, thus raising their cows to become 'multiple sucklers'.

Beefing-Up Britain's Cattle

Modern developments have taken account of the human desire for milk, and the slowness, cost and low yields of traditional methods; still more recently the demand for lean meat has increased for several reasons.

Traditional dairy breeds, such as Channel Islanders and Ayrshires, produce unappealing beef; therefore, the calves command a poor price – and indeed many of the males are knackered to feed packs of dogs kept for hunts. Dual-purpose breeds were developed in an attempt to reconcile 'beefy' and 'milky' traits and the arrival in Britain early in the century of the Friesian in this role, and of the Holstein as a crossing breed for dual-purpose rearing, transformed the picture. These changes aroused some misgivings: the 'unpatriotic' Holstein influence, for instance, was kept quiet during the First World War, and further introduction of 'exotic' breeds, mainly from the Continent, came with a rush as methods of artificial insemination (practicable for cows but not for sows or ewes) were improved. Health regulations were circumvented as semen was smuggled in during the rush to 'beef-up' Britain's dairy-herd, so that the male progeny and unwanted heifers could be reared to a finish that would attract subsidies and good prices from the butchers at the livestock marts.[1]

These changes, reinforced by a later disenchantment with creamy milk, saw the demise of Channel Island and Ayrshire herds and a change to the 'black-and-white' (a few red-and-whites were produced as well) – the milky Holsteins and dual-purpose Freisians, crossed with Charolais, Limousin, Simmental and Belgian Blue sires. Farmers could now buy from the Milk Marketing Board straws (of semen) of the 'bull of the day' or spend a bit more to nominate a sire chosen for the purposes of their own husbandry. Resort to a farm bull was a cost most farmers could thus avoid; consequently, the bull was dispensed with.

These breeding games were attended by problems. Friesian and Holstein cows have narrow birth canals; delivery of a beamy calf with good beef potential could therefore invite difficulties, the calf 'getting stuck' despite hefty tugs, even from a tractor, on the ropes secured to it. In a very recent development, used especially with Belgian Blue 'culards' (which, being double muscled, are very beamy and beefy), caesarean births have become the norm, although the cost of the vet has to be reckoned with. Suggestions have been mooted for rearing cows as 'disposables' – the caesarean 'birth' and slaughter of cow and calf being accomplished by the butcher. In another attempt at reducing the complications of calving (dystokia), the cow is induced with drugs to give birth prematurely, the slightly under-sized calf then being resuscitated with more drugs and special nutrient formulas devised for such 'prems'.

Enter the Animal Technicians

Embryo transfer has been hailed as the greatest advance in farming since Artificial Insemination (AI). It obviates the delay of gestation after a valuable cow has been inseminated by AI. At present, farmers must wait nearly ten months before a calf can be born (normally a singleton) and then another couple of months before they can bring the dam back to oestrus and readiness for the next conception. The 'calving index' is thus just over a year.

With embryo transfer, fertilized embryos are flushed from a donor cow who has by then contributed all the genetic material the farmer requires. She and a recipient cow (of nondescript genetic attributes) having been synchronized by drugs in their oestrus cycles, the embryos are implanted in the cheaper cow to act as the foster-mother in the prolonged period of reproduction; correspondingly, the valuable cow can be quickly reinseminated and, what's more, be induced by drugs to produce many embryos at each flushing, some of which may be kept in the freezer for subsequent use.

Drug manufacturers have been increasingly anxious to circumvent such controls as exist on the use of such artifices by vets and farmers; they promote sales of drugs that can be used on the farm, preferably without prescription or with the least participation by vets, whose attentions are costly. Although emasculation of livestock confers some advantages to the farmer, it is a cruel mutilation, causing setbacks in growth and adversely affecting feed conversion rates. Several drug-induced endocrinological possibilities therefore arise: castration-by-drug; drug-induced 'beefing-up' of heifers and cast cows, as well as of bullocks (and other castrates); and the suppression of the randiness of entires, while preserving their advantage as producers of lean meat. Some of the sex hormones used to achieve these ends are so powerful that workers in the factories manufacturing them were affected, men developing breasts and women suffering menstrual irregularities. While research laboratories and factories can take special precautions, control in farming conditions is far more difficult to exert.

The drug industry has availed itself of other discoveries. Certain drugs, originally used for purely therapeutic purposes, have also been shown to boost growth. This has led to the practice of plying such substances at sub-therapeutic doses continuously to apparently healthy stock. Pigs responded exceptionally to copper compounds, which were toxic to sheep – so much so that pastures treated with pig manure could intoxicate the sheep which were later run on them. Monogastric animals grow faster for a while if their rations are dosed with arsenical drugs like those once used in the treatment of syphilis. Further, these treatments can be administered to whole flocks by additions to the feed or drinking water; Salsbury Laboratories, for instance, advertised its Roxarsone with multipage advertisements, the first showing a tap with the text 'Medication on Tap'. Farmers have had easy access to these drugs, but under pressure from the EEC the British government has tardily imposed some restrictions on the use of copper boosters in pig-feed.

Drugs for Everything

During and just after the Second World War, the fermentation industry began producing the newly-discovered antibiotics and discarding spent liquors, broths and fungal felts into the pool of by-products that came to be included in animal feeds. Subsequent growth of stock was greater than could be accounted for by the content of nutrients or by the therapeutic benefits to sickly animals of residues of the anti-bacterial substances. For farmers the benefits were enormous: no longer did they need to segregate poor doers, treating and nursing them as necessary, possibly having to pay a vet's fee. Instead, they could both medicate whole flocks and herds and boost their growth by additions of antibiotics to the feed, often in conjunction with implantation of drugs with hormonal effects.[2]

The farmer has access to many powerful drugs, with little supervision, to deal with the ills rife in modern methods of husbandry. Some of the drugs, such as Levamisole (administered to combat infestations with parasites), should be used in medical treatments only with great caution, normally under a specialist's supervision, because they may affect the immune system. These drugs and vaccines are administered to livestock as boluses, drenches, injections, implantations, dips and sprays, with no systematic recording of the treatments. Injections with contaminated needles can start infections, which call for subsequent administrations of antibiotics. It is not unusual for ewe carcases to bear as any as 20 injection scars or abscesses when they are inspected in the slaughterhouses.

The treatments are mainly intended to counter bacterial, viral and fungal infections, and parasitic infestations (mange, lice and worms). External infestations – warbles in cattle, for instance, and scab in sheep – invite the use of dips and sprays, some manifestly toxic to the stock and to the operators applying them. The active compounds include organophosphorus and organochloro-insecticides. British farmers, at the MAFF's bidding, were using lindane (BHC, an organochloro-compound) so liberally that the smell hung over sheep at market and residues in the meat rose high enough for the French – admittedly politically anxious to pick a fault – to ban imports of British sheep and their meat. The British government, while protesting that the practice was safe, withdrew lindane from the battery of allowed insecticides.

The stress and crowding in modern farming regimens provokes ills other than blatant outbreaks of disease. Metabolic upsets (such as various types of 'staggers'), milk fever and ketosis invite treatment with nutritional supplements, as well as with steroidal drugs. Feeds (and even the litter for poultry) may contain more than growth-boosters and other drugs: they may have been treated with such additives as fungicides, preservatives and compounding aids. They may also contain toxic products of spoilage. Various chemicals are used in the production of silage, which is fed by the ton to farm animals. Routine tests cannot monitor adequately the residues conveyed in the meat and milk. Most of the residues concentrate in the excretory organs (in particular the liver and kidney) and fat.

Intensive farming creates a breeding ground for bugs, ranging from viruses and bacteria to flies and other insects. To counter infestation, chicken farmers are now administering insecticidal drugs in the poultry feeds, that pass through the birds' digestive tracts and thus impregnate their droppings. This practice is claimed to be safe, because the compounds are not absorbed into the birds' bloodstreams.

The barrage of artificial practices is maintained up to the minutes before slaughter. Pigs are notoriously difficult to handle in markets and slaughterhouses – many farmers and vets condemn the cruelty in these places – so farmers may tranquillize the stock by injecting Stresnil or Suicalm, preparations based on azaperone, a drug now little used in medicine. Some cattle are injected just before slaughter with papain, an enzyme that is distributed throughout the flesh and softens it, raising the proportion of tender cuts to stewing meat. Such treatment wreaks considerable injury to the animal's major organs. In 1984 the government-appointed Farm Animal Welfare Council called for a halt to this cruel practice.

Now It's Safe, Now It's Not

In 1981 the British government at last fell in with an EEC ban on the use of oestrogenic drugs of the stilbene type. These had been used in Britain for a quarter of a century, to assurances of their safety. Other substances, which affected thyroid function, were also outlawed for used on food animals. Illicit usage ensued in all areas after the bans.

Diethylstilbestrol (DES) is a stilbene. In Britain, its use was accompanied by that of hexestrol, a derivative with similar properties, although its pharmocology had been less investigated. The stilbenes are very potent synthetic oestrogens, with some carefully supervised applications in medicine in the treatment of malignancies (albeit with untoward side-effects), and they rate as carcinogens. They are especially dangerous to people with cardiovascular risks. They were used for a time in the USA for women with menstrual problems, and their sinister reputation was enhanced when a hitherto extremely rare vaginal cancer in baby girls was traced to previous treatments of the mothers with DES. Cellular abnormalities in male offspring were subsequently noticed.

Further evidence accumulated against the 'safe' use of stilbenes in farming. Medical reports from Italy and Puerto Rico incriminated DES as a cause of precocious sexual developments in children consuming meat from treated animals, probably cattle and poultry. Children evince early signs of trouble because they eat more food for their weight than adults. Residues of the stilbenes were also detected in meat and meat products, especially in some baby foods made from veal in western Europe. Some other evidence of adverse effects appeared to incriminate zeranol (Ralgro), another oestrogenic additive, which is still in use.

DES and hexestrol were being used in Britain to caponize poultry and as growth-boosters in other livestock, notably (with a testosterone derivative) in a product called Maxymin and sold by Elanco for use by pig rearers. Such substances should be withheld from the feed for some time before the animals are slaughtered, to allow residues to 'disappear' before the meat is consumed but the precautions are difficult to observe and monitor in farming practice. Elanco, the agricultural side of the Eli Lilly drug company, ceased advertising Maxymin directly to farmers and withdrew it six months before the EEC ban on such products took effect.

Other major drug firms trading in the euphemistically styled 'animal health products' markets have run into problems. ICI had to withdraw their Grofas, a drugged ('medicated') feed used for years to boost the growth of pigs and poultry, after toxic effects from the active substance proved embarrassing. ICI also had to withdraw a farrowing aid called Methallibure after its use had led to the births of deformed piglets from sows developing 'tight womb syndrome'. Merck, Sharpe and Dohm had problems with an anthelminthic worming substance that, notwithstanding the usual tests before marketing, had to be withdrawn shortly after its introduction owing to toxic effects in treated stock.

In 1985, the British Ministry of Agriculture, Fisheries and Food acknowledged a frequently overlooked hazard in the gross usage of drugs: the danger to farm workers. It was for this reason (rather than concern over the risks to consumers from residues in the meat) that the Ministry withheld permission to add carbadox (sold as Fortigro), a growth-booster for pigs, to feeds. Pfizer, the manufacturers of Fortigro, are contesting he estriction, citing the use of the booster in other countries.

No Harmonization over Hormonization

A decade ago, British farmers were guided by the drug companies and by the MAFF's Agricultural Development and Advisory Service (ADAS) into using growth-boosting, synthetic compounds with hormonal activity. Zeranol (Ralgro) and trenbolone acetate (Finaplix), together with hexestrol and antibiotics, were involved in these regimens. Hexestrol was discreetly omitted from subsequently recommended programmes.

Bullocks and heifers put on lean flesh faster if they are given implants of Finaplix, an anabolic steroid; the effect can be enhanced by the addition of other hormonal agents. Such practices fit in well with the nutritionists' innocent calls for leaner meat, since the costliest time for the farmer is the later phase of fattening when the beast lays down the marbling fat that the butcher esteems in a well-finished beast. The drugs also allow some 'beefing-up' of 'milky' breeds, and use of anabolics enables farmers to dry off cows culled from the dairy herd (then fit only to be slaughtered for manufacturing meat) and to implant them and rear them on concentrates for a few months to realize a better price at a market prepared to take them as 'butcher's beasts'. At least 70 per cent of Britain's beef originates in the dairy herd and cow meat comprises nearly 25 per cent of total consumption; beef from bulls accounts for less than 10 per cent. Bulls raised intensively in feed-lot systems (barley beef) can be sent for slaughter at less than a year old; they are therefore killed before puberty, and sex taints are unlikely. But such animals already manifest some masculine traits, being more restive and aggressive than bullocks (and therefore likelier to yield more condemned meat from 'damaged' carcases) and the excessively lean flesh is liable to dryness and 'dark cutting'. Administration of oestrogens exerts the expected endocrinological effects of toning the bulls down and inducing partial emasculation.

Recommended withdrawal periods between implantation and the earliest day of slaughter vary from nothing to two to three and a half months for the various products and stock. Confusion is likely to arise, especially as farmers may double up the implantations on the (mistaken) basis that 'two must be better than one'. Moreover, trading through markets, at which implantations are not declared, may lead to multiple treatments or to slaughter before the withdrawal period has elapsed. The rules expect a control out of keeping with the workings of the trade. A veal calf can go for slaughter aged three and a half to four months; the withdrawal period for the appropriate growth-booster is 100 days.

The implants should be inserted into the back of the animal's ear. Ears are supposed to be discarded at slaughterhouses (although the workers and inspectors will be unable to distinguish 'hormonized' animals) and meat from the heads is not much used for human consumption (most of it goes for pet food). These precautions are held to protect human consumers from a concentration of drugs in the meat they eat; indeed, it is argued by the trade that swallowing the pure drugs would do no harm (but this assertion does not run to chronic consumption). Producers connive at irregularities over the withdrawal times, pleading 'large margins of safety'. Results of experiments on laboratory animals are adduced; however, the results of the farmers' experiments on their animals are obvious, for the endocrinological effects can gain them £25 to £50 when the beasts are assessed at market, and the practices are worth at least £40 million a year to the British beef industry. Most of the steers and heifers are implanted and over half the carcases going to butchers have been so treated. Many farmers have misgivings and the NFU has had difficulty in establishing its stand against any ban on the use of the growth-promoters.[3]

Some farmers, with simple logic, argue that the implants would be more effective at the 'meaty' sites in the hindquarters and act accordingly. Farms and slaughterhouses are no temples of technological excellence.

Assurances over the safety of these growth-boosting practices abound. Nonetheless, misgivings over their health effects prompted the EEC to set up a committee of enquiry under Professor Eric Lamming of Nottingham University. After several postponements Professor Lamming's committee had not completed its report by the end of 1985. Meanwhile, the drug industry is busily selling the advantages of testosterone, progesterone, and 17-beta-oestradiol – the so-called 'natural' hormones (in reality, synthetic compounds identical to the sex hormones common to mammals of many species) permitted under EEC regulations.

In October 1985, the European Parliament called, by a large majority, for a ban on the use of Ralgo, Finaplix and the three 'naturals'; the EEC's Commissioner for Agriculture changed tack accordingly and pulled the rug from under the Lamming Committee which, it seems from the tenor of leaks, was set to uphold continuing use of the five substances, with possible reservations over Ralgro and Finaplix. (In April 1985, Britain's MAFF had feebly anticipated the forthcoming restrictions by moving Ralgro from 'the merchant's list' to the 'prescription-only' category, entailing some involvement by vets.) In December 1985 the EEC's agricultural ministers, with fierce opposition and threats of obstruction from Mr Jopling, Britain's Minister of Agriculture, agreed to ban the use of all the hormones and like drugs by the end of 1988 (with an extension to 1989 for Britain)[4].

'Consumer reaction' rather than 'science' was given as the reason for the sudden changes of opinion, which left Britain almost isolated within the EEC and able to continue its practices only by using its special position to deploy delaying tactics. French consumers had staged effective boycotts after the doubts over veal. West Germany's rigorous food regulations – as well as its high consumption of beef from stalled bulls slaughtered young, a system gaining little from application of the hormones – created a climate unfavourable to the production of beef from castrates. The Irish Republic turned rapidly from complacency to stringency, primarily with an eye on its exports to West Germany. Denmark went along with Britain, who could also count on support from the USA, anxious that EEC stipulations might debar imports from countries permitting use of the growth-boosters. While European consumers' organizations were almost united in favour of the ban, Britain's Consumers' Association, publishers of *Which?*, participated in discussions in Brussels, but dallied; it published no information to its members, but the president of the National Union of Farmworkers (NFU), Sir Richard Butler, was able to cite CA in its support of the hormones against the consumer lobby. It seems that CA may have committed itself to current practices, possibly with reservations over Finaplix and Ralgro, but by the beginning of 1986 it had retreated to indecision and silence. British farmers reckon that the imposition of the EEC's proposed ban will increase the price of beef, on average, by 4p per pound. Nonetheless, many members of the NFU's livestock section embarrassed the union's leaders by voting in favour of a ban.

Applications of hormones do not end with growth promotion in fatstock; prostaglandins and gonadotrophins are injected or implanted into breeding stock to synchronize or time oestrus, or to control farrowing and other births at the farmers' convenience and to avoid overtime working. These practices are highly developed with sows, and many ewes are 'regulated' by sponges, impregnated with hormones (one isolated from the urine of pregnant mares farmed for the purpose) and pushed into their wombs. The drug industry vaunts its prowess, won after experimentation on laboratory animals, in its artifices: ICI, for instance, booked full-page advertisements to trumpet to farmers and vets its Estrumate, a synthetic prostaglandin, introduced as 'Just a little better than Mother Nature provided', with the text asserting that 'we can modestly claim that Estrumate is a genuine advance on the genuine article'.

Antibiotics by the Ton

In 1965, the Brambell committee deplored the cruelty of factory farming and, after a number of inconclusive official enquiries, the government accepted the recommendations of the Swann Committee, based on medical risks, to impose some curbs. Britain scored a first among the industrial nations in this regard, thanks both to persevering detective work by the Central Public Health Laboratory and to outbreaks of serious and resistant salmonella bugs among calves which, being pitched into markets after abrupt weaning at a very early age, are under severe stress. Mortality often runs at over 10 per cent among young calves, and in 1986 the trade was attracting sharp criticism from the government-appointed Farm Animal Welfare Council, many branches of the NFU, and some auctioneers.

Alarm arose when it was acknowledged that the misuse of antibiotics was resulting in resistance by pathogenic bacteria, which could thus develop immunity to several antibiotics, passing these properties on as they replicated – and even adding to their repertoire of resistance by genetic material acquired from harmless bacteria of other types. Adoption of the Swann Committee's recommendations was all the more urgent after human deaths due to bacterial infections immune to most of the drugs hitherto effective for such diseases.

The Swann recommendations were well meant, but did not anticipate the ease with which farmers and vets could circumvent them. It has sometimes been suggested that a more effective curb was imposed by the death in a road accident of a notorious calf dealer. Certain natural factors bred complacency over the flood of antibiotics used in 'farming-by-needle': the organisms causing mastitis, like those causing gonorrhea in people, developed resistance slowly if at all, so millions of shots of penicillin and like drugs are administered for mastitis and in 'dry cow therapy'; and *Salmonella dublin*, which is dangerous to cattle, seldom imperils people, although both species are sensitive to *S. typhimurium*, which does develop resistance.

Bowing to pressure from farmers to be allowed to use antibiotics as growth-boosters, the Swann Committee divided antibiotics into two categories: those for feed purposes and those to be administered, on vet's prescription, for 'therapeutic' purposes. The feed antibiotics comprised those seldom used by doctors, such as virginiamycin and bacitracin, bacterial resistance to which could develop with little prospect of danger to people. This list has been extended as more anti-bacterials and drugs used against parasites have been found to boost growth.

The therapeutic list comprises penicillin, tetracycline, and many drugs common to medical and veterinary practice, resistance to which posed a serious threat. The Swann Committee toyed with a ban on chloramphenicol for all veterinary practice, this drug having special importance in treating typhoid, as well as being associated with serious side-effects. In deference to the veterinary profession (most of whom treat small animals and horses – pets – rather than farm animals), the right to prescribe chloramphenicol was retained, but with special appeals for caution.

Disease is rife in intensive units in which treatment of individual animals is out of the question, so it was not long before farmers were appealing for therapeutic antibiotics to be added to feeds for both prophylactic and curative purposes. Farmers soon dispensed with those vets who refused to sanction such practices and quickly found others who were not so scrupulous. Some veterinary practices were dominated by a few prosperous farming clients whom the vets could not afford to disappoint. Other vets were employed by animal producing enterprises, and some by the manufacturers of drugs and feeds. Venality undermined the vets' authority. Some were caught (and struck off for a while) in flagrant violations, others connived at administrations sanctioned by their signatures, but over which they exerted little supervision. After an initial interruption in the sales of therapeutic antibiotics for veterinary purposes, the conspiracy of farmers and vets breached the spirit of the Swann recommendations, and their use in feeds, as well as in the usual ways for drugs, was widespread and almost out of control.[5]

It is difficult nowadays for a rearer to buy calf-feed unmedicated. Millers and compounders run batches of feeds through machinery from which remains of earlier products have not been cleared, so substances added to, say, pig-feed turn up in rations for dairy cows, for whom they may be unsuitable or toxic.

Most doctors in the NHS prescribe drugs but do not dispense them, and they stand to gain little from sales of drugs. Most vets on the other hand are private practitioners running small businesses selling the drugs they prescribe; those with farming practices have recently increased their sales of drugs more than those dealing with small animals (pets). Farmers can obtain many medicaments and medicated feeds from their merchants without a prescription, but most of the growth-boosters and antibiotics are on the list of prescription-only medicines (POM), which require a vet's signature, although vets will be selling to farmers bulk quantities of, say, Finaplix, which will be administered in their absence. Definition of 'the animals in their care' is tenuous: vets may plead knowledge of occasional visits to a farm and of the size of the herd, but they would have no proof of their own on adherence to correct procedures and withdrawal periods. A vet established in a court that his supervision over a supply of Finaplix was adequate; although neither he nor any partner in his practice had ever visited the farm, he had seen some of the animals bought by the farmer he was supplying as he happened to attend markets occasionally as inspector. Vets demurring at infringements lose clients to other vets who are less strict; sometimes the drug company's salesmen will make an arrangement for the firm's vet to iron out any difficulties.

Farmers jib at prices charged by vets, who are accused of abusing their monopoly. Whereas medical prescription drugs are not advertised direct to the public – only to doctors and nurses – drug companies advertise POM drugs direct to farmers. Thus, the cover of the *Farmers' Weekly*'s Royal Smithfield Show supplement was for years adorned by a full-page spread on Finaplix, with no indication that it should be used only under a vet's supervision. In 1985, the space was occupied by an advertisement for treatments of ailing calves. Ralgro is advertised to both farmers and vets, with special offers, such as a free gun for the injections. Salesmen carry their business direct to farmers, who therefore expect the vets to comply with their demands at competitive prices. While vets object to the huckstering, they recognize they are being undersold in a developing 'black market' now worth about £4 million a year in Britain. On the continent, illicit sales are rife: in Italy nearly all veterinary drugs are sold illicitly. Such products may be sub-standard or even dangerously labelled.

The veterinary profession is further compromised, for British vets pledge themselves on qualification 'to do the utmost for the well-being of the animals in my care'; yet the veterinary press frequently advertises growth-boosters with no claims at all to therapeutic properties – only the effect of 'finishing' the animals faster. And many vets acknowledge that the torrents of drugs are surrogates for the missing husbandry of the good stockmen's tender, loving care – their labour no long needed, they are dismissed into the ranks of the unemployed.

Recriminations between vets and doctors over misuse of antibiotics are muted in a professional way. Doctors (and dentists) have some grounds for precautionary prescription – and, seen in retrospect, over-prescription – of antibiotics, such as the penicillins, of low toxicity. A doctor cannot risk delays in treating suspected meningitis, pneumonia or an ear infection in a child. Waiting for full tests on the disease could cost a life or a faculty. The malpractices of vets, however, do not deserve lenience. They are party to abuses for the gain of profits from animals exploited in conditions they should condemn utterly. The profession's reputation has been tarnished.

Doctoring by Government

Although members of the cabinet are required to desist from business activities while in office, many retain and pursue farming interests. The Minister of Agriculture is usually an active farmer, and the National Farmers' Union generally reckons to deploy 'the old boy' network to achieve its aims despite its weak representation among MPs. Farming is the nation's biggest industry but employs only 2 per cent of the population. The NFU has opposed restraints on the use of drugs and 'agrochemicals', but has had difficulty in overcoming support from its livestock sections for the consumer lobby calls for a ban on the hormonal growth-boosters. Producers of traditional beef, with no need to beef-up dairy-crosses, would be the happiest to forgo the use of the hormones, and they support elements in the meat trade who call for 'real beef', produced by traditional methods. Sales of real beef would certainly devalue products ineligible for the description, just as real bread and real ale cast doubt on the worth of today's commercial bread and beer.

Mr Jopling's almost solitary resistance to the EEC's intended ban on the hormonal boosters may earn British farmers a respite, but he is having to rely on an unworthy recourse to the threat of an uncontrollable black market if attempts are made at implementing a ban. This is commentary enough on the standing of the farming industry and its accomplices among the vets.

Defenders of the hormonal growth-boosters cite the presence of sex hormones and of compounds with similar activity in meat, milk, other foods, and the human body; moreover, absorption of dietary doses is slight. Experiments on animals may be adduced as further proof of safety, although these – being on rats and mice – may be no more relevant than the manifest short-term effects exhibited by animals on the hoof. Thus a scientific case is built up to quash the 'gut feelings' of the consumer lobby and 'politicians'.

These arguments allude mainly to the 'natural' hormones, for which no withdrawal periods are stipulated: they may be used licitly up to the day of slaughter. Ralgro and Finaplix are artefacts, however, potent drugs plied with little supervision, and we have to take account of metabolites in both animal and consumer, all these compounds and residues being taken year in year out, with much less supervision than would be exercised over the use of such drugs administered in short courses by doctors. The manufacturers of the growth-boosters include some contra-indications in their small print; for instance, their use in breeding stock is dangerous (are not some human consumers of the meat 'breeding stock' also?). Implantation of the hormones elicits the predicated immediate effects on the animals, which behave in excited and peculiar ways. The argument that tries to justify additions of hormones already present in meat is flawed: consumption of animal products has increased sharply since we abandoned our Stone Age diets, and we have insufficient information to state that 'rich' diets are an unmitigated nutritional success. Moreover, scientists have in the past pontificated over the 'safety' of ingredients subsequently withdrawn because of doubts over their health effect or because they have ignored sub-lethal effects caused by 'intolerance', allergic reactions and genetic variations. Present intakes of hormones may be excessive; increases cannot therefore be regarded as 'safe'.

Any scientist must be prepared for challenges to his assertions. Do we know all the factors involved in the increasing cancers of the breast and reproductive organs? Can we explain fully the apparent increase recently in anorexia nervosa and premenstual tension? Can we rule out dietary influences? If the growth-boosters are 'safe', why do scientists continue to subject animals to further experiments in attempts to prove this safety? (In the USA, human vegetarians are being enlisted as controls in surveys of consumers of all types.) With the complex of nutritional, anti-nutritional, and physiological factors in the simplest foods and the possibilities of interactions and varying types of metabolism, the gut-feelings of caution over further artefacts, which extends to other food-additives, are a more measured assessment of the unnecessary risks than the so-called 'scientific' assurances. It is not, after all, as if the EEC and Britain face food shortage: the EEC's beef mountain had grown to 750,000 tons by the beginning of 1986 – the Eurocrats faced the task of restraining production rather than stimulating it. The 'political' arguments spring more from the producers' reluctance to forgo the bonus of £25 to £50 they gain from every 'hormonized' beast, which is worth £40 to £100 million to farmers in Britain alone.

Routine tests of meat for residues are slow and costly. Far fewer are carried out in Britain than in other EEC countries, and a research group developing rapid and cheap radioimmune assays was disbanded in 1985 by Britain's Minister of Agriculture. Again, producers seem to take precedence over consumers in the ministry's scheme of things.

The reckless use of antibiotics at therapeutic and sub-therapeutic levels, and the flouting of the Swann recommendations, indicts the evils of intensive farming and its retribution in the spread of resistant disease and death. Like the use of growth-boosters and the abuses of other drugs, it debases husbandry, compromises farmer and vet, and exposes the consumer to unnecessary hazards. Such negligence over potent agents fosters fecklessness in other matters. The British government's excuse that a black market in drugs would develop if action were taken to ban their use is unworthy: the government's refusal to govern effectively means that the consumer is being blackmailed into participating in a crudely designed and ill-supervised experiment for the profit of producers who would resort to peddling even DES if there was money to be made. 'If the EEC bans hormones I'll buy the last million pellets and retire,' jested Julian Best, marketing director for Hoechst, the drug company selling Finaplix, at the 1984 Conference of the British Veterinary Association. There is many a truth, even in a sick joke.

Notes

1. The development of poultry farming in the late 1950s from a farmer's wife sideline to the battery and broiler moloch entailed similiar flouting of hygiene regulations as new breeds were imported. Nemesis struck as exotic viral diseases swept through the flocks. Contraband and illicit vaccines were traded; these together with the routine use of growth boosters, were adopted as part of the regimens which brought crops of broilers to the oven-ready state within seven or eight weeks from hatching. It took 40 weeks for the free-ranging fowl of yore to be considered ready to be killed for the table.
2. The mode of the growth-boosting action of the antibiotics remains incompletely understood. Some, such as Monensin, change the ruminal flora so that the animals'

production of methane is curtailed. Methane is normally belched off and is therefore the product of a wasteful conversion as far as the farmer is concerned – a beast may in this way blow off up to 10 per cent of the calorie value of the feed – and use of antibiotics reduces the loss. The altered metabolism raises the feed-to-food efficiency a little by incorporating the retained fragments into components in the meat and milk, thus modifying their 'natural' composition even further (concentrates and low ratios of fibre, as well as breeding practices and yarding and restricted movement, had changed the flesh from the sinewy muscles of animals left to roam mainly on grass or used as beasts of draft).

3. At the 1985 Royal Smithfield Show producers and butchers mocked the Meat Promotion Executive's display of beef, and many deplored the use of hormonal drugs. They scorned the beef as being as succulent as sawdust and as limp as a sister's kiss, and 'barely better than knackermeat'; Dr David Allen, the MPE's beef specialist, admitted that it was not the sort of meat he would eat. Members of the Meat Promotion Executive, such as Mr Geoff Harrington, the marketing director, have muttered their misgivings at the developments and alarmed colleagues with appeals for 'real' meat, reared without additives and drugs. In 1986, an editorial in the *Meat Trades Journal*, alluding to the success of the Vegetarian Society's Campaign for Real Bread, stated unequivocally that 'the EEC is right to oppose the use of growth-promoters'.

4. On 19 August 1986, Mr. Jopling announced that Britain would comply with the EEC ban on growth-promoting hormones earlier than expected. Jopling said: 'I opposed the decision of the agricultural council last December. However, we must now respond to the fact that intra-community trade will be limited to meat from untreated animals as from January 1st 1988.'

5. The USA recognizes strong arguments for restricting the uses of antibiotics in farming, but the failure in Britain of the Swann recommendations has delayed promulgation of similar codes in the USA, although warnings over chloramphenicol have been uttered and – as in Britain – flouted by vets who have allowed preparations intended for dogs and cats to be used on farm animals.

11

Pesticide Controls

A History of Perfidy

Chris Rose

Britain's environmental record in controlling pesticides has been nothing less than deplorable. Official papers show that in the 1950s, Ministry of Agriculture civil servants:

- Stifled moves to inform all doctors in areas where spraying was underway with organophosphorous chemicals (nerve poisons).
- Reversed the recommendations of a government-appointed committee which had urged that legal controls be introduced over pesticide operators and over the chemicals themselves.
  - Introduced a system – later to become the Pesticides Safety Precautions Scheme – drawn up to comply with the industry's interests, under which research on the biological effects of pesticides was to be kept a secret between government and industry.
- Doctored official minutes so as to reverse the meaning of a report on dangerous packaging of pesticides sold in shops.

As a result of policy decisions taken at this early date, the stage was set for 30 years of runaway pesticide use in Britain. By the mid-1980s, when EEC pressure forced the government to resurrect the idea of a statutory control scheme (made law under the Food and Environment Protection Bill), the industry was, in the words of the Friends of the Earth campaign report, *An Industry Out of Control*.[1]

Some Secrets Revealed

Recent research in the Kew Public Records Office by Maurice Frankel of the Campaign for Freedom of Information,[2] has revealed how early attempts to bring pesticides under a comprehensive system of legal controls in order to protect farmworkers, the public and the environment, were undermined and finally defeated by concerted lobbying from within the civil service on behalf (it would appear) of the pesticide industry. Official papers from the early 1950s show how civil servants manoeuvered and steered 'expert' committees away from imposing legal controls – and even rewrote and reversed their findings.[3]

In the main, the papers which Frankel found at the Public Records Office discussed the Working Party on Precautionary Measures Against Toxic Chemicals Used In Agriculture, chaired by Professor Solly Zuckerman, the Conference on Toxic Chemicals in Agriculture, and the Proprietary Pest Control Products Joint Panel on Retail Sales.[4]

The Zuckerman Report

The Zuckerman Committee started meeting in 1950 with instructions from the Ministry of Agriculture to comment on the recommendations of a previous working party – the Gower Committee – which had advised that protective clothing should be made compulsory for any farmworkers using toxic chemicals.

Zuckerman's group found that since 1946 'at least seven' agricultural workers had died of poisoning from DNC, a selective weedkiller brought into wide use against cereal weeds such as cleavers and marigold. In January 1951 the committee's draft report observed that in addition to those deaths: 'One man has died and a number of persons have suffered serious illness while engaged in the manufacture of DNC. Many agricultural workers have described symptoms suggestive of mild DNC poisoning occurring during the cereal crop spraying season.'

The committee heard that during 1950 some 300,000 acres were sprayed with DNC and DNBP aromatic dinitro weedkillers. In the same year, some 50–70,000 acres were treated with parathion: 20,000 with TEPP; and 60,000 with schradan – all organophosphates.

Zuckerman and his colleagues soon became convinced of the need for a coherent system of statutory controls over the manufacture and use of pesticides.[5] Correctly assuming that organophosphorous and dinitro pesticides would be 'used on an increasing scale', the committee determined that 'until alternatives can be found . . . our task is to recommend measures for the protection of workers who are handling these chemical compounds' . . . In its *draft* report, the committee wrote:

We endorse the suggestion that medical practitioners should be warned when spraying is to be undertaken on farms in their locality, and we think that the warning should include local hospitals;

In our view the labels of the containers should contain in large letters the words 'Deadly Poison';

Organophosphorous formulations should be coloured, preferably with one distinctive colour, during manufacture;

In the course of our enquiry, we have found that the public may be exposed to some risk to health arising from the agricultural use of dinitro and organophosphorous compounds . . . The chief danger lies in the chronic effects which result from frequent exposure to these chemicals, and this greatly reduces the hazard to the general public since *normally* [our emphasis] they are not so exposed. There are, however, certain precautions which should be taken in the public interest.

The committee went on to recommend that:

Spraying operations should be suspended in windy weather, particularly on land adjacent to public roads or footpaths;

Warning notices should be placed on gates giving access to fields that are being, or have recently been, sprayed;

Departments should take statutory powers to call for the registration and licensing of all chemicals that are introduced and offered for sale as substances which protect agricultural products from diseases and pests.

The Stifling of Controls

Faced with the prospect of pesticide manufacture and use being brought under a system of statutory controls, the Ministry of Agriculture soon produced a contrary 'Departmental View', which represented both the interests of civil servants (in the cause of administrative convenience) and those of the pesticide industry. Particularly active was the Infestation Control Division (ICD), which noted in a departmental memo:

Statutory control in one form or another over pest control is in force in some countries, including the United States, Canada, Belgium, Netherlands, Denmark, France, and Switzerland. It would seem, therefore, that there is a good deal to be said in its favour. It may be that in this country trade organizations have a highly developed sense of social responsibility, which enables us to achieve satisfactory control by voluntary means. The Industrial Pest Control Association would be opposed to the licensing of rodenticides and insecticides, but would support a voluntary approval scheme.

The ICD went on to comment:

> Infestation Control Division are constantly being pressed by local authorities to require the registration of pest control firms and operators, with the object of keeping a close watch on their activities . . . We have resisted such representations on the grounds (a) that it would be difficult to ensure that registration was complete . . . (b) that a new offence of 'failure to register' would be created, (c) that the work involved would not be justified, and (d) to impose the kind of conditions of registration that the local authorities desired would be an unwarranted interference with the freedom of commercial concerns.

There can be few clearer examples of the ability of the civil service to represent industrial interests; to reject reforms simply in order to avoid extra work; to draw its own selfish terms of reference; and to make judgements with no thought to the democratic process or the rights of the public. By its own admission, the rejection of the requests of elected local authorities (which are presumably in a good position to know the real problems of pesticide use in their area) was a regular event.

The Civil Service Wins Out

We can only guess at the lobbying required to bury the Zuckerman Committee's proposals for a legally enforceable system of controls over pesticides. We do know, however, that the lobbying was ultimately effective. The minutes of a meeting held at MAFF's Horseferry Road offices on 4 February 1953, for instance, clearly record that 'A *majority* of the Working Party felt that a clear case had been made for Statutory powers.' This view was opposed by Mr Barrah of the Ministry of Agriculture and, after discussion, the recommendation to implement statutory controls was quietly dropped. In its place, a formula drafted by Mr Barrah's senior, Mr Sutherland-Harris, was adopted. This origins of the Sutherland-Harris proposal lay in a memo of 19 January 1953 in which he noted, 'There seems to be too much emphasis on administrative control' in the Zuckerman recommendations: Sutherland-Harris suggested 'substituting' the word 'precautions' instead.

Sutherland-Harris proposed that the requirements set out in the report could very largely be met by voluntary arrangements. There then followed a description of the Pesticide Safety Precaution Scheme, which was to remain the mainstay of British pesticide control until external factors finally forced the government to introduce legal controls in the shape of the Food and Environment Protection Act in 1986.

The Third Revise of the Zuckerman Committee report simply recommended establishing an 'Advisory Committee' which could, for example, 'guide Departments on the scope of the initial investigations which should be undertaken by manufacturers and about the evidence which should accompany proposals to introduce new substances'.

Under the proposed scheme:

> Manufacturers and importers should agree that new chemicals or new formulations would not be introduced into practice until cleared with the Departments concerned . . . Manufacturers should be prepared to submit full information about the constitution of the preparation, about methods available for determining the extent of any contamination, about the toxic properties of residues.

In the event, this information remained a secret between manufacturers and the government: the public were excluded and although the committee referred to the need for general enabling powers, in practice this too was stifled by the civil service.

When the Zuckerman report, *Toxic Chemicals in Agriculture*, was finally published in 1951 even minor recommendations had been watered down or dropped. At a meeting in Carlton House Terrace on 11 February 1951, a Ministry of Agriculture committee agreed that the 'Warning to Local Doctors' recommendation 'was NOT a suitable recommendation for inclusion in the press notice'. Another item for which no publicity was thought desirable was the idea that spraying operations should be suspended in windy weather. Notices on farm gates were also played down. In this way, the ministry began to whitewash the effects and risks of pesticides in agriculture, gradually creating the widely accepted myth that all such farm chemicals are not only essential but safe.

Retail Sales: A Doctored Report

The ministry not only went out of its way to help manufacturers and commercial users of farm chemicals to fend off controls over their use, it also played a considerable role in undermining controls over the sale of pesticides in shops. The Proprietary Pest Control Products Joint Panel on Retail Sales was set up in 1951 to recommend safeguards on the sale of pesticides to the public and to pest control firms. With commendable initiative the panel sent out its secretary, Mr F. J. Snowdon, to buy samples from shops. In its draft report, the panel noted: 'We have seen sufficient evidence to convince us that (the problem of) bad packaging practice exist(s).'

The panel found highly toxic compounds (such as Barium carbonate) on sale in simple paper wrappings. It also noted that:

> Somewhat paradoxically, there seems to be a tendency on the part of some manufacturers of rodenticides and insecticides to convince the public of the harmlessness of their products . . . we are strongly of the opinion that emphasis should be laid on warning the public of the risks involved rather than on inducing a false sense of security.

Discussing advertising matter, the panel remarked: 'Misleading and exaggerated statements appear in advertisements and pamphlets issued in connection with pest control products.' The panel was also critical of 'excessive zeal for sales promotion'. It concluded that 'The only effective means of securing safeguards is by statutory control of the manufactured article.' Yet what happened? Were legal controls reintroduced? No.

The final report of the panel concluded merely that 'further safeguards are desirable'. A code of practice was suggested – this despite the fact that in its draft report the committee had noted that 'a voluntary scheme would leave uncovered an important part of the field with no guarantee of compliance over the remainder.'

The emasculation of the report was most evident in a remarkable passage on the dangers of unsatisfactory packaging. The draft report read:

> We have been shocked by some of the containers in which toxic chemicals have been packed and we consider that with notable exceptions the standard of packaging *generally is well below* [our emphasis] that necessary for toxic substances. Some preparations are wrapped in thin paper enclosed in flimsy, insecurely sealed envelopes. In one instance which has come to our knowledge the contents were seeping through a fabric bag displayed among consumable goods.

On the draft document, which is deposited in the Kew Public Records Office, Frankel found severe editing by an unknown civil service hand. In the final report, the passage simply reads:

> Although the standard of *packaging generally is satisfactory* [our emphasis], we deprecate the use of some of the containers in which toxic chemicals are packed. Some preparations are wrapped in thin paper enclosed in flimsy, insecurely sealed envelopes.

And with the exception of the few pesticides covered by the Poison Rules, the civil service was to out-manoeuvre doctors, biologists and farmworkers for another 30 years. Until 1986 (and the passing of the Food and Environment Protection Bill) Britain was to have no legal pesticide control system, and many obvious and badly needed safeguards were never introduced or implemented.

Notes

1. Friends of the Earth, *An Industry Out Of Conrol*, London: FOE, 1985.
2. Campaign for Freedom of Information, 3 Endsleigh Street, London, WC1.
3. The papers were released in 1985 under the '30 year rule'. Their release came in the middle of the debate on the 1985 Food and Environment Protection Bill. At the time, MPs were being asked to 'trust the ministry' and leave its civil servants to draw up detailed regulations to implement a broad enabling law.
4. MAFF Files 130/61, 130/58, 43/142.
5. It is ironic that until 31 March 1950, when the 1949 Prevention of Damage by Pests Act came into force, Britain had a system of legal controls over the manufacture and use of some pesticides. Under the Infestation Orders of 1943–45, the manufacture of 'preparations for the control of pests' was subject to licence. The scheme covered all pest control preparations, other than those intended for use on growing crops and those which the minister is empowered to control under the Diseases of Animals Acts.

12

Pesticides

An Industry Out of Control

Chris Rose

In 1980, Edward Goldsmith wrote in *The Ecologist*: 'In the last thirty years there has been a veritable explosion in the use of synthetic organic pesticides. Over 800 formulations are now used in the UK alone. They include nematocides, fungicides, herbicides and rodenticides . . . Each of us has in his body fats, traces of hundreds of different pesticides.' Today, just six years later, around 4,000 different proprietary products – made up of nearly 1,000 different pesticides – are in use in Britain. In 1982, according to the British Agrochemical Association (BAA), pesticide sales grew by 21 per cent, bringing the value of the home market to £329.3 million.

The agrochemical industry has been quick to exploit every market. Its sales force has been assiduous in seeking out every walk of life into which pesticides can be introduced – from insecticides in varnish and fungicides in wallpaper paste to aquatic herbicides and cosmetic glazes for fruits. But it is agriculture which provides the biggest market for pesticides. In the three years between 1974 and 1977, the area of cereals sprayed with aphicides increased 19 times. Between 1979 and 1982, the area of crops treated with insecticides doubled, whilst the area treated with fungicides more than doubled. BAA figures from 1979 to 1982 for the five major crops grown in Britain (cereals, potatoes, sugar beet, oil seed rape and peas) show a 29 per cent increase in the area sprayed with herbicides, a 37 per cent increase for insecticides, and a 106 per cent increase for fungicides. Yet, the actual cropped area only increased by 4 per cent.

By the early 1980s, 97–9 per cent of all main crops, cereals and vegetables were sprayed at least once. Official figures for 1983 show that one crop of lettuce was dosed 46 times with four different chemicals. Crops such as hops were receiving an average of 23 sprays a season;

orchards, 17; soft fruit and glasshouse vegetables, more than eight; and cereals, at least three; and arable crops, such as peas and potatoes, an average of just under five. A third of all fresh fruit and vegetables sampled by the Association of Public Analysts in 1983 was found to be contaminated with pesticide residues.

First Signs of Danger

From the very earliest days, government biologists were at least dimly aware of the ecological impacts of organochlorine pesticides. Dr V. B. Wigglesworth of the British Agricultural Research Council's Insect Physiology Unit believed the first reports of DDT's ability to kill insects were 'too good to be true'; soon after, in an *Atlantic Monthly* of 1945, he warned that DDT might act like 'a blunderbuss discharging shot in a manner so haphazard that friend and foe alike are killed'.[1] It was not until years later, however, that the insidious secondary effects of DDT and similar organochlorine insecticides were recognized and accepted by governments, and even then remedial action was slow and ineffective. In 1958, the California Department of Fish and Game tried to eliminate gnats from the 40,000 acre waters of Clear Lake, using DDT. A population of fish-eating grebes present on the lake was reduced from 1,000 pairs to just 25. Examination of phytoplankton, invertebrates, fish and birds, showed that DDT and its metabolites concentrated as they moved up the food chain.

Similar problems were occurring in the British countryside. In his spare time, Derek Ratcliffe, a young Nature Conservancy botanist (now Chief Scientist of the Nature Conservancy Council) was studying Britain's peregrine falcon population. He found that eggs were breaking in the nests. Eventually this lethal phenomenon was shown in sparrowhawks and other birds of prey, not just in Britain but in North America and much of the developed world. Populations crashed as a result. Bitter resistance from vested interests, ably represented by the Ministry of Agriculture, prevented DDT's complete withdrawal from the British market until October 1984, although it became official policy to phase it out from 1969.

Dieldrin, another fat soluble organochlorine insecticide, is still in use in Britain in 27 pesticides for use on wood and elsewhere. For decades it was employed in sheep dips and, as a result, caused widespread poisoning of golden eagles. The drastic decline in the British otter which had begun in the 1950s (and has led to the otter's extinction across almost all of England and Wales) is now known to have been triggered by dieldrin. And although the peregrine has now recovered much of its former abundance, other species such as the sparrowhawk have not.

Product	Manufacturer
Totril	Hortichem
Actrililawn 10	May & Baker
Mate	May & Baker
Topper	Union Carbide
Actril C[1]	Murphy Chemicals
Mylone[1]	Union Carbide, A. H. Marks
Topper 2+2[1]	Union Carbide
Iotox[1]	May & Baker
Malet[1]	A. H. Marks
Malet 50 EC[1]	Ciba Geigy
New Clovercide Extra[1]	Synchemicals
New Clovotox[1]	BASF, May & Baker
Synox[1]	Synchemicals
Marks Cubisol 580[1]	A. H. Marks
Deloxil[2]	Hoechst, A.H. Marks
Briotril Plus[2]	Alpha Trading Ltd
Novacorn[2]	Farmers Crop Chemicals Ltd
Hobane[2]	Farm Protection Ltd
Swipe 560 SCW[1,2]	Ciba Geigy
Brittox[1,2]	May & Baker
Actril Extra[1,2,3]	Murphy Chemicals
Certrol PA[3,4]	Marks
Atlas Feroxone[2,3,4]	Atlas Agrochemicals
Musketeer[1,5]	Hoechst
Post-Kite[1,5]	FBC Ltd
Doublet[1,5]	May & Baker
Assassin[1,5]	ICI
Belgran[1,5]	May & Baker
Twin-Tak[2,5]	May & Baker
Dictator T[1,2,5]	Ciba Geigy
EF 718[6]	Dow Chemical

Figure 7 Products known to contain ioxynil (21 June 1985)   (*continued*)

Figure 7 (*continued*)

Product	Manufacturer
Harrier[1,6]	ICI
Crusader[1,2,6]	Murphy Chemicals
Springclene 2[1,7]	FBC Ltd
Asset[2,7]	FBC Ltd
Super Verdane[8,9]	ICI
Super Verdane G[8,9]	ICI
Super Verdane Spot[8,9]	ICI
Bio Lawn Weedkiller[8,9]	Fisons
Glean TP[2,10]	DuPont
Certrol[11]	Hoechst

Notes:

1. With mecoprop
2. With bromoxynil
3. With MCPA
4. With dichlorprop
5. With isoproturon
6. With clopyralid
7. With benazolin
8. With 2,4D
9. With dicamba
10. With chlorsulfuron
11. With linuron

Source: Compiled by D. Buffin from Pesticide Safety Precaution Scheme 'Approved Products List', MAFF.

Suspect Pesticides

Trade secrecy and government secrecy laws have been deliberately and falsely used to allay or conveniently to quieten public fears over pesticides. In 1985, for example, just as the Food and Environment Protection Act was passing through its final stages, James Erlichman reported in the *Guardian* that both MAFF and May & Baker Ltd, had learnt that the herbicide ioxynil was teratogenic in some test animals – that is, it caused birth defects.[2] (Products officially listed as containing ioxynil are listed in figure 7.) Yet, instead of withdrawing the product and clearing it from shop shelves, the ministry had kept the discovery a secret and contented itself with an assurance from the company that trade would be phased out. At a moment when the government was keen to declare itself in favour of greater freedom of information, the ioxynil scandal was a clear reminder of how far the British system would have to be reformed if it were to serve the public rather than commercial interest.

The pesticides listed below are just some of those giving serious cause for concern. They include known carcinogens (cancer-inducing agents) in humans and animals, teratogens (agents causing birth defects) and mutagens (chemicals which cause genetic mutations). Some are banned in other countries or more restricted than under Britain's Pesticide Safety Precautions Scheme (PSPS). Most are marketed in many formulations under a variety of trade names. So long as the data on which the pesticides were cleared for use in Britain remain secret (and probably only summaries will become available under the Food and Environment Protection Act), we are simply not in a position to know on what evidence British authorities continue to allow their use:

- Paraquat: bipyridyl herbicide, garden and agricultural use in Britain (14 products). Severely restricted in Sweden, Finland, Denmark, The Phillipines, Turkey and Israel. Banned in West Germany. Extremely toxic to mammals if absorbed, kills by suffocation with no known antidote. Toxic to fish.

- Lindane/HCH: forms of HCH are severely restricted or banned in many countries as cancer agents. More than 130 products on sale in Britain. Wide variety of uses. Persistent.

- Dichlorvos: classified as 'highly hazardous' by the WHO, it is cleared by the PSPS in Britain for agriculture, horticulture, home garden, kitchen and larder, wood preservation, food trade and domestic use. It is marketed by 43 firms under 82 trade names, yet it is acutely toxic and animal tests have suggested teratogenic effects. American researchers have suggested it should be handled with caution by women of childbearing age. Other research suggests it interferes with the cholinesterase enzymes necessary for normal transmission of nerve impulses in the body.

- Thiram: a fungicide and bird repellant cleared by the PSPS for agriculture, horticulture and home garden use. Marketed by 15 firms under 27 trade names. Animal research indicated liver, kidney, and brain damage. In combination with alcohol, it produced violent nausea, vomiting and collapse. Evidence from animal tests of mutagenicity and teratogenicity.

- 2,4,5-T: a translocated herbicide banned in Sweden and very tightly restricted in the USA, it is cleared in Britain by the PSPS for agriculture, horticulture, home garden and forestry use. Yet the US EPA has stated 'the quality, quantity and variety of data demonstrating that the continued use of 2,4,5-T contaminated with dioxin

presents risks to human health is unprecedented and overwhelming.'
Sold in Britain by 19 manufacturers. The EEC, in 1982, urged
reductions in its dioxin content and restriction on its uses. Present
permitted dioxin level in Britain is double the EEC recommended
figure of 0.005mg/kg. Mutagenic effects and teratogenicity established by animal tests. Banned in India, USA, Sweden, Norway,
USSR and several other countries.

- Aldrin: an insecticide banned in Sweden, severely restricted in many
  other countries including the USA, and rated 'highly hazardous' by
  the WHO, it is cleared by the PSPS in Britain for agriculture,
  horticulture, and food storage uses. Eight manufacturers and
  product names. Numerous alternatives exist for most uses.
  Ingestion, inhalation or absorption into the body can cause
  irritability, convulsions and depression; continued exposure causes
  liver damage. Has produced mutagenicity in human cell tests and has
  proved carcinogenic in mice.

- Chlordane/Heptachlor: organochlorine insecticide. Twelve products
  marketed in Britain, despite EEC prohibition directive. Severely
  restricted in Israel and USA. Residues found in 99 per cent of
  people's blood. Suspected carcinogen. Earthworm and ant killer
  products in Britain.

- Pentachlorophenol: organochlorine insecticide little used in Britain.
  One product cleared for masonry and wood treatment, but others
  still on sale. Classified by WHO as 'highly hazardous'. Severely
  restricted in Canada, USA and New Zealand. Can cause liver nervous
  system and skin disorders.

- Parathion: organophosphorous insecticide (nerve poison).
  Withdrawn for commercial reasons in Britain in 1984 but existing
  stocks could still be used. British workers died when first used.
  Suspected of causing 50 per cent of all pesticide poisonings
  worldwide and 80 per cent in Central America. One teaspoonful is
  enough to kill through skin absorption. Banned in many countries.
  Classified as extremely hazardous by WHO.

- Aldicarb: an insecticide, is still cleared for some agricultural and
  horticultural uses in the UK although it is listed by the WHO as
  extremely hazardous. In Long Island mothers were concerned about
  a 46 per cent spontaneous abortion rate in areas with pollution of
  water supplies by aldicarb.

- Captan: a fungicide marketed in 29 products by 15 British firms, and cleared by the PSPS for agricultural, horticultural and home garden use, is restricted in some countries. Mutagenic effects have been demonstrated in bacterial, fungal, human cell, rat and mouse tests. It has been shown to be teratogenic in rats, rabbits and hamsters, and in the USA, the National Cancer Institute has proved carcinogenicity in mice. Russian studies suggest it causes sexual disorders in animals.

- Dieldrin: a persistent organochlorine insecticide, is still cleared by the PSPS for food storage practice (although little used for this in Britain), and wood preservation although they have withdrawn earlier clearance for agricultural and horticultural uses. Banned in Sweden and for most purposes in the USA and Japan. Carcinogenic in mice. Classified by the WHO as extremely hazardous.

These pesticides are by no means the only ones which are a cause for concern. There are perhaps 20 or 30 others (and consequently many more brand products) on sale in Britain which give rise to considerable concern. Figure 8 shows some of the pesticides reported to Friends of the Earth as having caused health or other problems in pesticide incidents which, of course, reflect the frequency of their use and misuse as well as their chemical properties.

Banned But Still in Use

There is extensive evidence that DDT and dieldrin are both still widely used by farmers, market gardeners and others despite official bans on certain products or uses. Indeed, so long as the manufacture and export of DDT is allowed, it seems impossible to prevent such abuses.

On Easter Sunday 1985, some young canoeists on the River Avon found two dead herons within a short distance of each other at Cleeve Prior, five miles north of Evesham in Worcestershire. They alerted Peter Riley of Vale of Evesham Friends of the Earth, who happens to be an organic farmer and one of Britain's most experienced local campaigners on pesticides. He sent the herons to the Institute of Terrestrial Ecology's Monks Wood Research Station for analysis. An appeal for information in the local newspaper also produced two further dead herons. As Riley records in *The Avon Herons*, FOE then again 'appealed through the local press for the public to report live or dead herons. The response was remarkable, over 50 telephone calls were received reporting dead birds and birds missing from their normal haunts, as well as living herons.'[3]

Pesticide
Aldrin
Azinphos-methyl/Demeton-S-methyl/Sulphane
Asulam
Amintriozole
Bisidin
Carbendazim
Cyhexalin
Dalapon
Demeton-S-Methyl
Dichlofluonid
Dicofol
DDT
Dieldrin
Dinoseb
Endrin
Glyphosphate
Lenacil
Maneb/Mancozeb/Carbendazim
Mecoprop
Mevinphos
Oxydeneton-methyl
Paraquat
Patafol plus
Permethrin
Propinconazole
Pyrethroid
Simazine
Simazine/Aminotriazole
Thiram
Triazophos
Tributyl tinoxide
Trimethyl benzene
2,4,5–T
2,4–D
3,6 Dichloropicolinic acid

Figure 8 Pesticides involved in incidents reported to FOE, 1984

Eventually 17 herons were found dead – all with lethal levels of dieldrin, a pesticide which had supposedly been withdrawn from agricultural use ten years earlier. Some birds also had lethal doses of DDT and PCBs in their fat.

The explanation, accepted by many conservation groups (for example, the Royal Society for the Protection of Birds), was that such contamination by banned pesticides is the result of leakages from buried cans of old chemicals, similar escapes from old sheep dips, or accidental spillages of wood preservation products. Farmers in many areas will privately admit, however, that they know of 'a shed or two' where banned chemicals are stockpiled: in the Evesham area, one farmer was said to have ten years supply of dieldrin stacked in a barn.[4]

In May 1985, Friends of the Earth photographed DDT in active use on an Evesham farm. The next month, a partly used can of DDT was produced when the Pesticides Action Network launched its worldwide 'Dirty Dozen' campaign in London. Complacent MAFF officials announced confidently that this was 'a one-off' and that it must be old stock 'because it is impossible to buy DDT now'. On 7 June, Evesham FOE members visited the local firm of Craven Chemicals Ltd, a company especially active in the British Agrochemicals Association. Posing as farmers, they found no trouble in buying a drum of DDT – in fact 20 litres of Croptex DDT concentrate – for £32.20. Only hours later, they purchased a second drum – 20 litres of Farmon DDT 25 for £26.04 – at Velmark Chemicals, a subsidiary (now sold off) of the agribusiness company, Velcourt. With no inspections of stocks and no checks on users,[5] it is anyone's guess how much more DDT is in use. Residues on crops suggest that market gardeners and fruit growers are the largest users, although Joyce Tait, a researcher at the Open University, found that most cabbage growers admit to using it. There are also persistent reports that dieldrin is on sale in south-west England.

Crop Diseases and Modern Farming Practices

The use of pesticides has dramatically transformed the ecology of farm crops. One of the most striking examples involves a virus known as Barley Yellow Dwarf Virus, or simply BYDV. Before the introduction of intensive farming practices, BYDV was little known outside parts of southern England; where it did occur, it did little damage. The disease is transmitted by the bird cherry aphid, itself restricted by the scarcity of its over winter host, the bird cherry tree. This changed with the introduction of autumn sown cereals. The lure of EEC subsidies prompted an ever greater rush to resow. When this finally closed the gap between crops to a critical few weeks, a few straggling weeds were enough to provide a 'green bridge' for the much commoner grain aphid, which had previously perished for want of a winter host. Now it could walk from one crop to the next and, as it did so, it carried BYDV. Within a few years, the disease had spread to the north of England and the midlands, where it had never been seen before.

When, in 1984, BYDV reached as far north as the Vale of York, the Ministry of Agriculture's Agricultural Development Advisory Service was prompted to issue a warning notice. In areas where BYDV had caused 'significant damage in recent years' (which included 'coastal areas of east Norfolk, most of east Suffolk, estuarine valleys of East Essex, the Epping area of Essex westward into Hertfordshire and the area of Brandon (Suffolk) south west to the edge of the Cambridge fens'), and depending on the sowing date, ADAS advised spraying with a synthetic pyrethroid. Spraying was also advised in areas with a 'high risk' warning, or if aphids were seen in the crop. The recommended list for anti-BYDV sprays included organophosphorous as well as pyrethroid insecticides. In effect, ADAS was recommending farmers to open up a second front – a 'winter campaign' – for major field applications of insecticides in large areas of Britain.[6] No thought appears to have been given to the root-cause of the BYDV epidemic – namely, modern agricultural practices.

Little Research on Crop Ecology

The effects of such twists in the ratchet of intensification are not always as obvious as in the case of BYDV. There is hardly any field research on the impact of pesticides on crop ecology and wildlife in Britain, and none of any consequence which is published by the government.

Nonetheless, anecdotal accounts abound. In the summer of 1985, for example, John Cherrington, the agricultural correspondent of the *Financial Times*, wrote, 'We have learned to kill the charlock . . . but a peculiarity of the situation is that (it has been) replaced by other weeds.' Cherrington reported that pansies, cleavers and white mustard were the principal invaders. He should not have been so surprised: it is a cardinal principle of ecology that if one removes a dominant species, then competitors which were previously scarce, will flourish. But ecologists are few and far between amongst the scientists who advise the government on farming policy: it is chemists who rule the roost.

British farmers are far behind their contemporaries in the USA and in much of Europe in trying first to reduce, then progressively to eliminate, pesticide use under a programme of Integrated Pest Management (IPM). A European Commission report points out that while psychological and institutional factors still impede the development of IPM in Europe, such methods can routinely reduce pesticide applications 'by between 20 per cent and 70 per cent depending on crop and local factors'.[7] As Lord Northbourne pointed out in the House of Lords during the debate on the Food and Environment Protection Bill, the Ministry of Agriculture cannot or does not provide any practical advice to farmers. Yet the ministry's own research demonstrated that large reductions can be achieved. For example Cranham showed that applications to the 2,500 acres of apple orchards which he studied could be reduced by 35 per cent.[8]

Other researchers have studied the ecological and economic consequences of intensified farm systems. A recent study shows how oil seed rape, when grown in massive monocultures (a practice encouraged through EEC subsidies) gradually acquires new pests, in the same way as new islands are colonized.[9] Consequently, between 1980 and 1982, the pesticide costs involved in growing oil seed rape rose from 8.6 per cent of total costs to 26 per cent. It was also noted that low levels of infestation meant that farms are using pesticides on rape when it is 'rarely justified'.

Pesticides Create Pests

Each summer, oil seed rape growers (like other farmers) are subject to intensive, and effectively uncontrolled, advertising campaigns from pesticide manufacturers. Hostathion (the organophosphorous insecticide triazophos) features heavily on the front page of several farming magazines for weeks on end. In a study of agrochemical advertising in Britain D. Leeks and J. D. Mumford found that between 1960 and 1975 'advertising of pesticides increased considerably, peak monthly advertising occurred earlier in the year, and the emphasis was less on safety and profitability, and more on ease of use and multiple action.'[10] In other words, contrary to the pesticide industry's claims, pesticides are becoming *less* discriminating in what they hit, and the ability to kill a wide range of species has become a selling point.

Modern wide-spectrum (or multiple-action) pesticides are a threat to wildlife in a way which the older long-lasting and accumulative organochlorines were not. Organophosphorous and pyrethroid insecticides are often ten or a hundred times more powerful per unit weight than DDT and similar wide-acting organochlorines. They are highly toxic to many species: indeed, 88 per cent of modern insecticides are officially acknowledged to be harmful or dangerous to fish. Moreover, drift of even minute quantities can wreak havoc with 'non-target' populations. Volatile and drifting herbicides can have a similar effect on flora. Birmingham University ecologist Graham Martin has pointed out that wild roses have probably disappeared from hedgerows in the Vale of Evesham as a result of herbicide pollution.

Both the elimination of food plants (violets, for instance) and direct 'knock-down' are likely to be involved in the massive losses in butterfly populations now being revealed by the Game Conservancy in studies of sprayed and unsprayed fields. In a unique research programme on Hampshire cereal farmland, Dr Nick Sotherton of the Game Conservancy found that both the abundance and number of butterfly species are dramatically reduced on sprayed as opposed to unsprayed 'headlands' (strips at field ends). In one sample, sprayed headlands produced only 300 butterfly sightings while the unsprayed area had 800. The research also shows that with increased insect food available to young birds on unsprayed headlands, the grey partridge population increased from 60 to 140 in two years. The 30 per cent production loss incurred in these headlands created an overall farm loss of only 0.7 per cent (on 2 per cent of the area), which the farmer regarded as insignificant.

In the course of its studies the Game Conservancy also discovered that a fungicide sold under the trade name Missile (containing the active ingredient pyrazophos – an organophosphate), is strongly insecticidal. Farmers using this against fungi in cereal crops actually risk increasing their pest problems by unwittingly killing off beneficial insects, particularly predatory beetles. The reduced numbers of beetles allow aphid populations to 'escape', proliferating to the point where they exceed the 'economic' threshold and justify the use of an extra spray.

Meanwhile, another problem for the farmer is emerging. Since the 1960s when leaf spot disease on oats first became resistant to organo-mercury fungicides, the chemical control of cereal fungal disease has undergone a cascade of failures as resistance has increased. After organomercury resistance appeared, there was a rapid spread of mildews resistant to the major ethirimol group of chemicals and, since 1981, control of eyespot disease with a third type of fungicides (MBC) compounds) has also begun to break down. By 1983, 60 per cent of all main cereal crops in Britain supported resistant fungi. There are also increasing reports that control of barley mildew fungus with triazole fungicides has become ineffective. Faced with growing resistance and cross-resistance, farmers have resorted to multiple dosing, or mixing fungicide cocktails. The health, safety and environmental risks of these cocktails are often not fully tested or known.

With no effective advice from the Ministry of Agriculture on crop ecology, no warning about ecological side-effects on pesticide labels, and no official strategy for reducing pesticide use, farmers are left guessing how many more problems they are creating for themselves and for wildlife.

Other Costs of Intensification

The effects of pesticide intensification can be alarming in themselves. The organophosphorous compounds recommended against BYDV, for example, are acutely toxic to people and wildlife at high concentrations. They also have serious sub-lethal effects, including loss of motor control, vision disturbances, depression, inflammation, flu symptoms and disorientation. In winter conditions, when birds require every minute of daylight for feeding, such behavioural changes caused by low-level poisoning may prove fatal. Perhaps significantly, it has been reported that MAFF discontinued research into such effects after they were uncovered in experiments on starlings.[11]

The impact of pesticides on human health has been systematically overlooked by successive governments. After the initial flush of concern in the 1950s, when farm workers dropped dead in the fields from using DNOC, the official view, repeated, reiterated, and amplified throughout the farming and agrochemical industry until it has become an article of faith, is: 'Pesticides are safe if used according to the instructions.'

There are several objections to this. First, as the Agricultural Workers Union has repeatedly pointed out, in practice it is often impossible to use pesticides 'according to the instructions'. The difficulties were vividly summed up by John Home-Robertson, an MP and a farmer, who told the House of Commons on 18 December 1979,

> I doubt whether any of the [Advisory Committee on Pesticides] has had the practical problem of, for example, having to mix a supposedly soluble chemical in precise quantities into cold water while balancing on a ladder leaning against a sprayer in the corner of a muddy field. Have they ever had to eat a sandwich meal in a tractor cab with their hands covered in DDT and miles from the nearest tap let alone proper washing facilities?

Secondly, the government system broadly assumes people do not come into accidental contact with pesticides as non-users. But they do. Thirdly, we are asked to accept the official view on trust as the results of safety tests are kept secret. Yet the industry and the government do not act in a trustworthy manner.

Officially there has been no major health, safety or environmental problem with pesticides over the last 20 or 30 years. Introducing the 1985 Food and Environment Protection Act (an act forced upon the government by the EEC but welcomed by the British chemical industry as a mechanism to keep out cheap imports), Agriculture Minister John

MacGregor stated 'existing arrangements were excellent in themselves.'[12] His views echoed those of the industry itself. Attacking parliamentary and public attempts to impose stricter controls over pesticides, R. F. Norman, chairman of the British Agrochemicals Association, wrote in the newsletter of the National Association of Agricultural Contractors that, despite the efforts of 'Lord Melchett and some other ecofreaks, we have a proud record of 27 years without any fatalities from pesticides and no unreasonable incidents.'

Yet within 18 months of starting its first campaign on pesticides, Friends of the Earth received hundreds of accounts of children, parents, farm workers, farm secretaries, farmers, vets, teachers, firemen, policemen, drivers and others, together with pets and domestic stock, being adversely affected by drift or other accidental exposure to pesticides. Many of these were included in FOE's *The First Incidents Report* and others are due for publication in a second report.[13] Some examples are given below.

FOE found that where one incident was reported, further inquiry often revealed many more in the same village or even street. In some areas, it was no exaggeration to say that pesticide spray created a local epidemic. Villages surrounded by oil seed rape or cereals, houses abutting potato fields subject to aerial spraying, and schools in intensive arable areas are all examples of places at special risk. In north Hertfordshire, for example, it has become routine for some schools to send children home with a note to parents warning them that 'as it is now the spraying season', pets and children should be kept off footpaths. Thus the countryside can become a no-go area if you value your health. As one consultant surgeon put it to me: 'I would rather live in the smogs of Los Angeles than in the chemical soup of rural East Anglia.'

Human Victims: Some Examples

The following case histories are just a few of those presented by FOE in *The First Incidents Report*:

- In 1984 J. N. B. Richardson of Grendon near Northampton wrote to Friends of the Earth in Rushden:

  Until March this year I was a perfectly healthy 45 year old with no history whatsoever of either asthma, hay fever or other allergy. Grendon, an agricultural village, has this year been surrounded by extensive crops of oil seed rape and I distinctly remember one particular occasion in March [when there was] a strange chemical taste in the air. A drive later in the afternoon revealed crop spraying to be underway on a wide scale and the prevailing winds were blowing towards Grendon. . . .

On the following day I developed what I thought to be a heavy cold which continued throughout the rest of March, April, and into May. All this time I was sneezing, gasping for breath, and coughing – all to an alarming extent. My doctor prescribed all sorts of remedies, indeed I went through the contents of Boots' shelves trying to find some relief. In desperation I was referred to the Ear, Nose and Throat specialist of Kettering General Hospital. To no avail, however. The discomforts eased in May, except I was convulsed by a racking cough for a further month and it is only now, the spraying season finished, and the rape gathered in, that I have found any comfort at all. The final diagnosis by my medical advisers is a type of asthma/hay fever from which it appears I am doomed to suffer for the rest of my life. I can only conclude that spraying was the original cause.

- On Tuesday 21 May 1985, a chemical company (H. L. Chemicals of Holbeach, Spalding) sprayed a field of rape in mid-flower with Fenvalerate, a synthetic pyrethroid – despite a label recommending that spraying be avoided at flowering because of possible damage to bees. Local beekeeper, V. M. Burnam, says this is typical of incidents in which a number of hives have been lost.

- Mrs Rosemary Reeve's mother believes that her 25 year old daughter, who has been in Colchester Mental Hospital for the past five years, is still hospitalized as a result of exposure to farm chemicals while working with cattle at the age of 16. A medical report for the disablement board stated her illness was due to poisoning with agricultural chemicals. Her case has been put to her MP, John Wakeham. The mental hospital is surrounded by intensive farmland and spraying may, Mrs Reeve thinks, have caused relapses.

- On 17 June 1985, Brian Jackson's smallholding near Goole, Humberside, was enveloped in spray drift causing fish to die and poisoning chickens. In 1984 his donkey died as a result of spray drift.

- On 13 June 1985, dozens of schoolchildren at North Ockenden near Upminster, Essex, suffered severe sore throats and running noses when Metasystox insecticide spray drifted through the school windows. The NFU told a local mother that the chemical was safe: in fact, it is a potent nerve agent which can have prolonged effects.

- In the period 31 July to 13 August 1984, at least four and possibly six dogs were found poisoned in the Bricket Wood area of St Albans.

Patricia Hughes contacted FOE London in October and reported that a 'rice like substance' had been found on nearby common land, and taken to London Colney Police Station for analysis. At least one dead dog turned out on post-mortem to have a large quantity of metaldehyde (slug poison) in its stomach. In another case 'seeds and spray poison' were reported to have been found in the stomach. A field ajacent to the Meads, the area in which the dogs died, was sprayed with paraquat on 7 August.

- In September 1984 a wood treatment company applied a variety of chemicals to a house in London SW16. These later turned out to include tributyl tin oxide (against dry rot) and dieldrin (a persistent organochlorine insecticide). The fumes from this treatment entered the part of the house used by Barbara Kearns and her baby son Jake. No written warning was made available at this time. After her son became ill, and she also felt ill, Barbara Kearns contacted the company. They were initially unable to tell her what chemicals had been involved, providing different accounts of the episode on several occasions.

On being admitted to hospital, baby Jake was found to be suffering from severe bronchitis. Staff had initially agreed with Barbara Kearns that her son was probably suffering symptoms brought on by spray fumes but once a journalist started questioning them, senior medical staff insisted that there was no proven link. The Health and Safety Executive which had initially said that chest problems resulting from spraying were 'quite common', later wrote in a letter to Barbara Kearns, that 'whilst TBTO is a poison, neither the contractor nor specialist colleagues have heard of any previous health problems arising from the vapours of applied dilute solutions' and 'Dieldrin is an organochlorine insecticide which is known to cause poisoning but the reference information available to me does not refer to possible chest conditions.'

Barabara Kearns told FOE in a letter in October that 'preventative precautions were virtually non-existent. When the workmen had *finished* spraying, *after* they had filled two rooms with vapour; *after* they had had lunch, replastered, cleared up and gone home at the end of the day – then they left a note saying to keep the place ventilated (etc.). No mention was made to me whilst I was actually talking to the workmen and they saw me with the small child. They then left this note in such a place that I didn't see it for a week.'

After being discharged from hospital, Jake's blood was sampled and sent for analysis at a special clinic. It was found to contain the following organochlorine pesticides: 0.50 ppb dieldrin, 2.90 ppb beta BHC, 0.90 ppb DDT, 22.20 ppb DDE (a metabolite of DDT), 0.50 ppb gamma chlordane, 0.50 ppb heptachlor expoxide, 0.10 ppb transmonachlor, 2.0 ppb hexachlorobenzene.

Lack of Epidemiological Studies

The long-term effects of pesticide contamination are difficult to disentangle without extensive and careful epidemiological study; yet, successive British governments of all political persuasions have shown themselves pointedly resistant to such studies. In the USA, where such studies are more advanced, it has been recently calculated that of four to five million agricultural workers, some two million are yearly at risk of exposure to organophosphorous compounds.[14] Assuming that only 1 per cent of cases are reported some 80,000 poisonings occur each year. The researchers also found that 80 per cent of the human urine samples they analysed contained detectable levels of PCP,[15] 99 per cent contained DDT residues, and 2–4 per cent carbamate residues. Of the human milk samples tested 80 per cent contained some trace of dieldrin and 63 per cent had traces of heptachlor. Another study reports:

> Chlorinated hydrocarbon residues in the serum of non-occupationally exposed pregnant women in rural Mississippi and from the umbilical cord of their infants at delivery were comparable to mean levels reported in occupationally exposed chemical company employees, and two to five times higher for both mothers and newborns than the average levels for the total US population.[16]

Recent (and still controversial) research from Canada suggests a link between pesticide use and the incidence of neurological illness. Professor André Barbeau of the Montreal Clinical Research Unit found a very high correlation between the amount of pesticides used across nine Quebec provinces and the occurrence of Parkinson's disease.[17] Further, a chemical involved in the disease – MPTP – is structurally very similar to the active agent in paraquat.

It is difficult to see how such findings have not prompted similar investigations in Britain. Yet apart from the work of doctors such as Jean Monro of the BUPA Nightingale Hospital,[18] who specializes in the identification and treatment of chemically induced allergies and sensitivities (and so finds herself working at the fringes of accepted medicine), there are few studies which are not kept government secrets.

Keeping the Data Under Wraps

The adverse health effects of pesticides on the public do not show up in statistics because the information is either not collected or is kept secret. The Health and Safety Executive's Agricultural Inspectorate frequently use the secrecy clause (28(2)) of the Health and Safety at Work Act to deny the public even such basic information as the name of the chemical involved in a given spray incident. This also stymies independent research. The complete absence of figures for the amounts of different pesticides used on farms makes the detailed study of pesticide related cancers, nervous disorders or other potential effects extremely difficult, if not impossible.

In 1984, only one government study was in progress to examine possible links between disease and the intensive use of agrochemicals.[19] Base-line environmental monitoring has been studiously ignored, or even reduced, so ministers can safely reply that there is 'no evidence' of problems. For example, in a parliamentary reply to a question by John Cartwright MP on airborne pesticide pollution, environment minister William Waldegrave stated that 'Routine general monitoring of ambient levels of pesticides in the atmosphere is not thought to be necessary or justified.' So too, the government has little idea of the extent to which drinking water is contaminated since 'Data on pesticides detected in rivers and groundwater are not held centrally.'

Despite the official patina of bland reassurances, local doctors have become extremely concerned at the levels of cancers in small villages, and at the incidence of nervous disorders, allergies, 'flu' (the latter especially during spraying of organophosphorous compounds) and other health effects apparently linked to pesticide use. In Lincolnshire, for example, general practitioner Peter Mansfield managed to get the tacit support of the Health and Safety Executive for a sampling programme to study the changes in cholinesterase in the blood of rural residents exposed to spray drift.[20] Interest in the problem has since prompted an unannounced but major reorganization of data collection within the HSE. Nonetheless, the government has taken no steps to encourage the public to report incidents and most observers are convinced that it is frightened of taking the lid off what would be a very controversial 'can of worms'.

Pesticides in Food

Although pesticides can present an acute hazard to users and spray victims, almost everyone receives small, repeated doses in their food. Although the levels found in British food are almost always far below those liable to cause short-term signs of acute poisoning, there is the concern that they may induce chronic disorders including allergies, hypersensitivity, birth defects or cancers. The official position is that there is 'no cause for concern'. Yet this reassurance is based on some very slim evidence. Pesticides are not tested for their effects on chemically sensitive people. Food sampling is extremely limited and the results are often kept secret. No national data are compiled.[21]

Whilst 87,000 samples of food were tested in 1981 for bacterial contamination only 338 were tested for pesticides. The fullest studies have been carried out by the Association of Public Analysts, whose laboratories are attached to county councils. A 1983 survey by the Association of fruit and vegetables bought in retail shops showed a third to be contaminated with at least one pesticide.[22] One fruit sample in ten and one vegetable in five had residues exceeding the 'reporting limits' set by the ministry. Such limits are themselves scaled down from levels known to cause acute toxic reactions (hence the well-worn argument 'it would take a tonne to poison you'), but do not take any account of carcinogenicity, teratogenicity or other chronic effects for which there is no safe lower level.

The study also showed that DDT, ostensibly withdrawn for use on such crops, was present on ten per cent of apples, and on other fruit and vegetables bought in shops in November 1983. In some cases the levels were high. The apples tested included Cox's Orange Pippins, Bramleys, Russet, Lambourne and Worcester: all English varieties used by the marketing industry to promote an image of health-giving countryside freshness. Yet one in ten contained a banned pesticide.

The consumer of gin and tonic is also unlikely to escape accidental ingestion of pesticides. Dr Bob Poller of Queen Elizabeth College, London, has found up to 20 times the recommended maximum limit of fungicides such as biphenyls in lemons. In fact, this is no accident as the fungicide is deliberately added in a cosmetic glaze based on bees' wax. And the glaze is necessary in order to promote sales, because so much insecticide is used in orchards that the fruit has to be scrubbed clean after harvest, so making it unattractive to consumers. There are a number of other studies showing how consumers may sometimes receive quite large doses of pesticide. For example, in a MAFF study which deliberately added fungicide to potatoes, residues of up to 218 times the maximum recommended limits for tecnazene were detected.[23]

Of course, the consumer cannot tell if a lemon, potato or apple is contaminated. The important question here is whether it is acceptable for people to be exposed at all to chemicals which may induce a malignancy or a birth defect, especially when the risk need never arise. The fact that one cannot single out a particular death and say 'Ingesting pesticide 12 years ago, killed this person' is a legal nicety which may enable companies to escape court actions but which is not a valid excuse for government to abrogate its responsibility to the public.

Pesticide contamination of food is just one end result of Britain's failure to control the use of such chemicals. It is primarily a failure of central government, which has given no resources or real guidance to other consumer protection agencies – a mere 50 samples are believed to have been analysed by Kent County Council in 1983 and less than 200 in Avon in 1984. It was partly the need to enforce an EEC regulation (Directive 76/895) which led the government to introduce the Food and Environment Protection Act; yet without a reversed policy, the act will protect neither food nor the environment. While British cabbage contaminated with captan may still be served up in British school dinners,[24] and while butchers still sell Welsh lambs' kidneys that carry dieldrin,[25] there seems no doubt that our pesticide policy is neither effective nor enforced.

Conclusion

It has been said only half jokingly that ICI has been known to receive cabinet papers before some cabinet ministers. The capacity of public interest groups to investigate, probe, question and challenge such companies, is puny. Moreover, the agrochemical industry enjoys the closest co-operation with government. In the case of companies like ICI this dates back to the time when as developers of phenoxy acid herbicides, their Jeallotts Hill Research Station was treated as a top secret, almost military, establishment. This is not to say that a company like ICI is deliberately malicious: there is little if any evidence of that, and ICI itself is probably one of 'the better' companies from a consumer, health and environment viewpoint. But the whole system is geared to accommodating and nurturing the industry's interest first, and anything else second.

In an era when we realize that the capacity of the environment to absorb and deal with toxic chemicals and pollutants is more and more exhausted, and as disease is increasingly linked to pollutants and diet, such a gross failure of policy is little short of catastrophic.

Notes

1. John Sheail, *Pesticides and Nature Conservation: The British Experience 1950 – 1975*, Oxford: Clarendon Press, 1985.
2. James Erlichman, *Guardian*, 25 June 1985.
3. Friends of the Earth, *The Avon Herons*, Vale of Evesham, 1985.
4. The attitude of MAFF has encouraged this practice: it has routinely allowed farmers to

'use up existing stocks' when products have been withdrawn under the PSPS, and one ecologist recalls that although MAFF finally agreed DDT was 'too dangerous' to wildlife to be used, the ministry men suggested the only 'safe way' to dispose of existing stocks was to allow them to be used up!

5.   Although such checks could be introduced under the Food and Environment Protection Act, they will be limited at best, as the government refuses to allow Environmental Health Officers to become involved.

6.   Whilst working for Friends of the Earth, I put this view (which is not contested by numerous other independent observers) to the Ministry of Agriculture. 'You seek to damage the reputation of my Department' shrilled an outraged Agriculture Minister, Michael Jopling, in a letter to FOE. But he went on to admit the cause of concern: government advice on BYDV meant that in some cases even where there was no pest present, the crop would be dosed with organophosphorous or pyrethroid insecticides 'just in case'. More important still, MAFF could not deny that BYDV was a disease problem actually created by over-intensive agricultural practices which the ministry itself had aided and abetted.

7.   J. Bassino and A. Mouchart, *DG XI Contract*, 6613/03, unpublished, December 1983.

8.   J. E. Cranham, *Agriculture and Environment*, 1982, pp. 63 – 71.

9.   Lane and Norton, 'Agronomy, physiology, plant breeding of oil seed rape', *Aspects of Applied Biology*, no.6, 1984.

10.   D. Leeks and J. D. Mumford, *Protection Ecology*, no.4, 1982, pp. 59–65.

11.   Gail Vines, *New Scientist*, vol. 97, January 1983, p. 1340

12.   *Hansard*, col. 795.

13.   C. Rose, *The First Incidents Report*, FOE, 1984.

14.   Coye, Lowe, Maddy, 'Biological Monitoring of Agricultural Workers Exposed to Pesticides', *Proceedings of Conference on Medical Screening and Biological Monitoring for the Effects of Exposure in the Workplace*, California Department of Food and Agriculture, July 1984.

15.   In 1984, for example, the US EPA cancelled almost all remaining uses of PCP – pentachlorophenol – after an accumulation of data showing birth defects in laboratory animals and cancers arising from impurities. Yet in Britain, PCP is still employed in many products and – in defiance of a supposed withdrawal – it is widely available over the shop counter in wood preservatives.

16.   A. J. D'Ercole, R. D. Arthur, J. D. Cain, B. E. Barrentine, *Paediatrics*, vol. 57, no. 6, 1976, pp. 869–74.

17.   Andre Barbeau, *Science*, vol. 229, 19 July 1985.

18.   BUPA Nightingale Hospital, Lisson Grove, London NW1.

19.   *Hansard*, col. 572, 20 November 1984.

20.   Recently, researchers discovered that less than 1 per cent of many pesticide sprays ever reseach the target organism and between 20 and 30 per cent often drifts in minute droplets for hundreds or even tens of thousands of metres, becoming more concentrated as the droplets evaporate. British studies are few but in the USA a report for the Environmental Protection Agency found that between 10 and 60 per cent of pesticides sprayed onto crops from aircraft drifted more than 1,000 feet from the target. For years, the British 'safe distance' from houses was just 50 feet. Under the proposed regulations of the Food and Environment Protection Act, it was increased to 200 feet. In Belgium the distance is 2,000 metres.

21.   At the time of writing several important publications on pesticide contamination of food are in press including a major report by the London Food Commission and a book by James Erlichman to be published by Penguin.

22.   R. S. Nicholson, *Journal of the Association of Public Analysts*, no.22, 1984, pp. 51–7.

23.   J. Watts, *An Investigation into the Use and Effects of Pesticides in the United Kingdom*, FOE, London, 1985.

24.   Banned in Norway as a potential carcinogen but detected in the 1983 APA survey on cabbage and cauliflower and sold freely to gardeners in Britain.

25.   Officially banned as a sheep dip and widely regarded as a probable carcinogen.

"It's perfectly safe Obergurgh, I've notified the Genetic Manipulations Control Board!"

13

2,4,5–T

Britain Out on a Limb

Chris Kaufman

2,4,5–T is a weedkiller used mainly for dealing with woody weeds of the bramble type. It is inevitably contaminated with dioxin – the most toxic synthetic chemical on earth – and for that reason its use is banned in a string of countries including Italy (remember the terrible effects of the dioxin cloud released in the Seveso explosion), Sweden, Norway, Denmark, USA (for most purposes), Japan and Holland. Yet in Britain, the Advisory Committee on Pesticides has repeatedly given the herbicide a clean bill of health. Although a number of brands have been withdrawn from sale, many can still be bought over the counter and are widely used.

The manufacture of 2,4,5–T has now ceased in every major country in the world except New Zealand, following a score of accidents. US GIs who were exposed to the defoliant Agent Orange (a mixture of 2,4,5–T and 2,4–D, another herbicide) in Vietnam have suffered cancers and there has been a high incidence of birth defects amongst their children. Recently, the affected veterans have won the largest ever out-of-court settlement to be awarded in the USA, amounting to 180 million dollars, from manufacturers of Agent Orange, such as Dow Chemical. Those who are at risk from the herbicide include such diverse groups as country ramblers, amateur gardeners, farm, forestry, horticultural, public parks and garden workers, shop workers, lorry drivers, chemical and dock workers and those who wash the protective clothing used by workers who use the herbicide. Over 100 local authorities in Britain and many major employers (like British Rail, the CEGB, and the Forestry Commission) no longer use 2,4,5–T and many trade unions (notably the Transport and General Workers Union, the Association of Scientific, Technical and Managerial Staff, and the Fire Brigades Union) refuse to handle the weedkiller. The TUC, the Labour Party and many others are committed to a total ban on the use of 2,4,5–T.

Given such overwhelming opposition to its use how does the ACP justify its continued support for 2,4,5–T? How have its members managed to convince themselves that they are right and that everyone else is wrong?

'Safe if Used According to the Instructions'

The ACP insists that the use of 2,4,5–T (and indeed many other 'problem' pesticides) is safe 'if the instructions are followed'. One of these instructions is to keep the chemical off the skin. I have yet to meet a farm, forestry or horticultural worker, however, who has not at some time or other been drenched in spray or choked in chemical dust whilst using pesticides. How many ACP members mix with the people who will be using the chemicals they pass as safe to use? How many have experienced the joys of breaking open a frozen pesticide container in a sub-zero dawn; having to wash their hair five times to remove chemical dust; feeling the splash of concentrates whilst pouring from an unco-operative can, perhaps whilst they are poised precariously on a wobbly ladder at the bottom of a muddy field; feeling the chemical on their skin or in their throat from a leaky back-pack sprayer; attempting to unblock the nozzles on a tractor sprayboom; or smelling the fumes in the kitchen as they strip off their protective clothing – clothing which, if provided at all, often offers little protection?

The answer is 'none'. There is not a single workers' representative on the ACP – or an environmentalist for that matter – no one, in fact, who could inject a note of practicality into the committee's decision. They would be out of place, it is intimated. With one voice, the British Agrochemicals Association and the Ministry of Agriculture declare, 'This is a forum for experts not for political arguments.'

Playing Down Incriminating Evidence

In the words of the old song, ACP reports and arguments 'accentuate the positive, eliminate the negative'. In assessing the safety of any pesticide (not just 2,4,5–T) the committee relies on data from animal experiments and other laboratory work, together with anything that is known of the particular pesticide's effects on people. It should be noted, in passing, that the committee does not initiate its own research. No research has been conducted on the levels of exposure to which pesticide users are subjected, save the crudest public relations stunts carried out by the agrochemicals industry.

In the case of 2,4,5–T the ACP admits that the chemical, with its contaminant, dioxin, will cause birth deformities and cancer in laboratory animals but argues that this does not necessarily point to the same effects in human beings. The committee omits to point out – or plays down – a number of other facts, including the following:

- Another well-known teratogen or birth deformer, called thalidomide, caused more harm to humans than animals.
- Dioxin is known to be a most potent initiator and promoter of cancer.
- Reputable studies of the effects of exposure to 2,4,5–T on Swedish workers, more recently on US process workers and on Danish workers, have shown that it greatly increases the incidence of a very rare form of soft tissue cancer over that among the public at large.

The ACP's reaction has been to 'rubbish' these reports, though they have been accepted as significant even by scientists employed by US chemical companies. Instead, the committee accentuates other studies of often dubious quality which give 2,4,5–T a clean bill of health. And this is the committee which tells the public that it has to 'Err on the side of caution'. When challenged about the medical effects of dioxin, the ACP claim that the levels of dioxin present in 2,4,5–T are safe. Yet the committee has failed to bring forward any convincing data to demonstrate what constitutes a 'safe' level of exposure. The level quoted is always that to which people are currently exposed.

Evidence of Health Effects: The ACP's Reaction

Because of the lack of any official research into the effects of 2,4,5–T on humans, the Transport and General Worker's Union compiled a dossier of incidents which had been reported to it by members of the public or the union. Among other cases, the dossier revealed how Mrs Hogben's child Kerry was born with terrible deformities from which she has since died; how Mrs Scheltinga miscarried after picking blackberries in an area recently sprayed; and how the children of farmworker David Thomas, forestry workers' wives Wendy Cobbledick and Mrs Chidgey, research chemist Dr Campbell and amateur gardener David Salt were all born with deformities after their parents had been exposed to 2,4,5–T.

The ACP reacted to the TGWU dossier by taking each case separately, accusing the union of being alarmist and attempting, by every means, to absolve 2,4,5–T from any blame in the matter. The British Agrochemical Association was also ready to join (and often lead) the chorus against the evidence the union had collected. The ministry too was unwilling to accept the implications of the union's dossier. When Jack Boddy (general secretary of the National Union of Agriculture and Allied Workers) and I challenged the then Minister of Agriculture, Peter Walker, he responded that if the union would go round campaigning against 2,4,5–T, then parents would inevitably seek to attribute to it any miscarriages or birth defects. If that were so, we replied, let him show that they were wrong.

Whilst the link between exposure to 2,4,5–T and subsequent health effects in any one of the 100 cases we have collected cannot be proved beyond *all* doubt (and therefore to the satisfaction of the ACP) the overall picture they paint, together with the very similar experience in many other countries, and all the important experimental evidence already referred to, provide a massive indictment of this herbicide which no government that is truly concerned with the health and welfare of its electorate can conceivably ignore.

Distorting the Evidence

The calibre of the ACP's investigation into the case studies we presented was disturbing. The committee had purported to investigate the circumstances in which the victims had come into contact with 2,4,5–T and the links with subsequent medical effects, like cancer, deformed births, liver damage and skin diseases. The final reports of the committee were severely at variance with important details (such as dates and subsequent medical history) given to them by the people whose cases were under scrutiny. For example the ACP's 1980 review said of Mrs Hogben: 'It is highly improbable that the malformations were the result of genetic mutation. The subsequent birth of a healthy child also argues against any long term genetic effects.' In fact Mrs Hogben carried her next baby for just 27 weeks and it lived for only a few hours.

One wonders what an ACP investigation might make of the death of Mrs Fiedler's son. The Coventry horticultural worker died of a rare cancer aged 24 after working with 2,4,5–T for five years, mixing it for back-pack sprays and then using it himself. He asked at the hospital whether 2,4,5–T could have been the cause of his cancer and was told, 'The Ministry of Agriculture says it is harmless.' Would the committee ever back down and admit to letting a killer onto the market?

Getting the Facts Wrong

When the Agricultural Worker's Union took up the cudgels against 2,4,5–T in earnest in 1979, we were frequently told by manufacturers and government spokesmen that it was a stupid issue to get worked up about because so little 2,4,5–T was being used. In fact, the ACP stated in its *Eighth Review* that only three tonnes of the herbicide had been imported into Britain in 1978. The figure seemed odd to us because so many of our members seemed to be using 2,4,5–T. Were they the only ones?

It was only by dint of tenacious probing that Dr Roger Thomas MP discovered that 164 tonnes had in fact been imported in 1978, of which 108 tonnes was re-exported (no doubt to Third World countries whose workers are even less protected or aware of the problems than British workers). 'How come,' we asked a clearly embarrassed ACP, 'you got the facts so wrong?' 'The figures we were given underestimated the amount used on grassland' came back the lame reply. But it left the union delegation wondering: if the committee was so inept in handling questions of fact, how reliable would it be on questions of judgement?

Peter Walker told us that as Minister of Agriculture, he had to reply on the judgement of his expert advisers – the ACP. But where had the ACP got its figures from? From the Ministry of Agriculture, of course.

There was an 'other wordly quality' about the whole debate and particularly our meeting with the ACP. It seemed so wrong that the potential, and probably real, victims were having to argue against a status quo which clearly threatened further casualties. Why was the onus of proof of safety not on those who allowed the chemical onto the market?

A Question of Proof

Here a crucial difference between our two approaches emerges. 'We will rescind its clearance if the union can prove to us that 2,4,5–T is harmful', was in effect what the ACP told the union delegation. 'No', we responded, we cannot supply proof 'beyond all reasonable doubt.' Our yardstick is to estimate the hazard on the basis of what we know, and if 'on the balance of probabilities', the substance appears dangerous, then it should clearly be taken off the market.

In short, our argument was and still is, that we are not prepared to supply the sort of evidence that they seem to insist is necessary – namely a sufficiently large number of proven victims, a high enough body count. We have learnt the lessons of asbestos, thalidomode, vinyl chloride and a string of other public health scandals. We will not have workers used as human guinea-pigs any more.

The government assures us that 'evidence' is something that only experts can provide. But experts can differ. We looked at the evidence and concluded that the use of 2,4,5–T presented such high risks that we had no alternative but to tell our members not to use it. The committee, needless to say, looking at the same evidence, concluded that it was safe.

A Collusory System

The system for adjudicating the safety of pesticides – despite proposals for a few minor changes going through parliament – remains essentially closed and secretive. It is also a collusory system to the extent that advice as to the sort of data which manufacturers should submit to the ACP when seeking the committee's approval for a new product is supplied by the Ministry of Agriculture. MAFF officials would not be human if they did not want to see their advice shown to be sound.

Then again, great play is made of the independent status of the committee. It consists of academic and government scientists, and toxicologists. How many of these are themselves doing research that is funded by the very industry whose products they are called upon to assess? How many may be running courses funded by these same chemical companies? How many blindly accept the basic assumption of the agrochemical industry and the Ministry of Agriculture that pesticides are a 'good thing', whose value to society must far outweigh their potential hazards to workers and wildlife, countryside and consumers? The answer is surely most if not all of them.

Overhauling the System

Although it is now intended to include union representatives at one or two new ACP subcommittees established under the Food and Environment Act, any co-opting of a union or consumer voice on to the ACP would fail to tackle the crux of the problem. We can have no faith in a decision-making body which comes under the Ministry of Agriculture's umbrella because of the close links already referred to between MAFF and the chemical manufacturers and also because of the ministry's overriding goal of maximizing productivity at any cost – a policy that is irreconcilable with its declared goal of assuring that the safety of human beings, animals and the environment be paramount. The two do not mix. It must be one or the other. Yet the Food and Environment Act goes further than even the existing arrangements in spelling out that this is exactly the role envisaged for the ACP.

By all means let MAFF and the ACP continue to consider the efficacy of pesticides. But let questions of health and safety be dealt with by the appropriate body: the Health and Safety Executive. This would make it possible for workers to sit together with employers and government nominees on the HSE tripartite committees. The existing HSE Advisory Committee on Toxic Substances, perhaps with the addition of environmental representatives, could well serve as the model.

That would be an important first step in absorbing the lessons of the 2,4,5–T campaign, which exposed, in the words of a successful TUC resolution, that 'the system for allowing safe clearance to pesticides is a complete shambles.' It would add to the advances we have made through the impetus of the campaign: the setting-up of the Chemicals in Agriculture Subcommittee of the HSE's Agricultural Industry Advisory Committee (a tripartite body which does useful work on 'user problems', such as container design and the disposal of chemicals) and the flourishing committees on both pesticides and reproductive hazards which include trade unionists, concerned scientists, medical people, environmentalists and others concerned with such issues.

And it would put into proper perspective the remark by Dr David Hessayon on his retirement as chairman of the British Agrochemicals Association in May 1981 when he said: 'If we give way on 2,4,5–T, the unions will then go on to campaign against another chemical and then another. We are engaged in a power struggle for control over the industry which we cannot afford to lose.'

Wrong. We are engaged in a power struggle to protect the health and safety of those who work on the land in whatever guise, or who are in any way affected by the unregulated flow of dangerous chemicals into our environment – be they wild animals, or local people, amateur gardeners, or city dwellers using public parks or school playing fields.

That's our vested interest, Dr Hessayon. What's yours?

14

Nitrates in Food

A. H. Walters[1]

It has long been government policy to encourage farmers – both by direct subsidies and via advice provided by ADAS (the Agriculture Development and Advisory Service) – to make the maximum use of nitrogen fertilizer so as to maximize short-term returns. No broadly based research has been undertaken in Britain to establish the effects of the resulting massive increase in fertilizer use either on the quality of food produced or on the health of those who eat it. Worse still, evidence, largely from abroad, establishing that such food may contain a high level of nitrates (constituting a potential health hazard) has either been ignored or systematically played down.

Neither the discovery that nitrate levels in certain vegetables grown in Britain may be up to four times higher than the maximum permitted in Holland, nor the recommendation of the Royal Commission of Environmental Pollution that a study be undertaken on the nitrate content of foodstuffs in Britain, have persuaded the government or its scientific advisers to change their policy of encouraging nitrate use. Indeed, a recent report by the Royal Society actually urges a further increase in the use of nitrogen fertilizers.[2] Can the government's policy be justified?

The Royal Society's View

Annual fertilizer usage in Britain over the period 1928–80 is shown in figure 9. From 1981–4, the trend of increasing fertilizer use has continued. There were 24 members of the Royal Society study group which issued the report, *The Nitrogen Cycle of the United Kingdom*, in 1983, only five of whom were Fellows of the Royal Society. All the members were employed at universities or government-funded institutions, except for one member who worked at the Fertilizer Industry Advisory Committee of the UN Food and Agriculture Organization. The report quotes 510 references to relevant scientific work. Not one is presented which sounds any hint of criticism of British government-sponsored research policies. The report is in effect the work of a closed shop.[3]

The study group's conclusion reflects the composition and interests of its members. The report states:

The increased use of fertilizer N (artificials) in UK agriculture, in conjunction with the introduction of better crop varieties, livestock breeding methods, crop protection and pharmaceutical products, has played a major part in improving the quality and quantity of UK crops and livestock in the last 30 years. Despite advances in agricultural productivity, there is still substantial potential for greater intensification of UK agricultural production. This would require, along with other inputs, the use of more fertilizer N, particularly on grassland. (p. 43)

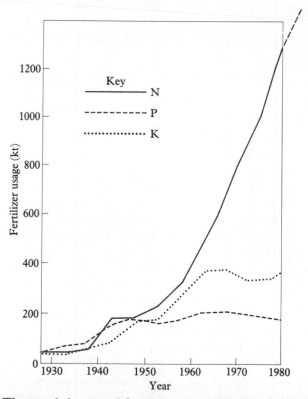

Figure 9 The trends in annual fertilizer usage in Britain during the period 1928–80

This commitment to the ever-increasing use of artificial fertilizer is re-stated, time and again. For instance:

> In the foreseeable future, this increased agricultural production will require the addition of more fertilizer than is applied at present. (p. 53)

> There is still scope for increased use of fertilizer N on grassland, and it is calculated that in England and Wales by the year 2000 an additional 350 kilotonnes of N may be applied annually to grassland; that is, about 40 per cent more than is applied at present. (p. 66)

> We conclude that fertilizer will remain the principal source of nitrogen in the foreseeable future, and that the annual rate of application is likely to continue to rise if maximum crop yields are to be achieved with existing crop varieties. (p. 19)

It is difficult enough to understand how the study group can support an increase in nitrogen use when British agriculture is already over-producing. It is still more difficult, however, to justify that support in the face of the growing evidence that nitrate is a potent health hazard.

Uptake of Nitrates: European Restrictions

When nitrate is used to excess, the food produced becomes loaded with the chemical. The extent to which excess nitrate uptake can occur was amply demonstrated over 15 years ago in laboratories in many different parts of the world. Many references to such work were given in a paper which I published in 1970.[4] In 1972, I was invited by WHO/IRC to lecture on the subject in Switzerland.[5] Serious notice of this work was taken in Switzerland and Holland. Both these countries now have legal limits in force for nitrate levels in certain vegetables. No official notice appears to have been taken of this work in Britain, however, and even in the Royal Society report there is no mention of it. In view of the British government's policy, that is hardly surprising.

For a long time it has been known that nitrate in plants can, after harvesting, break down to nitrite which is toxic. In this respect, the recent work of Claude Aubert in France, 'Les nitrates dans les legumes', is very revealing.[6] Not only does Aubert's work show how the uptake of nitrate into plants rises in relation to the amount of fertilizer applied to the soil in which the plant is grown, but also how, during transport and storage, it may break down to nitrite. The rate of breakdown will vary according to the type of plant, method of cultivation, weather, method of harvesting, and type and time of storage before sale.

In Switzerland and Holland, the permitted levels for nitrate in some common salad vegetables must not exceed 4,000 mg NO_3 /kg in the former, and 5,000 mg/kg in the latter. In lettuce on sale to the public in Britain, Dr J. R. Fletcher of Farnborough College of Technology, has found levels up to 22,000 mg/kg nitrate. In the face of such data, it is little wonder that the recent Royal Commission on Environmental Pollution considered that: 'Better information was needed on the nitrate content in foodstuffs as consumed, and they accordingly recommended that a study should be undertaken. The Government agree.'[7] When set against the Royal Society report, this makes very strange reading.

Human Intake of Nitrate

Since the amount of nitrate (and possibly nitrite) appears to be increasing in the food the farmer grows, one needs to ask what might be the human intake of these chemicals? Table 7 gives figures for the calculated human intake of nitrate from various sources. All such intake figures must be viewed with considerable caution, however. In practice, they will vary according to age, diet, local customs and local geology. In places where a high average concentration of nitrate is known to exist in drinking water, for example, high levels of nitrate intake have been found in the local population, as confirmed by urinary analysis. Such a situation was found in Worksop where the excess nitrate was blamed on local agricultural practices. The Royal Society report acknowledges that levels of nitrate in drinking water are increasing worldwide.

Table 7 Calculated human intake of nitrate from various sources

Foodstuffs	Weekly intake of foodstuff	Nitrate content in foodstuffs (as NO_3)	Human weekly nitrate intake as NO_3 (mg)
Meat products	370.00g	100mg/kg	37
Milk	2.89 litres	30mg/l	87
Cheese	110.00g	45mg/kg	5
Vegetables	1.12kg	200mg/kg	224
Potatoes	1.05kg	60mg/kg	60
Water (variable)	7.0 litres	15mg/kg	105
		Total	518

Source: Agriculture and Pollution, 7th Report, London: HMSO, 1979.

Health Effects

The effects of continued ingestion of increasing levels of nitrate in vegetables and other foodstuffs on the health of susceptible children and adults are unknown. What is known is that some humans are more sensitive to toxic effects than others. In some exceptional instances, *ingested nitrite may in some way break down further to form carcinogenic nitrosamines.*[8] It is also known that excess nitrate in plants may result in an imbalance of micro-nutrients in food. This may contribute, either directly or indirectly, to the development of such conditions as dyslexia or mental retardation in some children, as pointed out by Dr E. Lodge Rees.[9]

One thing is certain. If food contains excess nitrate and is constantly eaten, some people are potentially at risk in some way or another.

Reactions to the Nitrate Controversy

Despite the evidence that nitrate can cause serious health problems, there has been considerable resistance on the part of farmers and other fertilizer users to the imposition of controls on nitrate levels in food. A recent editorial in *The Grower* had this to say: 'For a long time now Dutch growers have had to deal with statutory limits for maximum nitrate contents of some leaf vegetables; these currently apply to spinach, endive and lettuce. But in Holland the list is about to be extended.'[10]

There are similar but more extensive regulations in force in Switzerland. *The Grower* editorial concludes, 'Every grower in the UK should take careful note of what is happening and be ready to fight off any attack when it comes. The nitrate threat could be one of the most serious the industry has yet faced.'

For its part, MAFF has tried hard to avoid becoming embroiled in the controversy. According to a ministry spokesman: 'We are aiming at a policy of agriculture for a plentiful and varied supply of food. The health of the individual is a matter for that individual. Health issues are the concern of the DHSS, not us.' Should we take such a crass statement to mean that no links are recognized between good health and good food? According to Drs Berry and Buss: 'In much of the work relating patterns of morbidity and mortality to diet, the principal source of information about eating habits has been the national Food Survey conducted by the MAFF.'[11] So it rather looks as though the right-hand side of the ministry does not know what the left hand-side is saying.

Alternative Research Needs

Historically, the development of government-sponsored research in agriculture and the commercial development of artificial fertilizers has gone hand in hand. That close collaboration, based on meeting the need for more home-grown food, has produced a one-sided research policy which has been essentially geared to short-term economic priorities.

The pursuit of such a policy, driven on by the powerful commercial thrust of the chemical fertilizer industry, has made the farmer the victim of galloping nitrate consumption. This has led to the uptake of increasing quantities of nitrate and nitrite in food. Warnings that this would happen have consistently been sounded for many years by independent scientists to whom the public are now beginning to listen. Only now has the government at last conceded that it requires better information of the nitrate content of foods, but there are no signs yet that legal limits will be enforced, as in Holland and Switzerland.

Had successive governments heeded the clear early warnings, then research aimed at monitoring the effects of excessive nitrate use could have been instituted long ago. Had that happened, further research would have been necessary to look into complementary or alternative methods of nitrogen supplementation in order to obtain the required yields. It is farmers who should be demanding such research, because it is they who will pay the price for the short-term policies which have precipitated the present problems of over-production, environmental pollution and potential hazards to health. In order to conform to growing consumer trends, every British Farm Institute should have an acreage devoted to alternative methods of farming suitable for the location.

Since these problems are not confined to Britain but affect all the EEC, perhaps the growing importance of public opinion should also be considered. I will leave the last word to Monsieur Jean-Paul Civet, Secretary-General de la Union Fédéral des Consommateurs, who writes: 'We need to make a fundamental reorientation which only organised consumers can bring about. These are questions that can be asked only from the outside, from another viewpoint, and it is there that the discussion and the debate is both useful and necessary.'[12]

Notes

1. This chapter is based on a lecture given to the Andover Branch, NFU, October 1984.
2. The Royal Society, *The Nitrogen Cycle of the United Kingdom – A Study Group Report*, London: The Royal Society, 1983.
3. This is in sharp contrast to a report by the US National Academy of Sciences, the equivalent of the Royal Society. For the compilation of this report, 24 named persons from research laboratories, industry and independent organizations, including myself, were invited to provide information. Thus the US report was very much the work of an open forum. The report was entitled *Accumulation of Nitrate* and was published in 1972.
4. A. H. Walters, 'Nitrate in soil, plants and animals', *Journal of the Soil Association*, vol. 16, no. 149, 1970.
5. A. H. Walters, 'Nitrate in water, soil, plants and animals', *Journal of Environmental Studies*, vol. 5, no. 105, 1973.
6. Claude Aubert, *Nature et Progrès*, Les nitrates dans les légumes, vol. 76, no. 18, 1972.
7. Royal Commission on Environmental Pollution, *Agriculture and Pollution*, paper no. 21, London: HMSO, 1983.
8. *The Environmentalist*, vol. 13, no. 219, 1983.
9. E. Lodge Rees, 'The Concept of Preconceptual Care', *International Journal of Environment Studies*, vol. 17, pp. 37–42, 1981.
10. *The Grower*, 12 January 1984.
11. *British Medical Journal*, 10 March 1984, p. 765. The National Food Survey to which Berry and Buss refer is published as: *MAFF Household Food Consumption and Expenditure in 1981*, London: HMSO, 1983.
12. Le Consumerisme Institut de Recherches de l'Études Publicitaves, Paris, 1978.

15

Food Additives

What are we Really Eating?

Erik Millstone

In the past decade we have witnessed an enormous growth of interest in the possible risks to human health of chemicals in our industrial environment. Yet in Britain few scientists outside the closed worlds of government and industry have scrutinized the use of additives in food. In the USA, France and Scandinavia, for example, far more public and professional attention has been paid to food additives, and these countries have far stricter regulations than prevail in Britain. Industrial chemistry has provided food technologists with a powerful array of tools with which to create their products, but we know surprisingly little about the effects of this 'new food' on human health.

The Use of Additives

Food additives are chemicals which are deliberately used by industrial processors, and they are used if and only if they make the products containing them easier to market and sell. We should be concerned about the economic and ecological significance of these additives, but we should not make the mistake of thinking that agricultural chemicals such as pesticides, hormones and antibiotics are less important or less problematic.

It is primarily highly refined and processed industrial food products which contain chemical additives; indeed many rely on such chemicals for their marketability. The use, consumption and growth of food additives can therefore only be understood as part of a complex process by which the food system and our diets have been changed. These changes have occurred at an accelerating pace particularly since the Second World War.

Currently I estimate that the industrial market for food additives in Britain is worth about £225 million (give or take £25 million) a year.[1] About 1 per cent of that total buys chemicals which protect consumers against acute bacterial hazards and extend shelf lives of products. Another 1 per cent buys 'nutrients' which 'enrich' products which have been impoverished by processing, such as white bread flour, or margarine which never had them in the first place. About 10 per cent buys chemicals which relieve some technical problems encountered in food handling and processing, and the remaining 88 per cent buys cosmetics such as flavourings, colourings and emulsifiers. On average the food industry is buying the equivalent of four kilogrammes of additives per person per year. Many of us eat far less than that, but some people undoubtedly consume significantly more.

In terms of value, approximately 75 per cent of our food is processed and contains additives. This may have at least two potential consequences for consumer health. In the first place processing may lead to a loss of nutrients. In recognition of this problem there are regulations in force requiring the partial 'enrichment' of white bread, flour and margarine. The Food Advisory Committee – the expert committee which advises government on these matters – considers this nutritional reinforcement to be at best a token. Some of our diets have deteriorated to such an extent that the use of these additives probably is valuable to many of the poor, to large families and to the elderly. For most of the rest of us, the *over-consumption* of fats, calories and additives is more likely to be a problem than their under-consumption. Chemical additives which are so ubiquitous in modern food products may also constitute a possible toxic hazard to some or all of the community.

Major Types of Additives

There are essentially three main groups of food additives. The first group consists of consumer protectors. This group is dominated by the preservatives, but sequestrants and antioxidants can sometimes be included under this heading. In Britain, food distribution and retailing is highly centralized, so that as few as 45 major distribution centres handle some 80 per cent of all food marketed. Hence, the ability of preservatives to give food products a long 'shelf life' is of enormous economic value to the food industry. Preservatives are, strictly speaking, those chemicals which have an anti-bacterial action, and which can therefore protect consumers from risks of salmonella and botulism. Chemical preservation is, however, just one way of protecting us from acute bacterial food poisoning, and freezing, canning, dehydration may be technologically, toxicologically and economically preferable. Other chemicals, mainly the antioxidants, are used to extend the shelf lives of food products.

'Consumer protectors' account for no more than 1 per cent of the total value of additives used in Britain. They provide substantial economic benefits to the food industry. For example, in Britain, meat processors spend just a bit less than £40,000 to buy some 150 tonnes of the preservative sodium nitrite, which is then used in the manufacture of approximately one million tonnes of nitrite-cured meats. As a result pig farmers are earning almost £1 billion, and rather less is being earned by the meat processing industry. As for antioxidants, the total British demand for the two most commonly used antioxidants, namely BHA and BHT, is about 100 tonnes, at a combined cost of about £100,000. The use of these antioxidants sustains, amongst others, a fried snack foods market valued at about £660 millions a year.

The largest single group of additives is made up of cosmetics – principally, sweeteners, colourings and flavourings. Cosmetics account for more than 90 per cent of the total number of additives in use and no less than 88 per cent of the total cost of additives. If all cosmetic additives were removed from our diet, many processed food products would become bland, pale, tasteless and flaccid – but they would be no less nutritious, just cheaper. In spite of the commercial and official secrecy which cloaks these matters, I estimate that there are at least 3,500 different flavourings available to industry. Flavourings are not regulated in Britain, and so we do not know which chemicals are being used, either overall or in particular products. Since they are unregulated, they have also not been subjected to systematic safety testing.

The final group of additives consists of processing aids. These can be used, for example, to reduce the extent to which mixtures stick to the walls of the processing equipment, which can mean that the equipment does not have to be washed so often. They may also be used to reduce processing times or labour costs. An important group of processing aids are the polyphosphates. These are liberally introduced into many processed meat, fish and cheese products. Their effect on meat is to cause it to absorb water and swell up. Polyphosphates are supposed to 'tenderize' frozen chickens and pork joints, but more importantly they enable meat processors to sell water at the price which meat can command.

Making the Most of Surpluses

The use of food additives has grown rapidly since the Second World War, doubling over the past decade alone. To appreciate why this is happening we need to understand something about the dynamics of the market for food. The market for food is in some ways quite special because, in the jargon of economics, the demand for food is notoriously 'price inelastic'. If the price of records were halved many of us would buy twice as many albums, but if the price of potatoes were halved few of us would eat twice as many potatoes. This is significant because there are technological and political reasons why the relative and absolute prices for foods have been declining.

Despite the expectations of Malthus and the Malthusians, the effect of technological changes in agriculture over the past 200 years has been to raise the productivity of labour in farming more rapidly than population growth, and more rapidly than in almost any other productive sector. But the farming lobby has been politically successful throughout the western world in persuading governments to subsidize agriculture so as to keep marginal farmers in production. Thus the problem for the food industries of the industrialized countries is not one of scarcity, but of a surplus of cheap products. This has provided an opportunity for, rather than a threat to, the food-processing industry, which generates profits by taking cheap and plentiful foods and synthetic ingredients and transforming them into 'value-added' (and hence higher priced and more profitable) food products.

Take the example of potatoes and their transformation into crisps. When we spend 13p to buy a packet of crisps, we are buying 1p's worth of potatoes which have been peeled, sliced, fried, flavoured, preserved, packaged, distributed and advertised into a highly profitable product, instead of a simple but relatively unprofitable spud. The production and sale of crisps and similar products reduces or eliminates potential or actual surplus of fresh foods and expands the total value of the market for foods. If the scale of the surpluses was not being reduced by industrial processing then agricultural prices and farm incomes would be significantly lower, or subsidies might be significantly higher, or both.

Industry's Defence of Additives

When the food industry seeks to defend its use of additives it does so primarily by reference to three considerations. By using additives the industry can expand the range of products available to consumers particularly in cities; it can extend the life of the product on the shelves of a warehouse, shop, fridge or larder; and it can make products appear and taste more attractive than they otherwise would.

These arguments are not, however, entirely convincing. The weakest argument is the supposed needs of urban life. Given the commercial organization of food distribution it is in fact far easier to obtain a wide range of fresh foods in city centres than in remote rural areas. Preservatives do extend the shelf lives of processed foods, but worryingly so, because they may be accumulating in our flesh, for they appear to extend the shelf lives of human corpses on the shelves of undertakers. More importantly, the shelf-life argument applies only to preservatives, antioxidants and anti-fungal agents, which together account for less than 1 per cent by value and a few per cent by weight of additives used. Certainly, cosmetic additives may enable materials to appear and taste more attractive, and this may reduce wastage, but it may also permit manufacturers to use inferior materials and poorer processing techniques and to sell the result at higher prices than they would otherwise command.

Inadequate Controls

The food industry insists that the use of additives is strictly controlled by the government and subjected to thorough and exhaustive testing, and that there is no evidence that additives are causing any harm. Unfortunately, none of these claims withstands critical scrutiny.

By number and by value, the majority of additives simply are not subjected to specific regulations. Flavourings, as well as all modified starches and enzymes, are effectively unregulated in Britain. At the last count, the government was regulating the use of less than 350 food additives, and approximately 148 of these were also being regulated by the EEC. The regulated additives, however, almost certainly constitute less than 10 per cent of the total number in widespread use.

Effective regulation can be achieved only by the use of a system of positive lists, and groups such as preservatives, antioxidants and colours are covered by such lists. Preservatives and colours may thus only be used if and when their use has been explicitly permitted. Permission should only be granted after a committee of experts has satisfied itself that there is a need for an additive, and that it can be safely used.

Problems with Testing

If we confine our attention to those 350 additives which are covered by positive lists, we still have to face the fact that there are many problems with the ways in which these additives are tested and evaluated. A toxicological assessment of an additive must rely on only two or three different kinds of information.

In a few cases, epidemiology has something to say about the safety of additives but in most cases it has nothing useful to say. Epidemiology studies patterns of illness in human beings. This science will enable us to identify the causes of some acute hazards if the harmful effects are evident within a few hours of consuming the offending products. It is just possible that epidemiology may one day enable us to discover if saccharin is as carcinogenic to human tissues as it is to the bladders of rodents. In relation to other chronic hazards, however, epidemiology is entirely useless, for it cannot pronounce on either the safety or the toxicity of chemicals in use. Moreover, epidemiology can make no contribution to the evaluation of a new additive because it can only be a *post hoc* science. Harm must first be done before epidemiology can detect it, and so it cannot inform us about the safety of a chemical yet to be introduced and consumed. When industry says that there is no evidence of any chronic hazards from additives, this does not mean that it has looked for such hazards – or that if it had it would necessarily have found them. It simply means that even where additives are causing chronic harm, we are unable to establish this epidemiologically. Indeed, with 3,500 or more additives being used in millions of combinations, and often in minute quantities (some products contain up to 30 different additives) it is next to impossible for epidemiology to identify any long-term or chronic effects from using particular food additives.

Over the last 12 years some toxicologists have developed a battery of *in vitro* and short-term mutagenicity tests. These mostly involve the use of bacterial techniques, and are intended as a screen for chemicals which may have a potential to provoke mutations or cancers. At best, these tests may enable us to identify and exclude some harmful chemicals; but if a chemical passes the tests, this does not yet prove that it cannot contribute to mutations or cancers. Nor can such tests rule out the possibility that the chemical might cause other types of toxic harm.

For the most part, food additive toxicology relies upon the results of animal tests. When I asked an industrial additive toxicologist how much we can learn about human responses to chemicals by feeding those chemicals to rats and mice he replied: 'Your guess is as good as mine!' We do not test chemicals on animals because we know that the animals provide a useful or reliable model of human biology, but because these laboratory animals are relatively cheap and simple to handle. When it comes to assessing the adequacy of animal models for human toxicity, we know that there are many important differences between animals and humans.

In 1983 David Salsburg of Pfizer Central Research in the USA published the first quantitative estimate of the validity of extrapolating from standard animal carcinogenicity tests to humans. Salsburg estimated that for known human carcinogens, the animal tests got it right no more than 38 per cent of the time.[2] This means that the results of the tests were wrong more often than they were right, and that they were significantly worse than tossing a coin! Subsequently this estimate has been challenged, but the best alternative estimate is that, for known human carcinogens, the animal tests are reliable about 75 per cent of the time.[3] This is still a very poor basis for regulatory policy. We know, from occupational epidemiology, of only 18 types of chemicals which definitely cause cancer in humans. All but cigarette smoke have been removed from our environment. Hence no less than 99 per cent of cancers (other than those related to tobacco smoke) are caused by chemicals or processes which have not yet been identified. We know of some 700–800 chemicals which are carcinogenic to animals, but we do not yet have a way of knowing which of them, and which others, are human carcinogens.

Most human carcinogens remain unknown, and for them, the validity of animal tests is bound to be significantly lower than for known carcinogens. As a result we have to conclude that while toxicity tests on animals tell us a great deal about the susceptibility of the animals tested, they tell us very little about the susceptibility of humans. There is no evidence that animal tests are valid models, and some evidence that they are not much use.

On account of possible differences between animals and humans and on account of differences within human populations, a safety factor of 100 is usually introduced between the no-effect-level in animals and the acceptable daily intake for humans (calculated in milligrams per kilogram body weight). There is, however, wide-ranging disagreement on how to extrapolate from the results of tests on small groups of animals to large groups of human beings who do not live in laboratory conditions. There are at least 12 different competing statistical techniques for this extrapolation, and the results which they produce may disagree by up to four orders of magnitude. In other words, the uncertainty is approximately the square of the supposed margin of safety.

Secrecy

We are not in a position to say whether or not most additives are safe; our judgements about safety and toxicity are profoundly uncertain. In these circumstances consumers need to know whether they or the food companies are getting the benefit of the doubt. It is often hard to tell because all the deliberations of the Food Advisory Committee of MAFF, as well as all the technical data on which it gives its advice to ministers, are covered by the Official Secrets Act. Its agenda, discussions and minutes are all secret, and it is illegal for us to know precisely what information is available to the government, or how it is interpreted. As a result, we do not know directly who is getting the benefit of the doubt.

We can obtain some indication, however, by looking at information published in the USA and by the EEC. We find that some chemicals continue to be permitted in Britain despite the fact that there is prima facie evidence that they may be toxic, particularly if they are of great economic value to the food industry. This state of affairs can readily be explained when we realize that industrial interests have effective direct access to secretive regulatory agencies, while there is, until now, no organized consumer group which is privy to this information.

In practice, we find that quite a few additives are permitted in Britain but banned elsewhere. For example, the red colouring dye amaranth (E 123) is banned in the USA, USSR, Finland, Norway, Greece, Austria, and permitted in France and Italy only in caviar. The antioxidant BHA (E 320) was banned in Japan in 1983 following some work in that country which showed the chemical to be carcinogenic to laboratory animals. Since that time the USA, Canada, and the EEC countries have discreetly expended a great deal of energy trying to discover reasons for disregarding the implications of the Japanese work, but have not yet succeeded. In the meantime, however, no action is being taken.

Even British publications sometimes reveal that it is industry which is getting the benefit of official doubts. In 1979 the Food Additives and Contaminants Committee of MAFF published a report on colouring additives.[4] It identified 16 colouring additives for which adequate evidence to demonstrate that they could safely be used was not then available. Nonetheless the committee recommended that their use should be permitted, but specified that further information was required and should be provided within five years. The recommendations were accepted by the government and seven years later we have not heard another word on this matter, but the 16 colourings continue to be used widely.

Research on the toxicology of food additives is almost entirely confined to commercial research laboratories, or to the laboratories of industrial organizations such as the British Industrial Biological Research Association (BIBRA). The government does not conduct safety evaluations on particular additives, nor does it commission independent research. (This is just one of several respects in which the USA and EEC countries are ahead of us.) As a result, in Britain, as the former director of BIBRA has himself pointed out, food additive toxicology is not a science which seeks to understand the biological effects of chemicals upon humans, but merely a technology designed to produce animal test data sufficient to gain permission from governments for the use of additives.

Keeping Consumers in the Dark

If the actions of the food-processing industry and the use of additives is genuinely to serve consumer interests then several conditions must first be satisfied. First, consumers must have adequate information and choice; and secondly, the wants of consumers must be prior to and independent of industrial actions. Neither of these conditions is currently satisfied.

If the last of these conditions were fulfilled the food industry would not be spending over £300 million a year advertising its products. (Currently, the annual advertising budget of one major international company specializing in dairy based food products is greater than the total annual budget of the World Health Organization.) The well-stocked shelves of supermarkets may give us the superficial impression that we have plenty of choice, but it remains impossible to buy bacon without sodium nitrite or crisps without BHA. The lack of information is the primary problem for consumers. Commercial and official secrecy combine to deprive us of the information which we require and to which we are entitled. Furthermore, additives – especially the cosmetics – can themselves be used to deprive consumers of information about the products which they buy and eat.

The access which consumers have to information about the composition of foods and the toxicology of additives is far too slight for any real consumer sovereignty to obtain. I have been studying the food industry for several years from the relative privilege of a university and I have had great difficulty in discovering the identity of all the additives in use in food products on the British market. It is now possible to obtain from MAFF a document entitled *Look at the Label*, which lists the names of the regulated additives, and the numbers which can be used on labels to indicate their presence in a product. (Numbers prefaced by an 'E' indicate an EEC-wide identification.) It is entirely impossible to obtain any list of all the unregulated additives which are in use. Even if consumers knew what the foods contained, however, this would tell them little or nothing about their safety. All toxicological data are published (if at all) only in technical journals and books, and no efforts have been made to open up this field to public scrutiny or accountability.

Food labelling regulations also leave a very great deal to be desired. Consider, for example, the contrast between the meaning of the terms 'flavour' and 'flavoured'. A product labelled as being 'smokey bacon flavoured' must have a flavour which derives entirely or substantially from smoked bacon, but a 'smokey bacon flavour' product need only have a flavour resembling the real thing, it need contain neither bacon nor smoke. This labelling device is not only misleading, it is intended to mislead.

Consumers are entitled to much more information about the composition of products, and their safety and toxicity. If companies are willing to provide the government with evidence which purports to show their products are safe, then the public should be fully entitled to see that information. If they are not prepared to share it with the public, then it should not be admissable as evidence. In particular, we are entitled to know the magnitude of the uncertainties, and who is getting the benefit of the very considerable doubts.

Notes

1. This estimate is based upon a *Depth Study of the Food Additives Industry in the UK*, London: Industrial Aids Ltd, April 1980; supplemented by interviews with sales personnel of companies manufacturing food additives in Britain.
2. D. Salsburg, 'The Lifetime Feeding Study in Mice and Rats – An Examination of Its Validity as a Bioassay for Human Carcinogens', *Fundamental and Applied Toxicology*, vol. 3, 1983, pp. 63 – 7.
3. J. K. Haseman et al. and D. Salsburg, *Fundamental and Applied Toxicology*, May/June 1983, pp. 3a – 7a; D. Salsburg and J. K. Haseman, *Fundamental and Applied Toxicology*, 1984, pp. 288 – 92.
4. *Interim Report on the Review of the Colouring Matter in Food Regulations 1973*, report no: FAC/REP/29, London: HMSO, 1979.

16

Lead Astray

Brian Price

If one issue can be said to encapsulate all that is wrong with British environmental policy-making, that issue must surely be lead. Official complacency, indifference, dubious scientific reasoning and the power of vested interests are all apparent in the sorry saga of this pervasive pollutant. A ubiquitous environmental contaminant, lead approaches toxic levels in the general population more closely than any other pollutant. Yet successive governments have permitted its continued addition to petrol and paint, tolerated its presence in water supplies and ignored the risks which it poses to the health of the nation.

The Toxicology of Lead

Before considering the politics of lead pollution, it is important to discuss the harm that, it is prudent to assume, lead can cause. Lead is of no use to our bodies, unlike some other elements which are essential in small quantities but toxic in large amounts. Its gross effects on the human body, resulting from the intake of comparatively large doses, are well known. Outright lead poisoning and its often fatal consequences has been documented since Roman times; fortunately it is now uncommon, being confined mainly to workers in industries which use lead and to children who chew old high-lead paint.

Less obvious, but in the long term probably more serious, are the effects of lead at levels which do not produce overt symptoms – subclinical lead poisoning. The metal exerts some of its effects on the nervous system – technically, it is a neurotoxin – and it is thus logical to look for look for the effects of small doses, if they are to be found, on the human being's most precious collection of nerves, the brain.

It has been known since 1943 that a consequence of overt lead poisoning in children is brain damage and mental impairment.[1] But what of children who have absorbed excessive lead but not enough to cause clinical poisoning? Attempts have been made to answer this question since the early 1960s and dozens, if not hundreds, of studies have been carried out to determine whether or not small doses of lead affect the intelligence of children.

The results of many of these studies have been inconclusive, tentative, confusing and sometimes contradictory. But as more sophisticated techniques have been developed and better experiments have been designed, suspicions about lead's subclinical effects have become stronger. Even now it is not possible to be certain about the magnitude of lead's effect on children's health, but it does seem clear that levels of lead common in the general population have a deleterious effect on the brains of a significant proportion of our young children.[2]

Of course, lead is not the only factor affecting children's mental development – social, educational and other medical considerations also play a part. These complicating factors have made the adverse health effects of lead difficult to quantify – and have enabled successive governments, as well as much of the medical establishment, to bring in a verdict of not guilty or, at least, not proven.

Sources of Lead

The controversy over the effects of lead has been matched by a similar dispute over the relative magnitudes of different sources of lead in the environment. The substance is present in air, water, dust, soil and food but how it gets into each medium and how it is transferred from one to the other has been obscure. The topic was discussed in *Lead and Health*,[3] the now notorious report produced for the Department of Health and Social Security (DHSS) in 1980 by a committee chaired by Professor Patrick Lawther. The government commissioned the report in response to pressure from environmental groups calling for a ban on the addition of lead in petrol. In doing so, it bought time during which it did not have to act and it also hoped for a whitewash of sufficient opacity to enable it to rebuff the environmentalists for some considerable time.

The ploy backfired, however, since the Lawther report was quickly discredited. A key deficiency within its analysis was its discussion of sources and exposure routes. Lawther argued that the major part of our average lead intake comes from food, with water being the second most important source and air the third. This was obviously convenient for the lead and petrol industries, as well as the government, since lead from motor vehicle emissions – which makes up about 90 per cent of airborne lead – was thereby exonerated of any blame for elevated lead levels in people.

The Lawther report virtually ignored two important factors: the contribution which lead in the air makes to food lead (and to a lesser extent, water lead) and the intake of lead from vehicles via dust adhering to children's fingers, dropped sweets and cooking surfaces. Dr Robert Stephens, of Birmingham university, has recalculated the available data and his breakdown of lead exposure for a two year old child living in an area of high traffic density is shown in figure 10.[4]

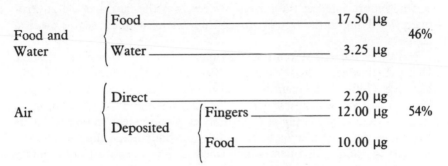

Food and Water	Food	17.50 µg	46%
	Water	3.25 µg	
Air	Direct	2.20 µg	54%
	Deposited — Fingers	12.00 µg	
	Deposited — Food	10.00 µg	

Source: R. Stephens, 'How lead finds its way into people', *CLEAR newspaper*, no. 1, 1982

Figure 10 Breakdown of Lead exposure of a two year old child

Clearly, lead in the air – and hence motor vehicle lead – contributes more to our lead body burdens than the Lawther Committee claimed. Further evidence of petrol lead's contribution has come from studies of the amounts of lead in the blood of Americans and Italians. Blood lead levels in New York children have been shown to fluctuate in line with variations in the quantity of petrol lead sold in the city,[5] while the levels of lead in the blood of a cross-section of Americans declined by 37 per cent after the amount of lead used in petrol was reduced by approximately half.[6] A study in Turin demonstrated that at least 30 per cent of the lead in the blood of people living there came from petrol.[7]

Government Reactions

The campaign for lead-free petrol – started by a subgroup of the Conservation Society and later taken up by Friends of the Earth and the highly efficient Campaign for Lead-free Air (CLEAR) – soon found out that demolishing the Lawther report was not enough. The government eventually yielded some ground by announcing in May 1981 that the limit for lead in petrol would be lowered from 0.4 grammes per litre to 0.15 grammes per litre in 1985 but refused to go further.

Many MPs and some of the general public were satisfied with this concession, although a number of environmentalists pointed out that, for several reasons, cutting the amount of lead in petrol by two-thirds would not automatically reduce the hazard from petrol lead by two-thirds. What was not appreciated at the time, however, was the fact that the government was flying in the face of advice from one of its most senior medical advisers.

In a letter, dated 6 March 1981, to senior officials of other ministries, Sir Henry Yellowlees, Chief Medical Officer at the Department of Health, effectively refuted the main tenets of the Lawther report. He stated that 'the simplest and quickest way of reducing general population exposure to lead is by reducing sharply or by entirely eliminating lead in petrol.'[8] This letter was, not surprisingly, kept confidential but when a copy was leaked to CLEAR, and subsequently published in *The Times*, a public outcry ensued.

No Threshold

Controversy in the political forum was matched by equally heated debate in the scientific arena. A key issue in this context was whether or not there exists a safe level, or threshold, below which lead has no toxic effects. The Lawther Committee's apparent view was that 35 microgrammes of lead per 100 millilitres of blood constituted a safe limit. Professor Michael Rutter, an eminent psychiatrist and member of the committee who was, in the end, reluctant to sign the report, commented in mid-1982 that 'it never could have been justified to assume that levels below 35 [units] were safe.' He pointed out that the report's conclusions in this respect could be interpreted as implying that there was a positive evidence of safety below this level – a scientific nonsense.[9]

Despite requests from members of his own committee and from CLEAR, Professor Lawther has refused to clear up this apparent confusion – which is most convenient for the government, since lead levels in the general public rarely exceed the 35 microgramme figure.[10] It should be noted, however, that some effects of lead on mental function have been described at levels as low as 13 microgrammes per 100 millilitres of blood, and it has proved impossible to demonstrate a level at which lead does not affect the electrical conductivity of nerves in the brain.[11]

The scientific controversy came to a head in May 1982 when the charitable trust set up by CLEAR organized a symposium in London to which the world's leading experts on lead were invited. The chairman of the symposium was Professor Rutter and in his summing-up speech, he stated: 'In my view the reduction of lead in petrol to an intermediate level is an unacceptable compromise without clear advantages and with definite disadvantages.'[12] Coming from a former member of a committee whose report was used as the justification for setting the intermediate level, this remark had profound significance.

The EEC Directive: A Red Herring?

The figure of 0.15 grammes per litre was not selected at random. It is a level which can be achieved by the oil refiners without undue disruption and which would not inconvenience the motor industry by requiring any modifications to vehicle engine design. It is also the figure which the EEC, in its Directive 78/611, sets as the minimum which member states may require of the petrol manufacturers. It must be emphasized that lead-free petrol is not illegal in the EEC, although it has sometimes been claimed that this is so, but that governments may not make a limit of less than 0.5 grammes per litre obligatory. This EEC Directive has been used by the British government as an excuse for failing to go the whole way and ban leaded petrol entirely but the argument is specious. Member states can exempt themselves from the Directive on public health grounds and a sufficiently determined nation could ignore the lower limit altogether.

The Royal Commission's Report

The opening chapter of the latest phase of the lead in petrol saga began in April 1983 when the Royal Commission on Environmental Pollution published its ninth report, *Lead in the Environment*.[13] The Commission, which had received a wealth of evidence from industry, government and environmentalists, concluded that it was imprudent to continue adding such a persistent and toxic pollutant as lead to the environment and called for – among other things – a ban on lead in petrol. Within hours of the report's release, the Secretary of State for the Environment announced to the House of Commons that the government had accepted the Commission's recommendations on petrol lead and that the matter would be pursued by Britain within the EEC.

This was clearly a victory in principle for the environmentalists but a timetable for the removal of lead was yet to be drawn up. The government considered a target date of 1990 to be reasonable but most critics felt this to be far too generous to the oil and motor industries. Exactly how lead-free petrol will be introduced is not yet clear – and there is still plenty of room for duplicity. The motor and oil industries are undoubtedly fighting a bitter campaign behind the scenes to stave off the evil day when lead-free fuel becomes the norm. The two industries are also, undoubtedly, squabbling amongst themselves about which of them is to make the most modifications to their products. The point at issue here is whether high-grade lead-free petrol should be produced for use in existing cars or whether a lower grade of lead-free fuel should be made for all new cars which would be specifically designed to run on it.

A Tough Stand in Europe

The one area in which the British government emerges with some credit concerns its stance in Europe. The memorandum produced by Britain for submission to the EEC was much stronger than environmentalists had expected and the British government appears to have held firm on the issue in the face of strong opposition from France and Italy. Lead-free petrol will be available in all member states from 1989, although individual countries will be encouraging its use earlier than that.

For once the big battalions of Whitehall, the oil and motor industries and, until recently, a significant proportion of the medical establishment, have yielded to the pressure from underfinanced and overworked pressure groups supported and inspired by a few courageous scientists (such as Professor Bryce-Smith from Reading University) and, it must be noted, the general public. Eighty-nine per cent of the population said that lead should be banned from petrol, in a poll conducted on behalf of CLEAR.[14] The cynical disregard for public health in the pursuit of profits exhibited by industrialists; the bureaucratic complacency which permitted this; and the unwillingness to rock the boat displayed by certain sections of the medical establishment, all reflect badly on Britain's environmental record. Those who might expect that the government would have learnt its lesson with regard to the issue of lead in petrol should, however, turn their attention to its performance on a related topic – lead in paint.

Lead in Paint: Why No Action?

It should be said from the outset that Britain lags decades behind other countries in the control of some types of leaded paint – for instance, products banned in France since 1939 are readily available in British DIY shops. The ingestion of only a small quantity of such paint by a child would not only impair intellect, it could prove fatal.

Incredibly, there is no law restricting the amount of lead which may be incorporated in paint for use in Britain. There are, now, legal requirements that paint containing large quantities of lead should be labelled as such with appropriate warnings concerning its use on surfaces accessible to children – but these regulations only put into law voluntary practices which have existed for years. US law restricts the amount of lead permitted in dry paint used on or in domestic buildings to 600 parts per million (ppm), yet in almost any DIY shop in Britain paint containing over ten times this amount may be purchased without even a warning appearing on the can. A survey carried out in 1984 on behalf of CLEAR showed that 82 per cent of primers, undercoats and gloss paints sampled contained more than 600 ppm of lead. Furthermore, shop assistants frequently had no idea about the lead levels contained in the paints or the hazards which they presented.[15]

Both the Royal Commission on Environmental Pollution and CLEAR have called for the US limit to be implemented in Britain. Predictably, the paint industry opposed this move virulently. For its part, the government has hedged, claiming that more information is needed before it can make a decision. The latest development is that lead additives to most domestic paints are now being phased out and by mid-1987 most should be clear. This move is a voluntary one, taken by the paint industry in response to public pressure. Legislation, however, is unlikely to be forthcoming.

The debate about lead in paint is, in essence, the same as the debate about lead in petrol. A persistent toxin is dispersed into the environment in an uncontrolled manner largely for the convenience of industry. As a result, levels of lead in people are elevated – in some cases greatly and in others only slightly. Once again the main argument is about the effects of low levels of lead but, in the case of paint, the issue is complicated by the fact that leaded paint occasionally kills children and certainly causes a significant number of acute poisoning cases each year.

The paint industry attempted to confine the debate to the question of acute poisoning which, nowadays, usually occurs only when old high-leaded paints are chewed or removed from surfaces carelessly. Modern paints, the industry points out, contain much less lead than pre-1939 products and are thus claimed to be safe. This, of course, completely ignores the point that the smaller – but still significant – amounts of lead still added to paint can increase the body burden of lead in children, which may already be undesirably high, without producing symptoms of poisoning.

Conclusion

Two questions emerge from the discussion on lead in the environment. The first is: 'What is the balance between risk and benefit?' There is no *absolute proof* that lead from petrol causes harm but ample evidence of the *risk* that it does. The benefits are clearer – slightly faster cars, marginally cheaper petrol, a little extra petrol per barrel of oil refined – and, in the case of paint, perhaps a slightly cheaper product. But are we, as a society, justified in risking the mental health of a significant proportion of our children for these advantages? Clearly we are not.

The second question is: 'Can we trust our official watchdogs, in government and the civil service, to balance the risks and benefits on our behalf?' The answer is again a resounding 'No', since they have signally failed to do so in the sorry history of British lead pollution. Indeed, the whole issue is succinctly summed up in the three words used by Des Wilson, the chairman of CLEAR, in the title of his recent book: *The Lead Scandal*.

Notes

1. R. K. Byers and E. E. Lord, 'Late effects of lead poisoning on mental development', *American Journal of Diseases in Childhood*, no. 66, 1943, pp. 471–94.
2. For a review of this topic see M. Rutter and R. Russell Jones (eds), *Lead Versus Health*, London: John Wiley, 1983.
3. Department of Health and Social Security, *Lead and Health*, London: HMSO, 1980.
4. R. Stephens, 'How lead finds its way to people', *CLEAR newspaper*, no. 1, 1982.
5. I. H. Billick, et al., 'Analysis of pediatric blood levels in New York City for 1970–1976', *Environmental Health Perspectives*, no. 31, 1979, pp. 183–90.
6. J. L. Annest, 'Trends in Blood Lead Levels of United States Population' in Rutter and Russell Jones *Lead Versus Health*.
7. R. Russell Jones and R. Stephens, 'Contribution of Lead in Petrol to Human Lead Intake' in Rutter and Russell Jones, *Lead Versus Health*.
8. H. Yellowlees, letter of 6 March 1981, published in *The Times*, February 1982.
9. D. Wilson, *The Lead Scandal*, London: Heinemann, 1983.
10. Wilson, *The Lead Scandal*.
11. D. Otto et al., 'Changes in CNS function at low to moderate blood lead levels in children' in Rutter and Russell Jones, *Lead Versus Health*.

12. M. Rutter, 'Low Level Lead Exposure: sources, effects and implications' in Rutter and Russell Jones, *Lead Versus Health*.
13. Royal Commission on Environmental Pollution, *Ninth Report: Lead in the Environment*, London: HMSO, 1983.
14. Wilson, *The Lead Scandal*.
15. B. J. Price, *Lead Levels in Decorative Paints. Report to CLEAR*, London: Campaign for Lead Free Air, 1984.

17

Asbestos

Britain's Record

Angela Singer

Goebbels, who should certainly have known, said it took the truth a long time to catch up with a good lie. The lie that only crocidolite (or 'blue' asbestos) is dangerous has been the most successful falsehood, in this country, since Father Christmas. Like Santa, the fantasy has generated millions of pounds worth of business. It has also claimed millions of lives, and will yet claim thousands more.[1]

The British government knew that asbestos – the genetic name for a group of minerals, consisting of fibrous silicates – caused asbestosis, a progressive and incurable lung disease, in 1906. It knew that asbestos caused lung cancer in 1934. The government, from 1964, was aware not only of the link between asbestos and mesothelioma, a fatal cancer, but also that members of the general public were at risk even from very slight exposure to asbestos.

Blaming it on Blue

Successive governments have done as little as they could get away with. In the late 1960s and 1970s, the families of dead asbestos victims (by no means all of whom had worked with asbestos; some had experienced much more casual exposure) went on television and gave interviews to the press about their experiences. The subsequent publicity forced the government and the asbestos industry to respond. With one voice, they pronounced that the deaths had been caused by exposure to crocidolite. This was ingenious. It was convenient for the asbestos industry because blue asbestos was only 3 per cent of production. The other types of asbestos – chrysotite (white asbestos) and amosite (brown asbestos) – were given a clean bill of health.

The industry put its own voluntary ban on the import of blue asbestos in 1970. Then it continued to expand the use of white asbestos, spending millions on publicity campaigns and handing out a few pounds a week in *ex gratia* payments to its ever increasing band of widows – bereaved, it could always claim (since the diseases take 20 to 30 years to develop), by earlier more lax methods of working and the former use of blue. This was also a face-saver for the government, since the outright ban of blue looked like tough action. The government, along with industry, still argues that white is less dangerous than brown or blue. In January 1986, brown and blue were at last banned officially. White, however, continues to be used, but with a stricter control limit.

Standards of Exposure

Regulations passed in 1969 permitted asbestos factory workers to inhale two fibres per millilitre of air – the equivalent of two million fibres per cubic metre of air, the amount normally breathed in an hour. In 1982, after Yorkshire Television screened a widely publicized documentary on asbestos and cancer, this limit was halved to one fibre; and in January 1986, the level was halved again to 0.5 of a fibre per millilitre, which is still 500,000 fibres per hour. Government regulations, however, have always followed industrial practice, they have never controlled it. Moreover, since the 1960s, any restrictions on the asbestos industry have been worked out hand-in-hand with the industry's directors. Nothing has been imposed without consent; no action has been taken which might harm business.

For the general consumer, there are no regulations. Anyone can buy any amount of white asbestos, take it home, drill through it, and shower the dust all over any nearby children. A compulsory label will be brought in during 1986, in line with an EEC Directive. This is to be a friendly little 'a' sign, however, known as 'Alphie Apple'. Alphie will be accompanied by the slogan: 'Take Care With Asbestos', as if the product were not a deadly carcinogen but something to be cherished like pure new wool. The sign has been derided by health and safety campaigners, who have pressed for a skull and crossbones symbol.

The industry has spent a great deal of money, in its own words, to 'influence' reports, legislation and government recommendations on new safety standards, both nationally and internationally. In 1976, fearing a major scandal over the health effects of asbestos, it spent £500,000 in six weeks. The decision to increase the publicity budget by this amount – from a steady £50,000 a year – was taken in ten minutes.

White Asbestos Kills

There is no doubt that white asbestos is just as deadly as blue or brown asbestos. There are well-documented case histories of people who died of mesothelioma (cancer of the chest and abdomen lining), the only known cause of which is asbestos, who have been exposed only to white asbestos – and only for short periods. Some of these cases are listed as government statistics on the mesothelioma register kept by the Employment Medical Advisory Service (EMAS), the medical section of the Health and Safety Commission. In one case, a man had been exposed to asbestos for one day only while sawing up asbestos sheets to build two sheds.

One reason why white asbestos has been deemed safe by the medical establishment arises from the use of the standard optical microscope. This is used to examine lung tissue at post mortem, but picks up only the larger fibres of blue asbestos. When examinations are made with a more powerful electron microscope, millions of the finer fibres of white asbestos become visible. Families who have not insisted on electron microscopy have frequently been refused compensation by pneumoconiosis panels. This keeps statistics low and perpetuates the deception.

A symptom – and a cause – of government weakness over asbestos is that much of the research on which the factory regulations have been based for the past 20 years has been carried out by the asbestos industry – principally, by Britain's largest asbestos company, Turner and Newall. The industrial hazards of asbestos have been drastically underplayed, as has the risk to the general public.

The First Regulations

In 1899, an asbestos spinner, aged 33, went to see a chest specialist, Dr Montague Murray, with what he thought was bronchitis. He was the last survivor of ten asbestos workers in his workshop, all of whom had died aged about 30. He died, too, the following year. His death, from asbestos, was reported to the 1906 British government inquiry into compensation for industrial diseases.

In 1928, Mr C. Leonard Williams, the medical officer of health for Barking, where there was a large asbestos factory (owned by Cape Asbestos), noted asbestos deaths. A year later, two factory inspectors, Dr E. R. A. Merewether and Mr Charles Price, confirmed an epidemic of asbestos diseases among asbestos workers. It was this which led, in 1931, to new regulations being drawn up for exposure to asbestos in the workplace: they laid down that asbestos workers should not be exposed to any dust. The regulations, which became law in 1933, have since been relaxed; 'permitted' levels of dust are now laid down for asbestos factories. Lung cancer among asbestos workers was reported in factory inspectorate reports in 1934.

The 1931 regulations were not enforced at the time – and never have been since. Indeed, the General, Municipal and Boilermakers Union (GMBU), which has 800 members at Turner and Newall's Rochdale factory, is currently pressing for the recommendations of the 1929 Merewether report to be implemented. Those recommendations stipulated improved exhaust ventilation; the enclosure of dusty processes; the use of conveyor belts, instead of individuals, to carry sacks of raw asbestos; and the substitution of wet methods for dry. In 1984, the factory inspectorate reported that it had visited Turner and Newall's factory 12 times in as many months, urging improvements (such as the enclosure of old carding machines, and the use of vacuum cleaners instead of oiled brooms to sweep up asbestos waste), but to no avail.

Mesothelioma and Asbestos

By 1964, it was clear that asbestos causes mesothelioma, the first cases having been diagnosed in the 1950s in miners in Cape's South African asbestos mines. Dr J. C. Wagner, who discovered the disease there, came to Cape's asbestos factory in Barking to see if there was mesothelioma among workers at the plant. Searching through the pathology records at the London Hospital in Whitechapel, he found several cases of the disease. His findings were later confirmed by British doctors – principally, Dr Muriel Newhouse at the London School of Hygiene and Tropical Medicine. From the start, cases appeared where the victims had suffered only brief exposure to asbestos. One victim was a nursemaid to the children of a gasworks manager. In those days gasworks contained large amounts of asbestos.

Publicity in Britain was confined to an article by Dr Newhouse in the *British Journal of Industrial Medicine*. In the USA, however, the link between asbestos and mesothelioma was discussed before a wider audience at a conference at the New York Academy of Science, at which British doctors gave papers. This led to fuller coverage in the British press.

A Suppressed Report

Subsequently, two government departments set up separate study groups to investigate the problem. The Ministry of Labour asked senior medical inspectors from the factory inspectorate to examine asbestos hazards to workers. The Department of Health asked its standing medical advisory committee, which had a standing subcommittee on cancer, to report on 'the cancer hazard due to asbestos to the general population'.

The Ministry of Labour's memorandum, *Problems Arising From The Use Of Asbestos*, was published in 1967. The Department of Health's report, however, was kept secret. When the BBC television programme, *Nationwide*, wanted a copy in 1976 (eight years after the report was finished), it had to be obtained in the USA under the US Freedom of Information Act. The report was not even placed in the House of Commons library until after the programme had been screened.

The story of how the report came to be placed in the library is illustrative of Whitehall's ability to deceive ministers and the public. The *Daily Mirror* was about to run a front page lead story on the report's suppression, and a question was asked in parliament as to whether the report would be published. The then Secretary of State for Social Services, Dr David Owen, said that the report was in the Commons' library, and his department deliberately gave the *Mirror*'s reporter, Roger Todd, the impression it had been there for months. The story was reduced to a few paragraphs inside the paper and lost all impact. In fact, Dr Owen's written answer had anticipated the report's arrival. A copy was rushed into the library on the evening after the answer was delivered.

The report made it clear that there was a cancer risk to the general public. It said: 'The risks associated with the exposure to asbestos are not limited to special industrial groups and the widespread use of asbestos may carry some risk of cancer to the general population.' It added: 'It has been possible to show that mesothelioma can be produced by slight exposure, and on present knowledge we must assume that no amount of exposure is completely free from risk.'

The report recommended that where materials containing asbestos are sold in DIY shops, they should be clearly marked with a warning to avoid inhaling the dust when drilling and sawing. The report said that tipping asbestos waste should be allowed only under strict control. Local authorities should measure the amount of asbestos in the air near asbestos dumps, building sites where asbestos was sprayed, and asbestos factories. It added that asbestos used in building and exposed to ambient air should be sealed to prevent particles flaking off. A balance should be struck, the report said, between the dust risk introduced by the use of asbestos and the corresponding reduction in the risk of fire.[2]

No action was taken on the 1968 report. There is still no warning on consumer products; asbestos roofs are never sealed, they are flaking away in every street.

The Regulations are Relaxed

By the early 1960s, it was obvious that the 1931 regulations were unenforceable, since it is impossible to work with asbestos without causing dust. New regulations were required. From 1965 to 1969, senior officials from the factory inspectorate negotiated with the asbestos industry. The industry argued for a permitted level of asbestos dust in factory air; in return, it was prepared to trade a ban on blue.

The inspectorate recognized that allowing a certain amount of dust in the atmosphere would make the 1969 regulations weaker than those of 1931 (which allowed no dust at all) but since the demand for no dust had been unenforceable it was hoped that at least a statutory limit would lead to a standard – thus making it easier to prosecute when the standard was not complied with. It was stressed in memos and minutes of meetings that the permitted level was not a safe level. It had no 'biological significance', the notes said. It was 'An empirical level which we have some hope of enforcing': or, 'merely what engineering controls could achieve'.

The new permitted dust level – called at various times a 'threshold limit value', a 'hygiene standard', and, recently, a 'control limit' – was thus set. The maximum for blue was so low that the industry brought in a voluntary ban in 1970. The new regulations required respirators and protective clothing when the concentration was over the limit for white. No level has ever been set for exposure outside the workplace, since there has been no official recognition of the environmental danger, identified by the DHSS in 1968.

The control limit set for white in 1969 was called the 'two-fibre standard'. This was two million fibres *per cubic metre* of air, the amount of air most people breathe in an hour. Despite the clear-cut statements from scientific bodies that this must not be presented as a safe level and that there is *no safe level*, the present government has twice used the device of halving the level as a public relations exercise – once, in July 1982, following the Yorkshire Television documentary on mesothelioma, *Alice – A Fight For Life*, and again in August 1984, to take effect in January 1986. The current limit still allows anyone working in a factory to inhale four million fibres in a working day. The levels are in any event a nonsense because there is an admitted 50 per cent error factor in measuring.

Protective Clothing?

Protective clothing is notoriously inadequate. Mr Norman Rhodes, the general manager at Turner and Newall's Rochdale factory, showing a visitor round in 1976, admitted the masks supplied were 'of cosmetic value, but they make some workers feel better'. These masks were still being supplied in 1982, when a box containing one was sent to the *Guardian*. A warning on it read: 'This product is not designed for use as a protection against asbestos, silica, cotton dust or any other toxic dusts, fumes, gases, and vapours.'

Inadequate protective clothing has serious environmental implications, because workers take fibres home on their clothes. There are well-documented cases of families of asbestos workers developing mesothelioma though they had never themselves worked with asbestos.

SPAID

The visitor to Turner and Newall's Rochdale factory in 1976 was Mrs Nancy Tait, who has since founded SPAID, the Society for Prevention of Asbestosis and Industrial Disease. Nancy Tait's husband, a post-office worker, had died of mesothelioma in 1968 after incidental contact with asbestos. She has spent every day since his illness collecting and disseminating information on the dangers of all types of asbestos.[3]

The government could have banned asbestos 30 years earlier, before it killed Ashton Tait and thousands of others. As it was, Mrs Tait, a retired civil servant, who could see through official euphemism and equivocation, tracked down medical reports and publicized obscure but critical documents on the hazards of asbestos. One was a report showing that Turner and Newall had been aware of the link between asbestos and lung cancer in 1943, but had agreed to keep the knowledge from their employees.

Nancy Tait has come to act as legal representative, researcher and counsellor representing families at medical tribunals and referring them to lawyers who can sue. SPAID was founded in 1978 and is the only charity which specifically helps the industrially disabled. SPAID has helped people to the benefits and compensation to which they are entitled, despite the obstacles of government bureaucracy. One man, who had worked with graphite, had appealed to pneumoconiosis panels seven times over 20 years. On the eighth occasion he arrived in an ambulance and needed oxygen to breathe. He received his money just before he died.

The Simpson Committee

In March 1976, the Ombudsman, Sir Alan Marre, criticized the factory inspectorate for lax conditions at Acre Mill, the Cape Asbestos plant at Hebden Bridge, Yorkshire. More than 200 employees and their families had contracted asbestosis and cancer. Faced with the biggest scandal so far, and 70 years after asbestos disease had first been diagnosed, the government responded by setting up yet another committee.

The government's Advisory Committee on Asbestos – known as the Simpson Committee after its chairman, Bill Simpson, the then chairman of the Health and Safety Commission – sat for two years. It took evidence from asbestos workers, experts, individuals including Mrs Tait, and the asbestos industry. It reported in October 1979 with 41 recommendations. By January 1986, only four of those recommendations had been implemented. Many of the proposed reforms (most of which have now been implemented) would have made little difference to industrial practices anyway. Indeed Simpson largely suggested making *de jure* what was already *de facto*. First on the list was to make official the ban on blue. Another recommendation was a ban on spraying asbestos, a practice dropped years earlier. Innovations such as labelling, though now law, should, according to Simpson, have been voluntary.

'Alice – A Fight for Life'

For three more years the Health and Safety Executive stalled on implementing even this limp package by saying it was waiting for an EEC Directive. Then, in July 1982, asbestos was brought to public attention again by the Yorkshire Television documentary, *Alice – A Fight for Life*. The programme was named after Mrs Alice Jefferson who worked at Acre Mill for nine months aged 17. She died at 47, distraught at leaving her 15 year old son and four year old daughter – and fighting to hang on to life until her last breath.

Viewers saw Alice and others in the last stages of excruciatingly painful cancer. They heard from the parents of a 12 year old boy, who died of mesothelioma after helping his father blow out asbestos dust from old brake linings. They listened to the widow of a 32 year old lawyer who had loaded asbestos cement sheets for two weeks as a student. He had died a few weeks before their first child was born.

They also saw copies of company documents showing that Turner and Newall, on whose statistics the factory standards had been based, had given the Simpson Committee underestimates of disease. The company had told Simpson that none of its mesothelioma cases resulted from slight exposure. In fact, one victim had been a telephonist, another was the office cleaner. On the company's own list of mesotheliomas, the exposure times included: 16 months, one year, seven months, five months, three years – and ten days.

There were demands for the evidence to be reviewed. Jack Ashley, the

Labour MP for Stoke on Trent, called for a Commons Select Committee which would hold public hearings and report within a year. The government acted rapidly to stop further discussion. It decided to implement four of the Simpson recommendations. The control limit was to be halved immediately. The bans on blue and on spraying were to be made law. And a licensing system was to be set up for asbestos removal companies.

Cowboy Operators

This appeared to be strong stuff. Indeed, the licensing system could have been a great step forward. Control is desperately needed over untrained, cowboy operators, who, in a declining building industry, are currently cashing in on people's fears, with no regard to the health of their employees or customers. The system contains no provision for vetting prospective licensees, however, or policing their performance.

The General, Municipal and Boilermakers Union, together with the Thermal Insulation Contractors Association (TICA), met the Health and Safety Executive for special talks in 1982, after the first proposals for the scheme had been circulated. The unions asked, in the strongest terms, for firms to be licensed only if they had been in business for a certain number of years and had a proven safety record, including the use of special decontamination units for their workers. They said the system should apply to all work with asbestos – not just lagging – and should include large companies stripping asbestos on their own premises, such as British Rail, who are currently exempt from the regulations.

None of these suggestions was accepted. As far as the GMBU is concerned the scheme is a licence to kill. Anyone can set up in business as an asbestos removal company and apply for a licence. By mid-1984, one thousand licences had been issued and no one had been refused. The licence is certain to give potential customers a false reassurance that the holder is an approved expert. The HSE refused to allow worker inspectors to supervise the work, or to appoint any more factory inspectors.

Alternatives to Asbestos

Since the beginning of the century the asbestos industry has used one of Margaret Thatcher's favourite arguments: 'There is no alternative.' In fact, there is – and in every case, it is a superior product. Asbestos brake pads on motor cycles, for instance, do not work in the wet. A stopping distance of 50 feet becomes 800. Metal disc brakes worked just as effectively, wet or dry, in tests in 1978 at the Road Research Laboratory even when they were doused with 30 gallons of water an hour. Asbestos-free brake pads have been fitted in police cars for some time because they are more effective.

The missing element is not the product, it is the demand. TAC Construction Materials, a Turner and Newall subsidiary, announced its first complete range of asbestos-free building materials in 1984. It said it was responding to public demand as a company of integrity. It was trying to stay in business. Turner's shares fell from 59p in July 1982 before the *Alice* programme to 23p by November. By the end of the year, the city had written the company off as dead.

The building industry had voted with its safety boots. As a last grab at a straw, TAC produced a booklet in October 1982 called *Living and Working with TAC's Asbestos Cement Products*. This booklet stated that the material could be safely cut, drilled and sawn with hand and power tools. The National Federation of Building Employers said it was 'absolutely disgusted' at this claim. Asbestos products should never be worked with power tools – and anyway there were already substitutes for 95 per cent of construction uses. The Union of Construction, Allied Trades and Technicians (UCATT), instructed its workers to refuse to work with any form of asbestos. The union produced its own leaflet for distribution to 300,000 members including thousands of carpenters and painters. Mr Doug Sanderson, the union's health and safety officer said: 'We've listened to experts long enough.'

Asbestos in the Environment: The Risk to the Public

The latest government report, *Effects on Health of Exposure to Asbestos*, written by Sir Richard Doll, emeritus professor of medicine at Oxford university, and Julian Peto, professor of epidemiology at the Institute of Cancer Research, London university, was published in April 1985. The authors say asbestos in the general environment is of 'extremely low risk'. They claim that the evidence of danger from asbestos in the environment is 'unquantifiable'. They say in their report:

> We have simplified our task by concentrating on those hazards that are liable to be met at work, as the evidence relating to non-occupational exposure is either too insubstantial to justify review, or, if it is clear, is unquantifiable, as in the case of the household contacts of asbestos workers who developed mesotheliomas of the plura from exposure to dust brought home on worker's clothes.

Doll and Peto made two adjustments in their report. They have taken measurements of asbestos dust in air taken by transmission electron microscope and adjusted them for an optical microscope. The calculation was made because it is for optical microscopes for that there is a risk measurement. For the optical microscope they halved the number of fibres found by the electron microscope. Secondly they extrapolated from workers' exposure to environmental exposure.

Doll and Peto calculated their risk estimates from figures provided by the Canadian government, which is candid about its wish for the use of asbestos to be continued. These figures are based on air in buildings containing asbestos, which is not crumbling or flaking but in good condition. Even so, the report says there will be one death per year among the general public. Three other sources of evidence are not taken into account: the government's own mesothelioma register, which contains lists of people who died of cancer after slight exposure to asbestos; the evidence gathered at the Mount Sinai Hospital in New York by Professor William Nicholson, who recorded cases of people who had died after household contact with asbestos; and, case histories collected by Mrs Nancy Tait of SPAID, offered to Doll and Peto, in 1982 and again in 1983.

One of the cases involved Mrs Patricia Kuflewski, who died of mesothelioma, aged 55, in 1983. She worked in the British Council's office in Oxford. In 1974, the office had brown asbestos sprayed on the ceiling. After she died, 31 million fibres were found in one gram of lung tissue examined under an electron microscope. The fibres matched those used in the office. In 1984, after an electrician had carried out repairs to the building, the London School of Hygiene examined debris from the staircase and found it contained 70 per cent brown asbestos.

Another victim was Mrs Julia Darlington, a hospital radiologist in South Wales, who died, aged 41, from mesothelioma in 1980. She had filed X-ray films on shelves made of asbestolux in a shed lined with asbestos. She did this work for three months in 1960 and for 30 days in 1975–6. The industry had stopped making asbestolux by 1981. The appeal tribunal which decided that Mrs Darlington had died of an industrial disease said it wished to record its 'strong view' that every mesothelioma death should be investigated with an electron microscope. It was clear, the tribunal chairman said, that many deaths due to asbestos were not being detected by the usual optical microscopic examination at post mortem.

The same feeling was expressed by the inspector reporting after a public inquiry in 1981. Langbaurgh Borough Council, near Middlesborough, had refused planning permission for private householders to build asbestos cement garages. The matter went to public inquiry and the council lost. The inspector said, however: 'It is clear to me that developments in the use of the electronic microscope, which post date the eminent advisory committee's final report, may call for changes in the recommended methods of handling and working with asbestos-cement sheets outside the place of manufacture; the pursuit of alternative reinforcement is an obvious priority.'

SPAID believes that it is the only group to insist on electron microscopy and fears it may have to stop if its Greater London Council grant ceases with the GLC's demise because each examination costs at least £500.

A Ban on Asbestos

Substitutes for asbestos became available in this country only after the media raised the alarm and the public refused to handle asbestos, or to work with it or buy it. The use of asbestos is, unfortunately, a barometer of poverty and ignorance. It is still being worked and mined in Africa and India, where it has been said cynically by asbestos producers that, 'Our workers do not live long enough to get cancer'. Not only should all asbestos products be banned immediately, British companies should be prohibited from having any interests in asbestos mines or factories overseas. Mesothelioma kills just as surely and just as cruelly wherever you are.

Notes

1. Official estimates in the USA put the potential death toll from asbestos exposure amongst workers alone at two million. In Britain it is possible that some 500,000 workers will die from asbestos related diseases over the next 30 years. If that proves the case then, as Alan Dalton of the British Society for Social Responsibility in Science remarks: 'Asbestos exposure will kill more people in Britain than were killed in the armed forces during the Second World War.' (Alan Dalton, *Asbestos: Killer Dust*, British Society for Social Responsibility in Science, 1979, p. 7.)
2. Ironically, asbestos is not and never has been fireproof. It has a degree of fire resistance, that is all. The mistaken belief that asbestos was fireproof contributed to the death of 50 people when the Summerland pleasure complex on the Isle of Man caught fire in 1973. In the original plan for the building the external wall, against which the fire started, was to have been made of reinforced concrete. To save money, however, it was built of a cheaper corrugated product called Gaybestos. This was a zinc-coated steel sheet with a layer of asbestos bonded to it. The architects thought it must be fireproof and would comply with a local by-law that external walls had to be made of non-combustible material with a fire resistance of two hours. In fact, the fire spread hidden behind the steel before anyone realized how far it had got. The inquiry into the cause of the fire found: 'The use of Gaybestos in the wall of this building was an error of judgement . . . the contribution of the combustible components of Gaybestos accelerated the growth of the fire in the early stages.' The death toll was so high because people had no time to escape.
3. Nancy Tait was awarded a Churchill Fellowship to continue her work. She has travelled the world collecting medical evidence and combed Britain meticulously recording case histories, particularly of low exposure. Her book, *Asbestos Kills*, was the first to present the cancer risks of asbestos to the general public.

18

Britain's Dirty Beaches

Fred Pearce

Britain has some of the dirtiest beaches in Europe. The government and the ten water authorities that discharge the sewage around those beaches are a bit touchy about this and in general refuse either to provide information about the state of those beaches or to spend money cleaning them up. From Brighton to Blackpool, Great Yarmouth to Hastings, the story is the same. The sewage is, in the words of one Cornish paper, being 'swept under the carpet'.

The EEC Directive

On this, as on many other pollution issues, Britain has been branded as the dirty man of Europe. A decade ago, the EEC published a Directive, binding on all members, that required all bathing beaches to meet certain basic standards of bacteriological cleanliness. It boiled down to requiring that all sewage dumped into the sea from holiday resorts either be treated to kill the bacteria or be discharged so far out to sea that, by the time it washed back onshore, it was safe.

Britain's game, it seemed, was up. But the EEC left a loophole. It would be up to each nation to come up with its own definition of what a bathing beach actually was. Britain decided on a definition so restrictive that it did not include Blackpool – by a long way the nation's favourite resort – as a bathing beach. This was fortuitous, since the North West Water Authority had reported that it would cost around £30 million to bring the resort up to European standards of cleanliness.

Of more than 600 beaches around the shores of England and Wales, the government 'designated' as bathing beaches just 27, at a mere 18 resorts. And Scotland was declared to have no bathing beaches at all. By comparison, France and Italy designated 3,000 beaches. The trick was to define a bathing beach as one containing more than 1,000 people per kilometre of beach. The government then left it up to the individual water authorities to decide how long a stretch of beach to measure at a time, when to count the bathers (would it be a wet Thursday in May or a sunny bank holiday in August?) and so on. Since the designation of bathing beaches was to be a once-for-all decision, with no adding to the list later, the great British beaches swindle appeared to have worked.

During 1985, the EEC finally realized that it was being bamboozled and a murky deal was done. It began with an innocent enough question asked in the European parliament late in 1984. Was the Commission happy with the way that Britain was applying the bathing waters Directive? The answer was harsh. 'In the Commission's view, the interpretation given by the UK government about what must be considered as bathing waters . . . is very restrictive . . . excluding all of the fresh water bathing areas as well as some popular beaches.' The answer then spoke of 'informal discussions' with the British government. 'The Commission may have to seek a common interpretation from the Court of Justice.'

As 1985 went on that threat evaporated. What had Britain done? The government did not designate any more beaches. But it did write a letter to the water authorities requiring them to conduct a major new survey of bathing beaches and warning them that the results would be published. In the way of such things, the government did not publicize its climb-down. It told the water authorities of its new approach in a private letter sent in April 1985. Nine of the ten water authorities hold their meetings in secret. Only the Welsh Water Authority still publishes minutes. But for those minutes the existence of the new survey would itself have remained a secret until the end of December 1985, when an announcement emerged from the Department of the Environment.

Whitehall had decided to abandon its absurd statistical definition of a bathing beach. From now on, it has told the water authorities, any beach having bathing huts, lifeguards and such will be regarded as a place where people swim. All such places, the ministry said in its letter, should now be surveyed and the results be published. The ministry says that this will lead to information on the bacteriological quality of 350 beaches being made public.

The Reaction of Water Authorities

This compromise may have got the government off the hook in Europe. But it has not pleased some of the water authorities. In November 1985, the North West Water Authority, custodian of Blackpool, Morecambe and other resorts, told ministers that the deal was 'unacceptable'. Publication of details about the bacteria off its bathing beaches would, the authority thought, increase public pressure for action at a time when it was more concerned about repairing its crumbling sewers and improving the quality of water supplies.

The authority said that it would cost £88 million to bring its bathing beaches up to the European standard. The authority's scientists were at that time planning to survey the state of bathing waters at four resorts: Blackpool, Morecambe, Southport and 'one somewhere in Cumbria'. Which of the dozens of polluted Cumbrian beaches would be chosen appeared to depend on the result of political discussions with local councils, rather than any assessments of the beaches' popularity. In the end, ministers have required the authority to survey 30 beaches including four at Blackpool and they have insisted that Blackpool's beaches must be cleared up, even if that delays other projects.

The Southern Water Authority has proved more helpful. It is surveying 65 beaches during 1986. It is prepared to be more candid because it is doing more to solve its problems, which are every bit as severe as those in the north west. The authority had completed seven new long sewage outfall pipes in its area (which extends from Herne Bay to Bognor Regis) in the seven years to 1985. Late that year, it hauled two new pipes out to sea at Margate and Ryde on the Isle of Wight. Broadstairs, Portsmouth (which spews much of its sewage onto the beaches of Hayling Island) and Ramsgate were all in line as part of a £125 million programme. The authority says that of its 81 sewage outfalls, only 25 are 'satisfactory'. In several towns, notably Hastings, Bexhill and Cowes on the Isle of Wight, there are embargoes on new building until the sewage problems are resolved.

A second authority willing to spend to clean up its beaches is the Welsh Water Authority. In October 1985, the authority unplugged a new three kilometre submarine pipeline at Tenby, designed to relieve the town's south beach of a notorious pollution problem. A press release for the occasion noted pertinently that the 'legacy of neglect is now seen as a potential threat to . . . the vital tourist industry.'

The authority also has pipeline schemes in hand near Porthmadog and at Barry, Porthcawl, Penarth and Llandudno. But there is plenty more work to be done. Only 38 of the 120 or so beaches tested by the authority meet the European standards. Some, in Swansea Bay, have been condemned by local doctors as unfit for bathing. Overloaded sewage tanks near Mumbles frequently dumped sewage onto the foreshore of the bay at low tide.

England's water authorities have traditionally said that a beach is satisfactory if there are no signs of sewage, such as sewer slicks or solids, in areas that people might bathe in. Even on these generous criteria, the failure rate is high. The Anglian Water Authority (Cleethorpes to Southend) says that only 45 per cent of its sea outfalls are satisfactory. The North West Water Authority will only vouch for 25 per cent of its outfalls, and the tiny Wessex Water Authority, which supervises the beaches of Hampshire, Dorset and Somerset, reckons 31 per cent are up to standard. The beaches of the north east, overseen by the Northumbrian Water Authority, have traditionally been polluted as much by coal slurry as by sewage, gushing from the rivers Tyne, Wear and Tees. But attempts are under way to stem the pollution that hits Redcar beach.

The EEC has set a scientifically testable definition of what constitutes a 'safe' Eurobeach. Samples of water should not contain more than 10,000 bacteria known as 'coliforms' per 100 millilitres, and no more than 2,000 faecal coliforms, which come from human excreta. According to ministers, preliminary findings from the national survey show that at least half of the 350 beaches surveyed do not meet these criteria.

Ministers promised the EEC in 1986 that more would be spent to bring beaches up to scratch. They told Parliament in late July that spending would double to £300 million in the next five years. But the figure is not what it seems. Further inquiries revealed that it was the total cost of all projects that were *already* under way, or which ministers hoped would start by 1991. Real spending, according to a less publicized parliamentary reply in April, would be £150 million in that period, a little *less* than in the previous five years. Even those figures include many projects – such as cleaning up the Mersey estuary – which may help beaches, but only incidentally.

Problems with Cleaning Up

Meanwhile, technical disasters have befallen some clean-up schemes. An ambitious tunnel, several kilometres long, designed to take Weymouth's sewage clear of local beaches, collapsed during construction. And a sewage scheme outside Brighton, which should be keeping the town's sewage away from the unlucky beaches of nearby smaller resorts, has already cost millions of pounds in repair as first the works itself, and then the outfall tunnel, collapsed under the strain of the resort's sewage. A scheme to save the beaches of Scarborough from the town's sewage (a problem worst after stormy weather) has been bedevilled by design problems and will be hard pressed to meet its 1990 completion date. This is unfortunate since the town's beaches are designated Eurobeaches. The Yorkshire Water Authority was given a special dispensation by the EEC to delay its compliance with the European rules for five years until 1990.

But at least Scarborough's beaches will eventually pass muster. The water authority's scientists say that they had wanted to designate neighbouring beaches, at Filey for example, but had been prevented from doing so by the government.

Do the Scots Not Swim?

What of Scotland? Unlike England and Wales, the Scots do not have water authorities that both create pollution and police it. The pollution from sewage is created by sewage works run by the regional councils. But the pollution watchdogs are the river purification boards. Without the English conflict of interest, one might expect greater vigilance about the bacteriological standards of the country's beaches. Yet there are no Eurobeaches designated in Scotland. Why not? Do the Scots not swim?

The answer comes from Desmond Hammerton, the director of the largest purification board, the Clyde River Purification Board. He wrote to *New Scientist* in July 1985:

> The government refused to designate a single bathing beach in Scotland on the grounds that no beach met the criterion of 1000 people per kilometre and that in any case the directive was really intended for the Mediterranean whereas in Scotland, because of the lower temperatures, few people naturally bathed and when they did so they only stayed in the water for very short periods . . . In 1976 and 1977 my board carried out a survey of bathing beaches along a 168 kilometres length of coast and showed that about half the beaches would comply with the directive. Since then, despite the fact that no beaches were designated and without any encouragement from the government we have published in our annual reports the results of bacteriological surveys of most of the important bathing beaches and the percentage compliance with the directive. My board's view has always been that the public are fully entitled to this information.

The Water Authorities: More Sinned against than Sinning?

Ever since their creation in 1974, the water authorities of England and Wales have been pressed by governments to cut their spending. For many, the task of cleaning up beaches has fallen off the end of the queue altogether. But by agreeing to the government's plans to sabotage Britain's implementation of the bathing waters Directive, and by indulging in an orgy of secrecy to cover the real state of affairs, many authorities have managed to portray themselves as more villain than victim. After a decade of telling water authorities to keep quiet about their beaches, ministers are now brazenly ordering them to 'come clean'. It is a sad, and rather typical, story.

19

Down in the Dumps

Britain and Hazardous Wastes

Nicholas Hildyard[1]

Incredibly, the Department of the Environment has no hard figures for the amount of hazardous waste generated every year in Britain. In 1981, witnesses from the Atomic Energy Authority's Environmental Safety Group told a Select Committee of the House of Lords: 'We do not know how much hazardous waste is produced in the UK, who produces it, what it is, and what happens to it.'

The Select Committee (known as the Gregson Committee after its chairman, Lord Gregson) agreed: 'Computation of the amount of hazardous waste for disposal is virtually impossible.' In order to establish a rough figure, the committee sent a questionnaire to all of Britain's local Waste Disposal Authorities (WDAs), the bodies which are responsible for overseeing and policing waste disposal. On the basis of that survey, the committee estimated that some 5.5 million tonnes of hazardous waste were disposed of in Britain in 1980. Of that total, '4.4 million tonnes arose in England, 435,000 tonnes in Wales, and 710,000 in Scotland and Northern Ireland.'

Those figures have been broadly confirmed in a 1985 report by the Hazardous Waste Inspectorate (HWI), a non-statutory body set up by the DoE in 1983 on the recommendation of the Gregson Committee. According to the HWI, 4.4 million tonnes of hazardous waste were generated in England and Wales in 1983. More recently, however, the Confederation of British Industry (CBI) told the 1985 Royal Commission on Environmental Pollution (which was investigating waste management practices in Britain) that some 12 per cent of the estimated 100 tonnes of industrial waste generated every year could be described as 'toxic or dangerous'. If the CBI is correct, then approximately 12 million tonnes of waste could properly be classed as hazardous.

The uncertainty over the figures is not a matter of purely academic concern. Commenting on the lack of firm data, the Royal Commission warns: 'We have concluded that there is insufficient reliable information available about industrial waste generation. Without this information, we do not consider that waste disposal authorities can properly prepare waste disposal plans for their areas.'

Past Sites: A Legacy of Destruction

The reckless, ignorant or indifferent disposal of hazardous waste has already resulted in gross pollution problems at numerous sites throughout the industrial world. As a result, massive sums of money are now having to be spent in order to clean up old sites. In addition to the *direct* economic costs incurred by cleaning up old dumps – costs which will place a heavy inflationary burden on the economies of many countries – the long-term 'hidden' costs to human health and environmental quality will be enormous. In the USA alone, it is estimated that it will cost *at least* $100 billion to clean up the 10,000 worst sites in the country. A more realistic figure would be in the region of $300 billion, equivalent to the entire annual defence budget of the US government.

No one knows how many 'problem' sites there are in Britain – *largely because no one has looked for them*. Indeed, the DoE does not even have a complete inventory of abandoned dumps – and, more incredibly, it has no intention of compiling one. The danger, as one member of the Gregson Committee pointed out, is that we are living 'in a minefield for which we have lost the chart'. Although a limited DoE survey, carried out in 1975, revealed 53 problem sites, several of those sites are still used for the disposal of highly toxic wastes. The most recent survey – a desk study carried out in 1985 by the independent consultancy firm ECOTEC – estimates that 'approximately 600 landfills may present a hazard'.[2]

Nonetheless, the DoE is insistent that the probability of Britain experiencing a major environmental disaster involving hazardous wastes is minimal. Although some past sites have required expensive clean-up operations before they could be redeveloped (with costs ranging from £40,000 per hectare to £110,000), it is confidently claimed that there are no sites in Britain on a par with the now infamous Love Canal dump in the USA, a site which caused the wholesale contamination of a housing estate near Niagara Falls and subsequently necessitated the permanent evacuation of several hundred local residents.

That view is shared by both the Gregson Committee and the Royal Commission on Environmental Pollution – the latter going so far as to state quite categorically that 'the probability that the UK may yet contain undiscovered horrors on the scale of Love Canal can be discounted altogether.' The Royal Commission argues that the scarcity of 'new' land for development has made planners in Britain aware of the importance of being able to re-use land. Consequently, suggests the Commission (though it does not state it explicitly), there has been a more responsible attitude towards land use in Britain than in the USA: waste dumps have been better located and planning controls more strictly enforced.

Likewise, witnesses from the DoE told the Gregson Committee:

> The United Kingdom is a small country and the chances of indiscriminate dumping having occurred seem likely to be small. For the rest, 130 years of public health legislation should have prevented the accumulation of waste becoming an environmental hazard . . . This contrasts sharply with the position in the USA where, until the passage of the Resource Conservation and Recovery Act (RCRA) in 1976, there was no federal legislation controlling the disposal of hazardous wastes.

On close analysis, none of the above arguments holds much water. It is spurious to suggest, for example, that because Britain is a small country, there is less likelihood of indiscriminate dumping having taken place. Experience shows that size is no safeguard against criminality. The state of New Jersey – where the bulk of the US's synthetic chemicals are produced – is pockmarked with unsafe (and frequently illegal) dumps. Yet, New Jersey is less than a tenth of the size of Great Britain, a mere 7,836 square miles as against 94,220 square miles. So too, thousands of potential Love Canals have been discovered in Holland, a country which is not only smaller and more densely populated than Britain but where pressure on land is also far greater. Indeed, Britain's size – and particularly her need to re-use land – makes it *more*, not less, likely that contaminated land will one day be built upon.

The remarks about US legislation are equally misleading. Until 1972, when the Deposit of Poisonous Waste Act (DPWA) was passed, there were no laws in Britain *specifically* designed to control the disposal of hazardous waste. British legislation on hazardous waste thus predates America's RCRA by only four years. Moreover, it was not until 1976 that a site licensing system was introduced in England and Wales under the 1974 Control of Pollution Act (COPA), whilst site licenses were not required in Scotland until 1978 and in Northern Ireland until 1980. Even in 1981 – when COPA's Special Waste Regulations were finally introduced and the DPWA was repealed – few waste disposal authorities had any facilities for disposing of hazardous wastes other than asbestos. The majority of WDAs were acknowledged to be 'understaffed or at full stretch', and most did not even employ a toxicologist. Indeed, to suggest that British controls have been superior to those operating on the continent or in the USA is no more than crude and unfounded propaganda.

With the exception of Britain, every major industrial country in the EEC has carried out a survey of its old dumps. Such surveys have quickly revealed numerous problem sites. In Denmark, 100 sites are thought to present an immediate threat to groundwaters or surface waters. In Germany, over 6,000 dumps have now been closed by the authorities as health risks: 800 are deemed a threat to water supplies. In Holland, a national suvey was ordered in 1979 after 10,000 drums of toxic wastes were discovered buried beneath a rubbish dump in Amsterdam; within six months, 4,000 other illegal dumps had been discovered.

Commenting on the Dutch experience, Graham Bennett, a consultant for the Institute for European Environmental Policy, wrote recently in the *New Scientist*:

> Four thousand illegal toxic waste dumps in a small country might seem to be an excessive number. Clearly, the environmental control for the past decade was less than water-tight. But surprising as it may seem, Holland is undoubtedly one of the better countries when it comes to environmental protection. The extent of the toxic waste problem has come to light because the Dutch were plucky enough to 'fish around in the drain.' And that raises the question of what might lie waiting to be discovered elsewhere. The lesson seems to be clear enough: *seek and ye shall find*.

It is a lesson that Britain seems determined to learn the hard way.

Today's Dumps: 'All is not well'

In June 1985, the Hazardous Waste Inspectorate published a refreshingly frank report which dispelled once and for all the myth (previously fostered by the DoE and others) that the British waste disposal industry now has its house in order. Although the Inspectorate stressed that it had visited some waste disposal facilities 'which are conducted in accord with current best practice and which are a credit to their operators', it warned: 'A significant number of landfill operators appear ignorant of the technical and scientific research that underpins the controlled landfill disposal of hazardous wastes . . . disposal site licence conditions are, in some cases, wilfully breached without attracting effective enforcement action by the WDA's.'

In the course of its investigations, the HWI visited some 400 sites in England and Wales. During those visits (most of which were pre-arranged with the site operators) the Inspectorate witnessed 'a considerable number of undesirable disposal practices.' Acknowledging that the 600 hours which it spent on site visits represented a minute fraction of the five million hours during which English and Welsh disposal facilities are open for business each year, the Inspectorate went on to comment: 'It may be inferred that malpractice at hazardous waste disposal sites is not uncommon . . . The HWI is in no doubt that the standards are anything but consistent or satisfactory.'

Among the most flagrant malpractices it witnessed, the Inspectorate reported having seen 'fibrous asbestos waste loose on the surface of landfills' and asbestos waste being delivered 'in bags in mixed loads with other industrial/hazardous waste, such that the bags . . . ruptured on deposit.' At one asbestos waste landfill, there were 'no site furnishings other than a demounted railway truck'; indeed, the site had no running water supply, no telephone and no electricity. At several other sites (unfortunately no figures were given) the HWI found that no record had been kept of where asbestos wastes had been deposited – in direct contravention of Britain's Special Waste Regulations. Incredibly, two WDAs took the view that asbestos should not be classified as a 'special waste' since, in their opinion, 'even in occupationally exposed populations, mortalityrates ascribed to asbestos are a small proportion of [all deaths in] that population.'

Although in Britain (unlike in the USA) there is no direct ban on the landfilling of drums containing liquid wastes, the DoE recommends against the practice in several of its *Waste Management Papers*. To its obvious disquiet, the HWI found that the DoE's recommendations were regularly flouted at many landfills:

> HWI have repeatedly discovered evidence of an unacceptably casual attitude towards the reception and disposal of drummed waste at landfills. For example, HWI has witnessed the deposit of a trailer load of [circa 80] sealed 45-gallon drums of waste on a landfill site where the operator had no idea of the contents and made no attempt at inspection. They were deposited en masse, still sealed, with no attempt to avoid a localised concentration . . .

The HWI also reported that at two in-house disposal sites (approximately 40 per cent of all waste from the chemical industry is disposed of in-house and the majority of landfills are in-house sites) 'very large numbers of drums were observed floating on lagoons.' Both sites were 'disposal outlets for major companies' and in one case a fire subsequently broke out.

Discussing the policing of hazardous waste disposal sites, the HWI records that standards vary widely across the country but are generally low. 'It is the HWI's opinion that the majority of disposal authorities barely discharge their statutory duties with respect to hazardous waste: few go much beyond this imposed minimum.' As a result, 'major operators, who generally try to adopt current best practice, are steadily losing business to other operators who are able to exploit variations in standards and enforcement.' Indeed, the Inspectorate reports that the last few years have seen a 'proliferation' of new hazardous landfills in precisely those areas where WDAs are failing most conspicuously to enforce adequate standards of management. In such areas, unscrupulous operators can clearly gain a commercial advantage over their more conscientious colleagues.

In effect, says the HWI, market forces – combined with official complacency – are allowing operators to employ the 'cheapest tolerable means' of waste disposal. Already, cut-throat competition within the waste disposal industry is leading to more work going to cowboy operators. In some areas, such cowboys have attacked inspectors. In Merseyside, for example, 'it has sometimes proved necessary to deploy inspectors in twos, and even threes, equipped with personal radios, when investigating the activities of local asbestos strippers and their "transfer" stations.' It is estimated that there are at least 150 incidents involving the illegal disposal of hazardous waste every year.

The Inspectorate's conclusions were unequivocal: 'All is not well with hazardous waste disposal. Though there is no evidence that hazardous waste disposal is posing unacceptable risk to public health, we are not convinced that the standards and practices widely adopted by the disposal industry provide a sufficient guarantee of protection of the environment.'

No National Standards

The disposal of waste on land is now subject to Part 1 of the Control of Pollution Act. Although COPA received royal assent in 1974, successive governments were slow to implement its key provisions on waste disposal 'due to the need for restraint on public expenditure'. Indeed, there are many who argue that were it not for pressure from the EEC, many sections of COPA Part 1 might still not be in force. Even today, Section 1 which requires all WDAs to ensure that they have adequate facilities for disposing of 'controlled waste' (that is, domestic, commercial and industrial waste) has yet to be implemented.

COPA's provisions do not lay down *mandatory, national* standards for the management of hazardous wastes. Under the act's site licensing system, it is left to local WDAs to specify what wastes may be deposited at a given site, in what quantities and under what operating conditions. Although the DoE – through its *Waste Management Papers* – sets out a series of 'recommended guidelines' for the disposal of hazardous wastes, those guidelines have no force in law. If a local authority fails to implement them, all the DoE can do (on its own admission) is to 'have an informal chat with the particular authority concerned'.

By contrast, in the USA, the Resource Conservation and Recovery Act (RCRA) and subsequent legislation not only specifies minimum standards for the design of waste disposal facilities but also bans certain waste disposal practices. In addition, an amendment to the act sets mandatory standards for the design of waste disposal facilities. All sites must also be monitored for groundwater pollution – although sites which do not contain liquid wastes may, under special circumstances, be exempted from monitoring by the EPA. Such monitoring is not a legal requirement in Britain.

Britain argues that voluntary controls are just as effective as the USA's use of the 'regulatory stick'. In particular, COPA's site licensing system is seen by authorities as offering more 'flexibility' than national standards, since site licences allow WDAs to take advantage of local conditions whilst, at the same time, ensuring 'an appropriately high level of site management'. The *proviso*, of course, is that WDAs adhere to the advice laid down in the DoE's guidelines. As we have seen, however, WDAs differ widely in the interpretation of the DoE's advice: in some cases, the department's guidelines are simply ignored.

Britain's faith in voluntary controls reflects both the cosy relationship that has built up between the DoE and industry, and the belief that 'good' legislation should be as 'flexible' as possible. Yet, all too often, experience shows that 'flexibility' simply means 'taking liberties' – particularly in an industry which still contains its fair share of cowboys. Moreover, it is a debatable point whether a coherent national policy on hazardous waste can ever emerge without national standards embodying national goals.

No Provisions for Aftercare

Under US law, the operator of a landfill is responsible for the safety of the site for 30 years after it has been closed. By contrast, in Britain, there is no law which requires the long-term management of closed sites. As soon as a landfill ceases to receive material for tipping, its site licence comes to an end. The restoration of the site is usually subject to conditions laid down by the planning authorities, who may be quite unaware of the technical problems involved.

In 1981, the Gregson Committee (which found that some sites were not even being monitored *during* their lifetime) recommended that 'site operators should be required to monitor sites during, and for a set period after, operation.' In 1985 – a full four years after the committee made that recommendation – the government replied that it would 'consider whether changes are needed in licensing legislation to facilitate imposition of requirements related to post-operation monitoring'.

Incredibly, under British law, there is no requirement to tell a potential buyer that a site has been used for the disposal of hazardous wastes. But, in the case of old dumps, the age-old principle of *caveat emptor* (let the buyer beware) takes on a new – and sinister – meaning. If housing estates, schools, or office buildings are built over contaminated sites – even in the distant future – the local population is almost certain to be exposed to a wide range of highly toxic chemicals, and heavily exposed at that.

Where an abandoned site is found to be polluting, the local authority often finds itself having to navigate a legal minefield in order to ensure that the site is cleaned up. Mr A. Q. Khan, the Waste Disposal Officer for South Yorkshire, illustrates the problem in his own county:

> A tip is on fire. Nobody works there. Nobody has the right to go there. And the tip has been burning for years. In three years it has gone down by about 30 feet. It is an old site. We have tried to look into all the legislation. Almost half a dozen pieces of legislation have some fringe involvement, but none to make the site owner do anything . . .

No Superfund

In the USA, a 'Superfund' – bankrolled by industry – has been set up for cleaning up those abandoned sites which are deemed to be 'imminent hazards'. Known formally as the Comprehensive Environmental Response, Compensation and Liability Act (CERCLA), the Superfund legislation was passed in 1980 – the last act that President Carter signed before leaving office. As originally conceived, the Superfund was to raise $4.2 billion – but this sum was reduced to $1.6 billion under pressure from industry, which is required to provide 80 per cent of the cash. It now seems clear that even the $1.6 billion raised by the Superfund is nowhere near sufficient to clean up all of the USA's 'toxic time bombs'.

Even so, Superfund is a huge step in the right direction. Britain, by contrast, has no fund to clean up polluting sites. Yet, as the HWI warns, 'it is doubtful that many waste disposal operators are making adequate financial provisions to pay for restoration or post-closure management.' How then will they pay if the sites prove polluting?

Equally important, the law does not make it clear who should pay for remedial action at polluting sites. If it can be *proved* that the site is polluted because the operator broke the conditions of his site licence, he is liable for any clean-up costs. If, however, the operator had complied with his site licence, it is left to the courts to decide whether he or the local WDA should pay for the clean-up: that means deciding who was negligent – a lengthy, and not always conclusive, business.

Licensed Anarchy

Laws are only effective if they are upheld. If they are continually broken – and openly so – then they not only fail to achieve their stated aims but eventually fall into direspect. The result is licensed anarchy. And that is precisely the state of waste disposal in Britain today. Many waste disposal operators openly abuse their site licences. 'We have a system at the moment where waste disposal authorities should enforce conditions on site licences,' John Newton of the Amey Roadstone Corporation told a 1983 conference on landfills. 'Quite honestly I think that if those conditions were enforced totally most of the sites in this country would close down.'

One problem facing WDAs is that the failure to comply with a site licence is not in itself an automatic offence under the Control of Pollution Act: *the site operator must be caught in the act of breaking his licence whilst depositing waste if he is to be prosecuted.* 'Some authorities like my own have been spending a lot of time trying to enforce the legislation and it is not easy,' says C. J. Tunaley of Yorkshire County Council.

> You've got to catch these people red-handed, infringing licence conditions and then you've got to take them to court and knock them out to get a draw – and they get a £10 fine. It's very demoralising, when having tried to enforce what is fairly 'holey' legislation to be always on the receiving end of a minimum derisory fine for the guilty party.

Although the maximum fine for all breaches of COPA was recently increased to £2,000, the HWI notes that 'many WDA staff complain of their frustration at the reluctance of their legal departments to mount enforcement actions when evidence of breaches of the site licences has been established by careful policing.' In 1985, the DoE recommended that 'consideration should be given to extending the scope of site licensing conditions to permit prosecutions where a breach of conditions is not related to the act of deposit itself.' If that recommendation is accepted by the government (and one hopes that it will be) waste disposers will not have to be caught red-handed in order to be brought to book. Until it is, however, WDAs will still have to cope with legislation that is generally accepted to be in dire need of strengthening.

Section 17: A Legal Minefield

Perhaps the most controversial provisions in Part 1 of the Control of Pollution Act are those dealing with so-called 'special wastes' – the Section 17 regulations. The regulations, which were introduced in March 1981, are intended to 'provide close control over the carriage of those wastes which might seriously threaten life itself if disposed of heedlessly'. To that end, Section 17 requires all 'special' wastes to be accompanied by a set of consignment notes which have to be signed at each stage in the waste's journey from its point of production to its final disposal site.

The controversy over the regulations centres not on this 'cradle to grave' manifest system (which is rightly welcomed) but on the criteria used to define 'special' wastes. One such criterion is whether or not a waste is 'dangerous to life'. Under the regulations, a waste is only deemed dangerous to life if:

(a)   A single dose of not more than 5 cubic centimetres would be likely to cause death or serious damage to tissue if ingested by a child of 20 kilograms body weight, or

(b)   Exposure to it for 15 minutes or less would be likely to cause damage to human tissue by inhalation, skin contact or eye contact.

That definition has variously been described as 'a scientific and legal minefield' and as 'virtually unenforcable in a court of law'. As the Association of County Councils put it to the Gregson Committee:

The regulations give considerable scope for uncertainty and disagreement with, inevitably, lengthy and costly litigation. Information on the toxic effects of chemical substances on humans (especially 20 kilogramme ones) and even on mammals is very scarce . . . Calculations of the effects on a 20 kilogramme child of the infinite variety of mixtures of chemicals of different concentrations and quantities would be so uncertain and open to challenge in courts that WDAs would have the greatest difficulty in proving that certain wastes are toxic or harmful and should be disposed of accordingly.

It is also pointed out that by defining 'dangerous to life' in terms of 'death or serious damage to tissue', the Section 17 regulations neatly side-step having to take account of the chronic, sub-lethal effects of hazardous wastes. Moreover, *the regulations take no account whatsoever of the damage that wastes may cause to the environment*, since 'special' wastes are defined primarily in terms of the hazards they pose to *human* health. It is thus quite possible for large quantities of environmentally damaging waste to be transported around the country (and possibly spilled or dumped illegally) without the authorities being any the wiser, since such waste is not subject to the manifest system.

In that respect, the regulations conspicuously fail to comply with the EEC's Directive on Toxic and Dangerous Waste, whose provisions Part 1 of COPA is supposed to implement. The Directive clearly stipulates: 'the essential objective of all provisions relating to the disposal of toxic or dangerous waste must be the protection of human health and *the safeguarding of the environment* against harmful effects caused by the collection of toxic and dangerous waste, as well as its carriage, treatment, storage and tipping' (my italics).

When the Special Waste Regulations were reviewed in 1985, the DoE's Joint Review Committee (which is composed of representatives from government, the local authorities and industry) concluded that there was no need to broaden the definition of special wastes.

Landfill: The Primitive 'Solution'?

No waste disposal technology arouses more controversy than landfill. The European Commission regards it as 'a temporary and second best' method of disposal and laments the reliance that many member states place upon it. In France, landfill is seen as 'a necessary and technically acceptable solution' – but only 'in the short-term' and only for wastes of low toxicity. In Holland, the dumping of all chemical wastes on land has been expressly forbidden since the early 1980s. In the USA, recent amendments to the Resource and Recovery Act require the EPA to limit severely the landfilling of hazardous wastes. Some states (Florida, for example) have already banned the landfilling of hazardous wastes, whilst others (notably California) do not permit the disposal of liquid hazardous wastes in landfills.

Britain, however, remains firmly committed to landfill as her primary method of waste disposal. In all, 85 per cent of the 4.4 million tonnes of hazardous waste generated every year in England and Wales is landfilled. In Scotland and Northern Ireland, the proportion is reported to be even higher. According to the Hazardous Waste Inspectorate, there are 4,202 landfills in England and Wales, of which an estimated 1,145 are licensed to take hazardous waste. In addition to those landfills, there are some 737 land based disposal facilities, including 99 storage sites, 53 lagoons, 5 'reception' pits and 14 mineshafts.

Britain justifies its reliance on landfill – which the DoE admits is 'heavier than some other countries' – by citing a single study, conducted in the mid-1970s, which concluded: 'Sensible landfill is realistic and an ultra-cautious approach to landfill of hazardous and other types of waste is unjustified.'

The study, known as the Co-Operative Programme, surveyed a tiny fraction of the total number of landfills in Britain. Just 19 sites out of the 5,000 landfills then operating were studied (that is, less than 0.4 per cent) and only 15 were examined in depth – hardly a 'statistically significant' sample. Of the sites surveyed, two were discovered to be causing water pollution problems. Although the Co-Operative Programme claimed that the 19 sites in the study were 'representative of the main geological types found in the United Kingdom', that claim is misleading: varying topography, differing rainfall patterns, differing soils and numerous other factors make each site unique. The truth is that we still know little about the geology of landfill.

But the most striking feature of the Co-Operative Programme's final report, published in 1978, is that its conclusions are diametrically opposed to those of every major study which has been carried out on landfills in the USA. In fact, US waste disposal experts are now generally agreed that the landfilling of hazardous waste is a singularly unsafe and grossly unreliable method of waste disposal. 'There is no such thing as a secure landfill,' commented Peter Montague, the project administrator of Princetown University's hazardous waste programme. 'There is no such thing as "permanent" diposal of hazardous waste in the ground. There is "long term storage" in the ground but sooner or later what we put there will move – and the place it is most likely to move to is someone's drinking water supply.'

Why then, does the DoE continue to insist 'in the particular circumstances of (Britain), landfill represents the best practicable environmental option'? Is the DoE party to information which is unavailable to US experts? Or are its standards less exacting than those of the USA?

Dilute and Disperse?

The majority of landfills in Britain are 'dilute and disperse' sites: that is, landfills which are built over permeable soils in order that their wastes can seep out of the site – on the assumption that they will be broken down and rendered harmless as they filter through the underlying soils.

Critics of the 'dilute and disperse' philosophy do not dispute that hazardous wastes undergo a variety of transformations when they migrate through the soils beneath landfills. What they question is that the resulting 'attenuation' of pollutants is as effective as the waste disposal industry – and the Co-Operative Programme – would have us believe. Certainly, some wastes are indeed rendered harmless by the physical and biochemical processes which take place beneath (and within) landfills. But, equally, other wastes can be transformed into new and more hazardous compounds. For example, tetrachloroethylene (TCE), and 1,1,1–trichloroethane (TCA) can all be converted by the action of microbes into vinyl chloride, a potent human carcinogen. Still other wastes (certain chlorinated hydrocarbons, for example) are effectively immortal: once in the soil, they do not break down but persist indefinitely. At present, however, scientists are quite unable to predict with any degree of certainty how the majority of wastes will behave.

Indeed, our understanding of how wastes behave once they are released into the environment is pitifully small. For that reason alone, the practice of purposively depositing wastes into leaking holes in the ground is, at best, questionable. We do not know how the majority are likely to react with each other; what effects they will have on underlying soils; whether or not they will be taken up by the soil organisms and thereby enter into the food chain; whether they are persistent or degrade easily; and at what concentrations they constitute a threat to human health and the environment. What we do know, however, is that 'dilute and disperse' sites throughout the industrialized world are now a major source of groundwater pollution.

Such uncertainties – and, in particular, the fear of groundwater pollution – have led many industrialized countries to ban (or phase out) the use of 'dilute and disperse' sites. In France, for example, only sites with a five metre thick base of impermeable material are permitted to take special wastes – and, even then, the government insists that 'high toxicity' wastes (cyanide salts, for example) be shipped out of France and stored in the disused salt mines of Herfa-Neurode in West Germany. According to the Ministry of the Environment, there is no landfill site on French soil where such wastes can be safely deposited. Permeable sites are only permitted to take inert wastes, such as building rubble.

Similarly, in the USA, all new landfills are now required to have 'a liner to prevent the migration of wastes to soils and surface waters'. In addition, a leachate collection system is obligatory (leachate is the polluting liquid that accumulates at the base of a landfill), and the dumping of liquid wastes in landfills is only permitted at 'facilities with liners and leachate collection and removal systems'.

Only 12 sites in Britain are lined. In effect, the vast majority of Britain's landfills would not be permitted to operate in either France or the USA.

Nor, for that matter, would many be countenanced in West Germany, where all new landfills must have an impermeable base.

The Myth of Containment?

But do liners work in the long term? Because such lining materials as clay and plastic are impermeable to water, it is often assumed that they will be equally impermeable to other liquids. It is now known, however, that some hazardous wastes – particularly industrial solvents – can transform even clay soils into sieves. Experiments in the USA reveal that clay exposed to the chemical naphtha is 730,000 times more permeable than when it is exposed to water.

Research has shown that synthetic liners are just as likely to leak as clay liners. In 1980, Peter Montague of Princetown university studied four supposedly 'secure' landfills in New Jersey. All the landfills were double-lined. At one site, where the liners consisted of hypalon (a tough polymer material, reinforced with nylon), the primary liner leaked 124 gallons a day within four months of being installed. All three other sites also leaked. The US Environmental Protection Agency has now said that all landfills should be regulated 'on the assumption that migration of hazardous wastes and their constituents and by-products . . . will inevitably occur.'

Groundwater Pollution

Leaking landfills have already led to the irreversible contamination of groundwaters in many parts of the industrial world. Much of that contamination results from the downward drainage of wastes dumped years ago. It is likely therefore that contaminated groundwaters will be discovered at an increasing rate as chemicals dumped in the 1940s and later finally enter aquifers and supply wells.

In the USA, where 50 per cent of the population relies on groundwaters for their drinking water, the EPA estimates that 1 per cent of all 'commercially exploitable' groundwater supplies is now contaminated. In many industrial and urban areas, the percentage is far higher. According to the EPA, 'improper waste disposal accounts for a substantial amount of groundwater contamination.' Over 40 per cent of impoundments – that is, pits, ponds and lagoons – are sited on 'thin or permeable soils' and lie over aquifers which are either currently supplying drinking water or which could be used to supply drinking water. A 1982 study examined 929 abandoned hazardous waste sites; one-third were found to be contaminating groundwaters: and another third were 'strongly suspected' of causing contamination.

Despite the US experience – and, indeed, that of continental European countries – Britain persists in denying that groundwater supplies in Britain are threatened by the landfilling of wastes. But can we be so sanguine? There is no record of any major landfill in Britain receiving a full hydrogeological survey prior to 1973 – some old landfills are therefore likely to have been sited (albeit unwittingly) over vulnerable aquifers. Even today, hydrogeological surveys are seldom carried out and many landfills are not regularly monitored for groundwater pollution. In the absence of such monitoring, what grounds have we for believing that Britain has avoided the problems now being encountered in *every* other country where landfill is being practised?

In fact, the evidence from abroad strongly suggests that groundwater pollution from landfills is all but inevitable. A 1977 survey, conducted in the USA, for example, revealed that of 50 sites investigated, 40 had contaminated groundwaters with organic chemicals. At 26 of the sites, inorganic chemical contamination of groundwaters exceeded the EPA safety limits – and, at 43 sites, the migration of at least one hazardous chemical was detected. Unless we are to assume that toxic wastes in the USA behave differently from those in Britain, it would seem inconceivable that the groundwaters beneath British landfills are not similarly polluted – or will not soon become so.

Significantly, the DoE's Working Group on Waste Disposal Legislation acknowledges that 'micro-organisms of a wide range of mainly chlorinated and brominated aliphatic compounds' have been found in groundwaters. The working group states that the sources of those materials 'have not been fully identified'. It does not, however, tell the reader that many chlorinated wastes are highly toxic even at 'micro' concentrations. Once polluted, groundwaters are difficult – if not impossible – to clean. The cool, dark, virtually lifeless nature of underground aquifers allows contaminants to be stored for 'hundreds of thousands of years, if not for geological time'. At Norwich, Norfolk, for instance, groundwaters contaminated with whale oil in 1815 still contained residual toxic compounds when wells were dug there in 1950 – almost 150 years later.

Conclusion

Britain's continuing commitment to landfill, her refusal to survey past sites and her failure to tighten existing laws on waste disposal are all symptomatic of the 'Out of sight, Out of mind' approach adopted by successive governments of all political persuasions to the problem of hazardous wastes. The result is not only a stupendous squandering of resources (much of the waste that is currently dumped into holes in the ground could be recycled) but a legacy of environmental destruction that will cost the taxpayer many millions of pounds to clean up – if indeed a clean-up is even possible. In addition, there are long-term, indirect costs which can never be ascribed a monetary value – the cancers incurred by those living near dump sites, for example, or the irreversible contamination caused to an underground aquifer. The bitter irony is that such destruction is not necessary: alternatives to landfill exist – and are being adopted by the more responsible industrial nations of the world. Why then does Britain continue to play Russian roulette with the health and safety of her citizens?

Notes

1. This chapter is drawn from Nicholas Hildyard and Samuel Epstein, *The Toxic Time-Bomb*, Oxford University Press (forthcoming).
2. The figure of 600 sites includes landfills which received household rubbish, as well as those which received industrial waste.

20

Dirty Water under the Bridge

Fred Pearce

Ten years ago Margaret Thatcher was the opposition spokeswoman on the environment. In one of her last acts of bipartisanship before her conversion to 'conviction politics', she supported a Labour bill to clean up the environment. The Control of Pollution Bill was, she said, 'likely to have a greater, more lasting impact on the quality of life in many parts of Britain than most other measures'.

Today the most important part of that legislation, on water pollution, is still meandering into force. And, far from boasting of the 'lasting impact' of this delayed enforcement of parliament's will, Margaret Thatcher's ministers now reassure industrialists that it will have little effect on their rights to pollute. They are right. Almost every polluting pipe or drain that the act was intended to bring within the law has been granted an exemption. It will, civil servants at the DoE agree, be well into the 1990s before these discharges are all licensed. And there is no indication, either from Whitehall or the regional water authorities that will police the licensing system, that they will do anything other than bend over backwards to meet the demands of industrialists. The destruction of one of the most important pieces of environmental legislation ever passed in Britain will be complete.

Pollution of inland rivers is already under the control of water authorities, thanks to laws passed in 1951 and 1961. The Control of Pollution Act Part II (usually known as COPA II) was intended both to bolster the powers and duties of the water authorities to clean up inland rivers, and to extend those laws for the first time to estuaries, tidal rivers and coastal waters. Ministers promised that, within a decade, filthy estuaries such as the Mersey and the Tees would be running with fish. But the zeal within Whitehall to clean up Britain evaporated almost as soon as the bill was passed. And, since most of the reforms in the new act required a ministerial decision to activate them, Whitehall's coolness has been fatal.

Delays in Implementation

Fred Lester the former director of scientific services at the Severn Trent Water Authority once described the act as a 'masterpiece in the art of flexibility'. He was not being complimentary. Below is the timetable of promises for the implementation of the sections dealing with water pollution:

- *July 1974*: royal assent for the act.
- *February 1974*: Gordon Oakes, a junior environment minister, promises that the water sections will be phased in from autumn 1975 to summer 1976.
- *August 1975*: a change of plan. Denis Howell, Minister of State for the Environment, announces another postponement – this time indefinite. He blames government spending cuts.
- *April 1978*: Howell announces a new timetable. There will be a two-phase implementation, he says, to be completed by the end of 1979.
- *May 1979*: the Conservative Party wins the general election and new ministers at the DoE, Michael Heseltine and Tom King, announce a new examination of the water provisions of the act. 'There is a feeling at ministerial level that it will cost a lot of money,' say civil servants.
- *February 1982*: after almost three years of review, King announces a five-part programme with the 'maximum use of transitional provisions'. The first parts will be 'brought into force in July 1983'. The rest will be in force by 1986. King reveals that, eight years after the act was passed, there is to be more consultation with industrialists about how the sections should be implemented.
- *August 1983*: Civil servants reveal more about the 'transitional provisions'. Most discharges that escape existing controls will be exempt from the new ones. The remainder, discharges to the Mersey estuary and of certain toxins, such as mercury and cadmium, that are blacklisted by the EEC, will be given 'deemed consents'. This, as we shall see, amounts to the same thing.

The Mersey

In the past 30 years, many of Britain's worst water polluters – the oil refineries, chemical works and steel and paper mills – have moved to the coasts and estuaries. They liked the wide open space and the deep berths for big ships – and they liked the absence of the pollution controls they faced upstream. The result of these moves on the ecology of the estuaries has been disastrous. The Mersey estuary, for example, offered a living to many fishermen in the 1930s. Today it stinks in summer and the fish are banished.

In 1983, Michael Heseltine, the then Secretary of State for the Environment, described the Mersey as 'an affront to the standards a civilised society should demand of its environment' and 'the single most deplorable feature of this critical part of England'. He claimed that cleaning up the Mersey would attract high-tech firms looking for a clean environment and would 'provide an incentive for the location of industry that needs clean water'. But Heseltine's successors have not given local water authorities the legal powers that they need – and which parliament thought it had granted ten years ago – to force industrialists to clean up their estuaries. While the Mersey estuary alone escapes formal exemption from the new rules (presumably because of Heseltine's concerns), the North West Water Authority has been told that industrialists should for the time being be given 'deemed consent' to carry on polluting as before. As Noreen Bovell, a civil servant at the DoE, told a seminar in late 1983, exemptions and deemed consents are regarded by her department 'as virtually interchangeable'. In mid-1984 Bovell's successor, Richard Mills, told another conference that industrialists on Merseyside need take 'little action . . . As the government has emphasised, environmental improvement must not inhibit wealth creation unreasonably.'

The North West Water Authority's scientists have taken advantage of Heseltine's interest in their problems to gain money from central government to clean up the authority's own pollution of the estuary from raw sewage from its sewage works. The authority plans to spend £170 million by 1995, and it wants firms that pump their own waste into the estuary to match that. If they did the river would no longer smell in summer and migratory fish could make their way upstream. But most of the big polluters are holding out. At Bromborough, Unilever, the multinational combine that makes everything from soap to margarine, refuses to discuss its response to the plan. Unilever disgorges some two tonnes of solid organic matter into the estuary every day. One unpleasant consequence is that large unsightly balls of fat are washed up on New Brighton beach.

Among the firms that pollute the Mersey in its worst reaches, around Widnes and Runcorn, are Albright and Wilson, UKF Fertilizers and ICI. The north region of the CBI, which represents these companies, says of the water authority's plans for the Mersey estuary: 'There is no point in spending this large sum of money . . . there are too many uncertainties for industry to be able to support such a programme.' CBI officials say the plan is 'not helpful to the region' and 'could dissuade new firms from moving to Merseyside'.

Industry's Influence on Water Pollution Policy

Ministers take these voices from the 'poachers' at least as seriously as they do the voices of the 'gamekeepers', the water authorities' scientists. In 1983, ministers abolished the National Water Council, a statutory body charged with working for higher standards in the water industry, especially on pollution, and with advising ministers on policy. Now industrialists provide Whitehall with the advice. One such official adviser has been Philip Edwards, who also works for ICI's Mond division, in Cheshire, one of the biggest polluters of the Mersey catchment. He says that the water authority's plans for the estuary are 'unrealistically optimistic' and 'more than is necessary'. He warns darkly of factory closures and lost jobs.

Ministers now value the opinions of the poachers so highly that they place them in ever increasing numbers on the boards of the water authorities. The 13 members of the North West Water Authority, for example, have included past and present senior officials of Shell, ICI, Unilever, a Cumbrian paper mill and the CBI. ICI has people on no less than four of Britain's ten water authorities.

Minimal Fines

The eating away at controls on water pollution goes beyond COPA II. The old legislation covering inland rivers has also fallen into disrepute. Licence consents to pollute rivers were widely rewritten in the late 1970s. In many cases the rules were slackened because the old consents had been widely flouted and lacked any credibility. The new consents were all said to be immediately achievable by the companies concerned. But, given an inch, many of the polluters have taken a mile. In the north west, where the consents were relaxed more than anywhere else, the water authority now reports that more than 30 per cent of the industrial discharges in its region 'do not reach the required standard'. The Welsh Water Authority says that less than 20 per cent of the effluents that it regularly monitors meet all the conditions of their consents. The National Coal Board's Phurnacite smokeless fuel plant near Mountain Ash in South Wales discharges 30,000 cubic metres of effluent into the River Cynon every day. In 1982, 8 per cent of samples met the consent condition. In 1981 and 1983 no samples met the consent. But the Welsh Water Authority, like others, has virtually given up prosecuting firms that fail their consents.

The fines are so minimal when they win a case that there is little incentive to prosecute. Take B & N Chemicals of Haverill in Suffolk. The firm has been a thorn in the side of the Anglian Water Authority for several years. After an especially serious pollution incident at the plant in 1981, the authority's pollution staff wrote: 'Once again acute pollution problems in the river Stow have been caused by chemical spillages at Haverill . . . The public water-supply abstraction (from the Stow) at Langham was closed on one occasion.' Chemical smells filled the town and fish flesh was tainted. The company was eventually convicted in court of discharging styrene and xylene, a suspected cancer agent, into the river. The fine was £325.

Keeping the Public in the Dark

One of the most important purposes of COPA II is to open up the system of licensing polluters to public scrutiny and participation. Before being granted, new consents must be advertised and a public inquiry may be held. Registers of samples taken by the water authority's scientists, who police the consent, are open to public inspection, and members of the public can, in theory prosecute polluters who do not meet their consents.

But again the Whitehall loophole specialists have been at work. As Richard Mills of the DoE puts it: 'Dischargers may view the prospect with some trepidation. I doubt if this is justified.' On advertising, he says: 'the request to advertise can, for instance, be waived where it is judged that the discharge will have no appreciable effect on the receiving waters.' This is a giant loophole. As Eric Harper, chief scientist for the North West Water Authority, points out: 'Any industrial discharge in the Mersey estuary would cause only an insignificant depletion of dissolved oxygen by itself, but the additive effect can result in total depletion of several kilometres.' So, according to the government's plan, any river could be entirely stripped of fish and left stinking without any chance of public participation in the decision. Without advertisements there will be few objections that might generate a formal inquiry. Even so, according to Mills, 'while there is provision for formal public inquiries, where appropriate, inspectors may proceed by informal hearings.' Another door closes.

Appeals: A Licence to Pollute

Industrialists will, of course, keep their right to appeal to the Secretary of State if they do not like the decisions; and they can carry on polluting while the appeal is considered. This buys valuable time. Two recently decided appeals took nine years.

In April 1984, Coalite, one of Britain's worst river polluters, appealed against a new consent set by the Yorkshire Water Authority for its works at Bolsover in Derbyshire. The plant dumps ammonia, phenols and cyanides into the Doe Lea, one of Britain's deadest rivers. The new consent had been under negotiation between the authority and the company for ten years – the water authority desperately wanted a deal (and believed it had got one) – precisely because the water authority's scientists feared that the company would drag them into an appeal.

Now it is anybody's guess how long it will be before the Yorkshire Water Authority can do anything about the Doe Lea. Certainly Coalite's chairman will be familiar with the meandering ways of Whitehall. He is Eric Varley, the former Labour Secretary of State for Energy.

"What Earth?"

21

Pollution on Tap

Brian Price

When we turn on a tap we take it for granted that the water issuing from it
will be clean, pure and safe. After all, are there not strict regulations
governing the quality of our drinking water? The short answer to this
question was, until recently, 'No' – or at least, 'Not yet'. Until a new EEC
Directive came into force in July 1985 the only legal requirement
governing drinking water quality in Britain was that the water should be
'wholesome' – a term of Victorian vagueness which has never been defined
precisely in terms of either chemical or microbiological parameters.

In general terms, the microbiological quality of Britain's drinking water
is excellent, although, as both sewers and water mains of elderly years
crumble under the onslaught of heavy traffic and corrosion, the possibility
of pathogenic bacteria from the former entering the latter increases. Of
more widespread concern, however, is the chemical contamination of water
before, during and after abstraction, treatment and supply. The chemicals
involved range from simple inorganic substances (such as nitrate and lead)
through mineral particles (such as asbestos) to complicated – and often
unidentified – organic molecules. Many of the materials entering our water
supply may be harmless but some definitely are not.

Drinking water may become contaminated long before it enters the
series of processes which culminate at its delivery to the tap. Underground
water supplies (aquifers) and rivers may receive loadings of pollutants
which, ultimately, can affect the quality of the final product. The sources
of these pollutants may be natural (spa waters in some areas, for instance,
may be undesirably radioactive) or may be industrial or agricultural
processes. In the USA, many groundwater suppliers have been irrevocably
polluted as a result of the careless disposal of hazardous wastes, while in
Britain at least one supply is known to have been polluted by potentially
carcinogenic solvents discarded on a US air base.

Nitrates in Drinking Water

By far the greatest pre-abstraction menace to drinking water in Britain is nitrate. A simple inorganic material, nitrate forms the basis of the bulk of the artificial fertilizers applied to the land. It occurs naturally in modest amounts but serious problems have developed – and more are inevitable – as human activities have grossly distorted parts of the nitrogen cycle so that high levels of the material are accumulating in food and water.

If excessive quantities of nitrate fertilizer or animal manure are spread on fields (and the quantities spread often are excessive) the growing crops are unable to absorb all the nitrate available. Even when absorbed, the nitrate may not contribute to the growth of edible plant tissue but may remain as unaltered nitrate – at alarmingly high levels.[1] The unabsorbed nitrate may either be washed off the fields into rivers and lakes or it may percolate downwards to an aquifer, with local geological and climatic conditions determining which process predominates.

Nitrate running off the land into rivers may be taken up by water plants – a process which can have dire effects if plant growth is stimulated excessively – but in many cases the capacity of river life to handle nitrate may already be overloaded. The result is a high level of nitrate in drinking water derived from such rivers since the usual water treatment processes do not remove nitrate to any great extent. The percolation of nitrate from fields to aquifers is a much slower process – nitrate released at the surface by the ploughing of grassland 20 years ago has not yet reached the groundwater in some areas – but there is little scope for nitrate removal on the way. This slow but inevitable assault on our aquifers has been dubbed the 'nitrate time-bomb'. When it goes off, the consequences could be devastating.

Health Risks of Nitrate

The health risks of nitrate are twofold. First, nitrate is converted in the digestive tract into nitrite, a process which is particularly efficient in young babies.[2] The resulting nitrite can combine with the red blood pigment which carries oxygen around the body, thereby blocking the uptake of this vital gas by the blood. This causes a condition known as methaemoglobinaemia in which the tissues are starved of oxygen. The condition only affects young babies on made-up bottle feeds and is rare in Britain, but as nitrate levels in drinking water rise the likelihood of its occurring increases. In several areas of Britain, bottled low-nitrate water has been supplied to the mothers of young babies on a number of occasions when nitrate levels in drinking water have risen to dangerous levels.[3]

The second health risk from nitrate is perhaps less certain but is potentially much more alarming. Nitrite, formed from nitrate in the gut, can combine with chemicals naturally present in food to produce substances called N-nitrosamines. These materials have been found to cause cancer in 39 species of animals.[4] Although no link between nitrate levels in water and human stomach cancer has been proved conclusively, there is some evidence to suggest that such a link exists. It would therefore seem prudent to keep the levels of nitrate in drinking water as low as possible, but medical advice to the government describes the risk of cancer as 'not proven' and sees no need for urgent action,[5] a *laissez-faire* approach which the government has seized on.

The gravity of the situation in Britain is best illustrated by means of a few statistics. A survey of 25 rivers, reported by the Standing Technical Advisory Committee on Water Quality, reveals that nitrate levels have, on average, doubled during the past 20 years.[6] Levels in the Stour, said to be a fairly typical East Anglian river, exceeded the World Health Organisation's recommended limit of 50mg nitrate per litre of water (sometimes expressed as 11.3mg nitrate nitrogen/litre) for 400 days in the three years from 1974 to 1977.[7] If present land-use practices continue, it is predicted that nitrate levels in the River Thames will breach this limit on an annual average basis by the mid-1990s – and that maximum levels will be well in excess of it.[8] A fifth of the groundwater supplies in East Anglia have exceeded the limit, on an average basis, while in east Yorkshire the limit is likely to be broken in many aquifers by the early 1990s.[9]

The EEC Directive: Britain's Response

In July 1985, the EEC's *Directive on the Quality of Water Intended for Human Consumption* came into force. The Directive lays down 'guide levels' (GLs) and 'maximum admissible concentrations' (MACs) for some 62 water quality parameters. The GL for nitrate is 25 mg per litre, whilst the MAC is set at 50mg per litre – the same as the WHO level. Member states are supposed to embody the MAC in law but the British government is intending to ignore the new limit, at least for the time being, by issuing 'derogations' (that is, exemptions) in the case of some 350 specified water sources.[10] Moreover, as David Wheeler, a research fellow in the department of microbiology at the University of Surrey, points out, 'The Government has elected to allow water suppliers to *average* their results over 3-month periods to help them meet the MAC; and to relax the standard to 80 mg per litre because its medical advisers cannot detect a health risk from nitrate at levels up to 100 mg per litre.'[11]

The attitude of some water authorities to the nitrate problem seems to be one of 'Ignore it and it will go away': but with more and more sources becoming polluted and nitrate fertilizer use growing by 4 to 5 per cent per annum the problem is certain to increase.[12] Predictably, the fertilizer manufacturers are resisting any attempts to curb the use of their products, while the government is unwilling to interfere with the powerful chemical industry. The result is a cynical buck-passing exercise with the water authorities claiming that the farmers should solve the problem while the agricultural lobby claims that the water authorities should treat the water to remove nitrate.

Pesticides

According to David Wheeler, there are some 10 to 20 pesticides which are 'quite common' in drinking waters at or about the 0.05 microgramme per litre level. Examples include atrizine and simazine. Yet,

> under direct instructions from Ministers, the Department of the Environment has asked Water Authorities effectively to ignore contamination of the public water supplies by pesticides. The interesting justification for this is that there are too many different types of pesticides in Britain to apply a simple standard! The few Authorities which actually analyse for pesticides may continue to do so, but those Authorities and other water suppliers (the vast majority in fact) which do not presently look for pesticides may continue in ignorance. No specific moves are recommended to take water sources with known excessive pesticide concentrations out of supply.[13]

Micropollutants: Another Threat

Public water supplies receive contaminants from many sources other than agriculture. The fall-out of materials from the atmosphere and the run-off of materials from the land all add toxic substances on a more or less continuous basis. These include lead and potentially carcinogenic materials from roads, as well as pesticides from farmland. The repeated re-use of water discharged by sewage works and industry can result in a build-up of many contaminants, particularly where industrial effluents are involved.[14] Accidental discharges can cause widespread contamination, as occurred in the River Dee in 1984 when phenols discharged by a factory were taken into public supplies and over 40 people subsequently complained of symptoms of phenol poisoning.[15]

The processes used to purify drinking water may remove many contaminants and will destroy pathogenic organisms but many materials –

such as certain drug residues and other organic compounds – will not be removed. Indeed, the treatment methods used may even aggravate problems or create new ones. For instance, aluminium compounds used to clarify water cause no harm to healthy people but the elevated metal levels resulting can cause dementia in people on kidney machines.[16]

Of more widespread concern, however, is the production of halogenated organic materials from contaminants by the action of chlorine, a gas used to kill bacteria. Organic materials, after chlorination, may cause taste and odour problems but they may also be dangerous to health. Apart from possible allergic reactions in sensitive people the main risk is one of cancer. Literally hundreds of these compounds – known as organic micropollutants – may be present in a water sample and because it is impossible to identify all of them predictions about the health risk can only be tentative. Nevertheless, it is known that chlorinating some water supplies makes them more mutagenic, thereby indicating a potential increase in the risk of contracting cancer after drinking the water.[17] One group of compounds which has been identified is the trihalomethanes which includes chloroform – a known animal carcinogen.[18]

Leaching Pipes

The risk of drinking water contamination does not end once the water leaves the treatment works and enters the public supply. The pipes used to convey the water may add chemicals to their contents while the surrounding soil may harbour penetrating pollutants. Vinyl chloride monomer (a known carcinogen) may be leached from PVC pipes under certain circumstance,[19] as can materials from epoxy resins and bitumen used to reline old cast iron mains.[20] Water mains made from asbestos cement release asbestos fibres into the water supply;[21] the official position in Britain is that this presents no hazard, but scientific and political concern about the issue is mounting elsewhere.[22] Plastic pipes passing through soil containing oil, petrol or phenols may be permeated by these materials, leading to foul tastes in the water carried.[23]

Perhaps the oldest – and in many cases the most serious – distribution contaminant of drinking water is lead, a powerful neurotoxin which has been shown to have an adverse effect on the intelligence of children (see pp. 189–95). Soft water running through lead pipes readily dissolves the metal and unquestionably harmful levels have been reached in the past. A survey carried out in 1975–6 revealed that 7.8 per cent of random daytime samples taken from taps in England exceeded the EEC limit for lead in drinking water, while in Scotland the corresponding figure was 34.4 per cent.[24] The report triggered a programme of remedial action in the form of grants for the replacement of lead plumbing in appropriate areas, while water hardening (which reduces the ability of water to dissolve lead) was commenced in a number of treatment plants. Although this action has certainly helped, it was long overdue and has not solved the problem for several reasons. First, the replacement programme is voluntary rather than compulsory; many houses have not therefore been replumbed. Secondly, grants are only payable for premises with rateable values below certain limits, thereby excluding many buildings where children may be affected. Thirdly, local government expenditure cuts have reduced the amount of money available for housing grants in general and lead plumbing replacement has not been immune from these constraints. The Royal Commission on Environmental Pollution has called for more vigorous action to control lead in water supplies and an official response is awaited.[25] Meanwhile, the government has simply ignored a recent research report which suggests that lead solder used in copper pipework might be harmful.

Drinking water supplies are not only contaminated by chemicals. Radioisotopes are ultimately discharged into the Thames from the atomic research establishments at Aldermaston and Harwell and the main material involved – tritium – is not removed by water treatment processes. The official argument is that the levels involved are trivial and well within international limits, adding little to doses of radiation received from other sources. It must be remembered however, that any increase in radiation exposure increases the risk of cancer.

Conclusion

For most of this century, the nation's drinking water has been largely free of the pathogenic organisms which plagued our forebears. With the exception of lead derived from plumbing, chemical parameters have also been broadly acceptable. But during the past two decades, two new threats have been identified. The insidious and inevitable build up of nitrates and the increasing cancer risk from the organic micropollutants are causes for serious concern and it is likely that extensive – and expensive – treatment will become necessary over the next 20 years. The government's view seems to be that everything in the reservoir is lovely but this view is being challenged with increasing frequency and urgency. Judging by the sales of overpackaged mineral waters in recent years, many people are turning away from mains water on grounds of taste. Unless action is taken soon, many will feel obliged to reject it on health grounds as well.

Notes

1. A. H. Walters, 'Nitrate and cancer – a broader view', *The Ecologist*, vol. 14, no. 1, 1984, pp. 32–7.
2. P. N. Magee, 'Nitrogen as a potential health hazard', *Philosophical Transactions Royal Society of London*, B 296, 1982, pp. 543–50.
3. The Royal Society, *The nitrogen cycle of the United Kingdom: a study group report*, London: The Royal Society, 1983.
4. Magee, 'Nitrogen as a potential health hazard'
5. Joint Committee on Medical Aspects of Water Quality, *Advice on Nitrate in Drinking Water in Relation to a Suggested Cancer Risk*, London: DoE/DHSS, 1984.
6. Anon., 'The rising curve of nitrate pollution: to prevent or cure?', *ENDS Report*, no. 97, 1983, pp. 9–12.
7. 'The rising curve of nitrate pollution'.
8. The Royal Society, *The Nitrogen Cycle of the United Kingdom*.
9. Anon., 'Nitrate report spurs action on farming practices', *ENDS Report*, no. 111, 1984, pp. 3–4.
10. Article 9 of the Directive allows such 'derogations' to be made provided that the exempted water source does not constitute a 'public health hazard'.
11. David Wheeler, 'Public Health Implications of the EC Directive on the Quality of Water Intended for Human Consumption (80/778/EC) 1980', unpublished MS, 1985.
12. The Royal Society, *The Nitrogen Cycle of the United Kingdom*.
13. Wheeler, 'Public Health Implications'.
14. M. Fielding and R. F. Packham, 'Organic compounds in drinking water and public health', *Ecologist Quarterly*, no. 2, 1978, pp. 149–62.
15. Anon., 'Health impact of Dee pollution incident', *ENDS Report*, no. 111, 1984, p. 7.
16. Safe Drinking Water Committee, *Drinking Water and Health*, vol. 4, Washington: National Academy Press, 1982.
17. Water Research Centre Open Day Exhibit, Leaflet to accompany Exhibit 6, 1984.
18. Safe Drinking Water Committee, *Drinking Water and Health*.
19. Water Research Centre, Abstract 84–1303, WRC Information, vol. 11, no. 173, 1984.
20. Water Research Centre Open Day Exhibition, Leaflet to accompany Exhibit 4, 1984.
21. Water Research Centre Open Day Exhibition, Leaflet to accompany Exhibit 8, 1984.
22. Anon., 'Could asbestos water pipes cause cancer?', *New Scientist*, 24 May 1984, p. 6.
23. Water Research Centre Open Day Leaflet 1984 (Exhibit 4).
24. Department of the Environment, *Lead in Drinking Water*, Pollution Paper no. 12, London: HMSO, 1977.
25. Royal Commission on Environmental Pollution, *Ninth Report: Lead in the Environment* London: HMSO, 1983.

22

The Hazards of High-Voltage Power Lines

Hilary Bacon

Section 37 of the 1957 Electricity Act, headed 'Preservation of Amenity', adjures the Electricity Boards, Electricity Council and Minister for Energy to have regard to 'the desirability of preserving natural beauty, of conserving flora, fauna and ecological or physiographical features of special interest, and of protecting buildings and other objects of architectural or historic interest'. Despite these instructions, the massive pylons and lines of the National Grid straddle our land, often disfiguring areas of outstanding natural beauty, often sited in unsuitable terrain, often towering close to houses or dwarfing the green belts around cities. Such is the respect shown by those in authority for parliamentary law and the environmental quality of life of the ordinary person.

But even apart from such overriding technological brutalization of the environment, there are other, more serious, hazards resulting from high-voltage (HV) transmission. These hazards are threefold. First, the often lethal accidents caused by direct contact or by non-contact earthing flashovers. Secondly, the newly discovered, subtle and cumulatively very damaging biological effects caused by the extremely-low-frequency (ELF) pulsed electromagnetic field created by the power lines. And, thirdly, the attitude of the authorities to the evident dangers of electromagnetic pollution.

Electromagnetic Fields: Their Known Health Effects

The electromagnetic (e/m) field generated by high-voltage transmission lines is just one form of 'electrical pollution' – a modern menace which results from the myriad electronic devices (from radar stations to television transmitters, microwave relay towers and microwave ovens) on which our industrial way of life is increasingly dependent. Those devices inevitably create waves of electric and magnetic radiation, whose strength and intensity is gauged along what physicists call the 'electromagnetic spectrum'. At the bottom end of the spectrum are extremely-low-frequency (ELF) waves, such as those generated by high-voltage power lines: at the opposite end are gamma rays. In between, in ascending order of frequency (a measure of how many times the waves vibrate each second) are microwaves, infra-red radiation, visible light, ultra-violet radiation and x-rays. Unlike gamma rays and x-rays, sources of low-frequency radiation are 'non-ionizing' – that is, they lack the ability to break up the material through which they pass into charged particles – and consequently they have long been considered benign. Indeed, in the early discovery of e/m fields, scientists believed that the lower, or non-ionizing, frequencies were biologically inactive because they produced no 'thermal effects' – that is, they could not heat body tissues.

But, by 1962, work from Moscow University had already been translated and published in the USA, showing that biological systems (which are tuned to receive the earth's natural, weak e/m field) suffer a wide range of adverse health effects when subject to a further, pulsed, e/m field, however low. Symptoms range from headaches, eyestrain, rashes, loss of appetite and exhaustion to cardio-vascular malfunctions, altered leukocyte counts, food allergies, disorientation, epilepsy, severe depression, even suicide.

A long-term study of these effects was undertaken. The investigators concluded that work at 500 and 750 kilovolt substations without protective measures resulted in 'shattering the dynamic state of the central nervous system, the heart and blood vessel system and in changing blood structure'. That initial finding has subsequently been confirmed in over 100 reports published in the Soviet Union. Other effects have also been documented: mice exposed to magnetic fields of 50 hertz quickly lose their ability to expel foreign matter from the liver, spleen and lungs; the function of the glands of rats exposed to fields of similar intensity is grossly impaired; and a survey of some 200 workers at 220, 330 and 500 kilovolt substations has shown a significant increase in the haemoglobin content of their blood. Soviet scientists now believe that electrical fields as low as 50 volts per centimetre can have an adverse effect on human health.

As a result of these findings, the Soviet authorities have imposed strict rules relating to the exposure of electricity workers and the general public to electrical fields of more than 250 volts/cm, and even in fields of 200 volts/cm, unprotected workers may only be exposed for ten minutes in any 24 hour period. A 360 foot zone centred on the line is restricted to certain authorized personnel. It may not be used for recreation; buildings, bus shelters and other places where people may congregate are forbidden in the area; no vehicle is allowed to stop or be refuelled under the line – for fear of a spark igniting its fuel; if a vehicle does break down it must be towed away before any repairs are done; and, finally, metal shields must be used over the seats of farm machinery.

By the late 1960s, research from Italy and the Soviet Union showing adverse effects on the cardio-vascular and nervous systems from ELF electromagnetic fields had been published in Britain. In 1973, the American Institute of Biological Sciences held a colloquium, *Biotic Management Along Power-Transmission Rights-of-Way*, at which bio-physicist Louise Young presented a cautionary paper which was immediately published in the *Sierra Club* Bulletin. A committee of the US Navy presented two papers to the White House, in 1973 and 1974, on the bio-effects of an ELF underground communications system, Project Sanguine. Both papers documented alterations in the blood serum triglycerides of volunteers subjected to ELF e/m fields. The implication of that and other findings – notably another study which revealed that those living near Sanguine were likely to suffer mental illness and other behavioural effects – was not lost on the committee, for the electrical fields created by Sanguine were one million times less intense than those under high-voltage power lines. Indeed, the committee was sufficiently alarmed to recommend that the US government 'be apprised of (these) positive findings and the possible significance should they be validated . . . to the large population at risk in the United States who are exposed to 60 hertz fields from power lines and other hertz sources'.

Despite the growing body of evidence from abroad that the ELF electromagnetic field below high-voltage lines can cause adverse health effects, no research into the problem was undertaken by the British authorities. Indeed, the clinical data, which have now been amassed on such adverse effects in Britain, have been collected thanks only to the efforts of the public and of concerned scientists and journalists. *The authorities have not lifted a finger to help*. On the contrary, whenever the CEGB has been presented with research data showing evidence of biological damage (data which it should have been aware of), it has persistently refused to admit their validity.

Fishpond: Britain's First Victims

In Britain, the issue came to a head when those living under the 400 kilovolt National Grid lines which straddle the village of Fishpond in

Dorset, where I then lived, began to complain of adverse health effects. In 1973, they approached the authorities for help – to no avail. The DHSS dismissed them with wrong information; the Electricity Council could only help consumers in cases up to 132 kilovolts; and the CEGB sent pleasant but completely unbriefed junior staff to reassure them. None of this relieved the villagers' symptoms; so through friends, and with the help of a sympathetic telephone operator, they obtained the US material from the USA, and the Soviet and Italian research from the Institute of Occupational Health and Safety in London. Finally, through media interest, they received the testimony that was then being presented at a public hearing held by the New York State Public Health Commission. This testimony came from two scientists employed by the US government, who had offered to testify on behalf of the public against major electrical companies which wanted to route 750 kilovolt HV lines across New York State to Canada. The two scientists were Dr Robert Becker, an orthopaedic surgeon, who was using ELF e/m fields to promote cell-growth in injured bones and who argued that this effect could be harmful in the normal body; and his colleague Dr Andrew Marino, a biophysicist who reported results of experiments on rats and mice in ELF e/m fields – stunted growth, lowered fertility, shortened life span.

Armed with this material, the MP for Fishpond, Jim Spicer, approached the CEGB who arranged a private meeting in July 1976 between his constituents and senior members of the CEGB, among them physicist Dr Toby Norris and Deputy Chief Medical Officer, Dr John Bonnell. The meeting was a disaster. The villagers were treated like kindergarten children; hardly any of the material they presented was considered – and Marino and Becker (a Nobel Prize candidate) were dismissed as troublemakers. The only positive outcome was that Jim Spicer asked the CEGB to measure the electric field under the Fishpond HV lines (as high as 6 kilovolts per metre).

New Evidence: The Innsworth Inquiry

It is now clear that even these measurements were misleading. The public discovered this thanks only to another independent and conscientious scientist, Dr David Smith, who offered to testify on their behalf at the 1978 Innsworth public inquiry – the first public inquiry in Britain where the question of ELF e/m bio-effects was raised. The inquiry took place because those present at the original Dorset meeting noticed that the CEGB were using survey maps that were 50 years out of date to support its case for moving 400 kilovolt lines closer to the village of Innsworth in Gloucester. The maps did not show post-war developments, such as a housing estate and a school located close to the new route. The testimony offered by Dr Smith was extremely revealing. After correspondence with the CEGB, he pointed out that the measurements taken at Fishpond omitted an important factor known to all physicists: namely, electric field 'enhancement'. This field, although composed of separate particles, behaves as a unified sheet. If it is 'perturbed' (for example, if it has an object within it), it drapes itself like a curtain in close folds over the top of the object. These folds can reach voltages a hundred times greater than the ambient field. Such an object might be a baby in a pram, a man standing next to a gate, a child's climbing frame, and so on. The CEGB had measured only an unperturbed field at ground level, using an instrument on a long handle.

This testimony could not of course be refuted by the CEGB, although it tried to do so in a 'rebuttal session' allowed to them by the Inspectors, against the opposition of the Innsworth objectors. Such a session is common practice in the USA but not in Britain. One of those who gave evidence was the CEGB's medical officer, Dr John Bonnell. In his original testimony he had quoted almost exclusively from research undertaken by an US electrical company employees to support his claim that there are no adverse health effects from ELF e/m fields. In cross-examination he dismissed research which proved otherwise because it had been done on animals and, therefore, he claimed, could not be extrapolated to humans. When it was pointed out to him that all the research he himself had quoted was done on animals, he said that important supporting evidence was even then on its way 'by jet-plane'. In fact, it did not arrive until a month later, and turned out to be papers presented at a Washington conference which offered far more support to the Innsworth objectors than to the CEGB.

Meanwhile, I presented all the earlier material on adverse health effects and more, together with signed statements from other sufferers and corroborative letters that I had received. After this I was cross-examined for a day and a half. The Inspectors took two years to reach a very equivocal decision, by which time the lines had already been uprated to 400 kilovolts and moved much closer to Innsworth, across wet meadows. The CEGB was merely requested to make annual measurements of air ionization under the HV lines at Fishpond – not, it should be stressed, at Innsworth. It took until 1983 to perfect an instrument for this purpose – but alas! it only worked in the CEGB laboratory.

The Aftermath of Innsworth

There were other interesting aspects of this inquiry. On the first day, Dr Bonnell was interviewed on BBC Television. He claimed that all research showing ELF e/m bio-effects had been proved faulty. At the end of the inquiry, an Innsworth resident demanded a signed statement of this claim – which took several months and the threat of legal proceedings to materialize. It is now safe in a Gloucester bank vault. Meanwhile, the villagers of Fishpond were fascinated to learn that six months earlier, when several of them had suffered blackouts, one circuit of their HV lines had been uprated from 132 kilovolts to 345 kilovolts – needless to say, without their knowledge. Despite CEGB protests, their case was taken to the Health and Safety Executive but HSE Inspectors found they had less information than the public, and were also hampered by cut-backs in staff and funds. In 1979, representatives from the Japanese government approached the CEGB on the subject of ELF e/m bio-effects; the CEGB refused to talk to them, so they went to the HSE – who sent them to me on the grounds that I was the most informed person in the country! Media interest grew, and produced letters of corroboration from all over the country.

The same year, the New York State Public Health Commission decision was published: the electrical companies were ordered to fund $5 million for further research by independent scientists (among them Dr Marino). That work is now duplicating Marino's earlier results. Meanwhile the Fishpond protesters were enlisting the help of more scientists, medical journalists and doctors. A GP from Staffordshire, who had observed a statistically significant number of severe depressive illness and suicide among his patients, discovered that these were clustered around HV lines – even those lines buried underground. It appears that the pulsed magnetic field, from which nothing can be shielded, has the most disruptive biological effects. Research by Dr David Melville of Southampton University on ELF e/m bio-effects produced findings similar to those of Dr Marino; pioneering work by Dr Cyril Smith of Salford University showed effects on production of DNA and RNA in *e. coli* from weak pulsed magnetic fields.

Meanwhile, a curious exchange was printed in the *British Medical Journal* in October 1980. A doctor wrote to ask if a patient's health might be affected by HV lines nearby. In reply, an anonymous 'expert' dismissed the idea, referring only to an 'overview' compiled by Dr Bonnell and other electrical company employees – but omitting any mention of this employment. A London consultant, worried about the adverse health effects of HV lines, wrote asking for further information, and received a letter from Harwell signed by someone who described himself as 'a colleague of Dr Bonnell'! After this the *British Medical Journal* announced that henceforth all 'expert' answers would be signed.

Another Inquiry

Growing public concern led to a second public inquiry, held in April 1982 in Haddington, East Lothian, at the demand of farmers and villagers who were alarmed at the proposed route for HV lines from Torness nuclear power station (when completed) to Edinburgh and Glasgow. There were many groups of protesters, one of which asked Dr Melville to testify for them; he also read out a statement from Professor Watson of the department of medical electronics at St Bartholomew's Hospital in London, who is supervising research on ELF e/m effects (including loss of calcium and nor-adrenaline-H) on the brain and nervous system. He warned of effects on the body's Circadian rhythms and also of the possible extent of factors as yet unknown. Another group asked me to testify for them. And oddly enough, the SSEB, whose province this properly was, found themselves supplemented by three senior members of the CEGB – among them Dr Bonnell – who had asked beforehand who was testifying for the opposition.

The Haddington inquiry presented other curiosities. The SSEB medical officer said first that there would be no ELF e/m bio-effects, then that there would be alterations in the leukocyte count. The CEGB witnesses invented a device even more ingenious and unusual than the Innsworth rebuttal session; they testified as a triumvirate, that is, a corporate body with three heads, and sat within whispering distance of each other.

During his cross-examination Bonnell was asked what he felt about public anxiety. He replied that the public was none of his brief, but he was planning a questionnaire for CEGB workers, also a volunteer study of three to five hour exposure to higher voltages than transmission line fields. Such short exposures are irrelevant to 24 hour exposure effects, which are time related and cumulative. Bonnell should surely have mentioned this. He should also have mentioned that in terms of bio-effects, exposure to lower e/m fields is far more clear-cut – especially when the fields are pulsed, which he did not specify. In 1982, Becker and Marino published a path-breaking book, *Electro-Magnetism and Life*,[1] in which they suggest a new understanding of life as originating not from an aqueous chemical solution but a crystalline lattice molecule which conducts information electronically as in a solid-state system, operating at various ELF frequencies. Inputs at biological frequencies can create 'trigger' effects at cellular or even molecular level which lead to misinformation and abnormalities. But these frequencies and voltages are so low, and so specific, that high-level power input can mask them, or drown them out. Therefore high-level exposures, although dramatic, can be both uninformative and misleading. Bonnell either did not know, or did not say, this.

After Haddington

In 1983, the Reporter at the Haddington inquiry ruled that the SSEB's preferred route over farms and houses should be rejected until further research proved ELF e/m fields to be harmless. But when similar testimony was presented at the Sizewell inquiry in 1984, the CEGB asked that the Haddington decision not be taken as a precedent. Yet in the intervening years, articles and letters on ELF e/m leukaemia effects had appeared in the *Lancet* and the *New England Journal of Medicine*; articles on childhood cancers resulting from exposure to ELF electromagnetic fields have appeared in *Health Physics*, which also published a paper by Dr Cyril Smith on ELF e/m production of brain endorphins.

Dr Smith's work caught the attention of Dr Jean Monro, who heads a clinic of environmental medicine in London. Together, in 1983, they presented a case history of severe blackouts caused by ELF e/m fields, at a conference in Texas. Early in 1985, they attended a similar conference in the USA with Dr Marino. Throughout all this history, the CEGB has demanded medical proof of adverse ELF e/m health effects; now it has been presented, their response is a resounding silence. Dr Monro, who believes these ELF e/m effects to be allergenic, is most anxious to continue her work, but her funds are limited. In early 1984, Jim Spicer MP proposed to the CEGB that they should fund Dr Monro's work. As yet, the CEGB has not replied to the request. Spicer also asked the DHSS to assess Dr Monro's treatment of some of his constituents; nothing has yet transpired. He had already approached the Department of Energy in 1981, asking for help, and been told 'This will take a long time.' For once the authorities appear to have been telling the truth.

Notes

1. R. O. Becker and A. Marino, *Electromagnetism and Life*, Albany, N.Y.: New York State University Press, 1982.

23

The Sellafield Discharges

Peter Bunyard

Windscale, now known as Sellafield, is a complex of nuclear installations sited on the western reaches of Cumbria between the Irish Sea and the Lake District. Sellafield's prime purpose is to extract out plutonium and uranium from spent reactor fuel rods and to make it available for civil or military nuclear programmes. Indeed, plutonium from Windscale's plutonium piles was used to make the bombs exploded over the Montebello Islands off the Australian coast in 1952, an act which heralded Britain's entry into the arms race, next in line after the USA and the Soviet Union.[1]

The Sellafield site therefore contains cooling ponds for storing spent fuel awaiting reprocessing, silos and special tanks for holding nuclear waste, a fuel fabrication plant for manufacturing plutonium oxide fuel for the prototype fast reactor at Dounreay in Scotland. There is also the B 205 magnox reprocessing plant and its B 204 predecessor, various derelict reactors including the twin plutonium piles and the steel-domed prototype advanced gas reactor, as well as an experimental vitrification plant for encapsulating high-activity waste and the excavated site for THORP, the next in line reprocessing plant. The four 50 megawatt Calder Hall reactors form part of the complex, and like the rest of the installations were taken over by BNFL in 1971 when the company was formed as an offshoot of the British Atomic Energy Authority (UKAEA).

An 'Organized and Deliberate Scientific Experiment'

The military origins of Sellafield have left a double legacy. Not only are operations there shrouded in secrecy, particularly with regard to re-processing and the quantities as well as quality of plutonium in storage,

"Here is the Mid-day News . . . Sir Walter Marshall, Chairman of the Atomic Energy Authority, knighted for his services to the Welfare of the People of Cumbria, has just been awarded a Life Peerage"

but little attempt has been made by British Nuclear Fuels Ltd (BNFL) or its predecessor, the UKAEA, to put its house in order with regard to radioactive discharges into the environment. On the contrary, as has now come to light some 30 years later, discharges of radioactive effluent were *purposely* increased during the 1950s as part of an 'organized and deliberate scientific experiment' in order to follow the pathway of various radionuclides through the environment. The man behind the experiments, Dr John Dunster, was at the time health physicist with the UKAEA, then in charge of the plant. Today Dunster is director of the National Radiological Protection Board (NRPB), the body responsible for recommending the maximum permissible radiation exposure standards for members of the public and for workers in the nuclear industry. In 1958, at the second United Nations conference on the peaceful uses of the atom, Dunster told delegates, 'The intention has been to discharge fairly substantial amounts of radioactivity . . . the aims of this experiment would have been defeated if the level of radioactivity discharged had been kept to a minimum.' He went on to say that the discharges were deliberately maintained at levels 'high enough to obtain detectable levels in samples of fish, seaweed and shore sand, and the experiment is still proceeding. In 1956 the rate of discharge of radioactivity was deliberately increased, partly to dispose of unwanted wastes, but principally to yield better experimental data.'

In fact, as a future disclosure in 1964 in *Health Physics* made clear, the experimental discharge of radioactivity into the Irish Sea began in May 1952, and was continued for several years. The stage was already set for the much greater discharges of radioactive waste that were to come more than a decade later with the operation of the B 205 reprocessing plant and the accumulation of spent magnox fuel from Britain's first nuclear power programme.

An Appalling Record

Sellafield's record has not been a good one. Various installations on the site have leaked, some for years before being detected, and that after many thousands of curies have run off into the soil. A silo containing the cladding from spent magnox fuel leaked for at least four years before the leak was discovered, by which time as much as 50,000 curies of radioactive waste had escaped, most of it caesium. Some pockets of soil were found to be giving off absorbed dose rates of up to 1,200 rads per hour, enough to cook an unsuspecting individual. Another leak from a building containing high-activity waste was discovered in 1978 by which time as much as

100,000 curies of waste might have escaped. That particular leak was the result of considerable bungling, a line for emptying a sump containing high-level waste being cut and capped when it should have been left intact. To compound the error the wrong size gauge was fitted which indicated that the sump was nowhere near full when in fact it was overflowing. For once the Nuclear Installations Inspectorate was openly critical of BNFL, stating that the management had been 'lacking in the level of judgement and safety consciousness expected'. Nevertheless, it did not prosecute the company.

Opting for the Cheapest Means of Disposal

As we know today, reprocessing spent reactor fuel is a relatively expensive item in the nuclear fuel cycle, especially when proper care is taken to control discharges and to dispose of higher activity waste. As Dunster himself remarked, the rationale behind the discharges of the 1950s was not only experimental curiosity but also to get rid of 'unwanted wastes'. To be able to dispose of wastes into the environment whether sea or atmosphere, was undeniably cheap and as long as the authorities in question gave their blessing, perfectly legitimate.

To date, authorizations imposed on rates of discharge from Sellafield have had more to do with what the industry has been willing to achieve than with targets set by authorizing bodies such as the Ministry for Agriculture, Food and Fisheries. In 1968, the UKAEA needed an increase in the authorization for discharge of highly toxic alpha-emitters such as plutonium-239 and americium-241. The limit was then 1,800 curies per year and it wanted 6,000 curies – a threefold increase of an already considerable amount (see figure 11). Two years later, in 1970, authorization was given; yet, the Fisheries Radiobiological Laboratory of MAFF did not start sampling for transuranics such as americium and plutonium until 1973 – fully three years later.

In 1974, BNFL discovered that corrosion of the spent magnox fuel in the cooling ponds was contaminating the pond storage water with radioactive caesium, particularly the 137 isotope. MAFF had originally called for a limit on discharges of 10,000 curies per quarter of caesium-137, but BNFL held out for 15,000 curies on the grounds that tighter control would lead to additional costs. In the event the authorization on caesium was never confirmed either way, which was just as well for BNFL, as by 1977 the quarterly discharge was up to 40,000 curies and the yearly discharge to more than 120,000 curies.

Figure 11 Radiological Units

Unit or Quantity	Symbol	Brief Description
Curie	Ci	3.7×10^{10} nuclear transformations (distintegrations) per second
Becquerel	Bq	1 nuclear transformation (disintegrations) per second
Rad	rad	0.01 Joules/kg (100 erg/g)
Gray	Gy	1 Joules/kg (=100 rad)
Dose equivalent	H	dose x Q x any other modifying factors
Quality factor	Q	Biological effectiveness of radiation
Rem	rem	rad x Q x any other modifying factors
Sievert	Sv	Gy x Q x any other modifying factors

By the early 1980s, BNFL had reduced annual discharges of beta activity – including caesium-137 – to just under half the peak value of 250,000 curies discharged into the Irish Sea during the mid-1970s, when the corrosion problem in the cooling ponds was at its worst. Alpha activity was also brought down, some 1,000 curies being discharged in 1980 compared to as much as 5,000 curies a year between 1973 and 1975. Even so, as the Radioactive Waste Management Advisory Committee made clear in its 1984 annual report, radioactive discharges from Sellafield remained the highest of any nuclear installation in Europe, with certain local fish-eating members of the public receiving up to 69 per cent of the maximum dose level as recommended by ICRP – the International Commission on Radiological Protection – and the NRPB.

Persistence in the Environment

Radionuclides such as plutonium-239, with long half-lives, will, once released into the environment, persist for tens of thousands of years. They are pollutants for all time, and are particularly dangerous on account of their intense radiotoxicity should they be taken up into the body. In the early days, scientists for the Atomic Energy Authority assumed that the plutonium and other toxic alpha-emitters discharged into the Irish Sea from Sellafield would be trapped and locked in the sediment offshore. Yet, as Dr V. Bowen, who had been an analytical chemist on the Manhattan Project and was later at the Wood's Hole Marine Biological Station in Massachusetts, pointed out at the 1977 Windscale inquiry, he himself had found that scallop caught off the Isle of Man – hence at least 35 miles away from Sellafield across the Irish Sea – had a plutonium burden 40 times and more higher than that found in plaice caught close to Sellafield off the Cumbrian coast. His finding was enough in itself, Bowen argued, to contradict official claims that transuranic nuclides which were incorporated into the sediments of the Irish Sea were not available biologically and therefore would not become part of the food chain. Should Isle of Man scallops be consumed at anything like the rates assumed for Windscale plaice, then, according to Bowen, individual consumers would get as much as 10 per cent of their maximum permissible levels of plutonium from that one source alone.

Six years later, on 6 October 1983, BNFL announced that it would spend £10 million on a programme to reduce plutonium and americium discharges from 1,000 curies to 200 curies. The factors involved in BNFL's decision, apart from public outcry at the discharges, were that the NRPB had come to the conclusion, after further research, that the absorption rate of plutonium by shellfish, including of scallops, was five times higher than it had maintained previously, that more shellfish was eaten than was previously thought, and that analytical methods for measuring plutonium contamination in shellfish had improved. In all, the NRPB decided that the contribution from plutonium to the critical group dose should be increased by a factor of 15.[2]

But Bowen was not only concerned with radioactive contamination of seafood. The main body of his evidence dealt with ways in which transuranics such as plutonium and americium might find their way ashore. He suggested three pathways: plutonium contaminated dusts that escaped the filter systems in the exhaust stacks of the reprocessing plant, waste storage and fuel fabrication plant; the sediments that washed up on the beaches, dried out and blew away; and the atomization of alpha-emitters by the action of waves along the shoreline. Bowen was amazed that the British authorities only began their examination of the inhalation pathway associated with the resuspension of radioactive substances from contaminated sediment in 1976, even though the Department of the Environment had considered such an eventuality in 1969, just at the time when it was considering BNFL's request for a threefold increase in alpha discharges. When the DoE carried out a few tests it found the mean concentration of plutonium particulates in the Ravenglass estuary, some 10 kilometres to the south of Sellafield, to be ten times higher than that found in the immediate vicinity of the Sellafield site.

In November 1983, the Environmental and Medical Services Division of the UKAEA at Harwell reported the results of experiments which mimicked the action of surf on plutonium contaminated sediment. It found the concentration of plutonium in the spray to be as much as 800 times greater than that of the seawater from which the droplets were formed. Americium was even more volatile, the concentration in spray being some 10,000 greater than in seawater. The mechanism by which plutonium and other alpha-emitters return to shore would therefore seem to have been confirmed.

Meanwhile, A. D. Horrill and his co-workers at the Institute of Terrestrial Ecology have found enormous variability in radionuclide levels within a distance of a few hundred metres in the surface silt and vegetation of a grazed saltmarsh in the Esk estuary, which itself leads into the Ravenglass estuary. Such variability, for instance between 20 and 460 pico-curies per gram dry weight for caesium-137 (one pico-curie is one trillionth of a curie), can lead to considerable errors in evaluating the extent to which the Cumbrian coast has become contaminated by the Sellafield discharges. Horrill, for instance, found americium levels to range from 4.59 to 232 pCi per gram and plutonium-239/240 to range from 11 to 240 pCi per gram dry weight.[3]

Tissues of ewes and lambs that had grazed over the contaminated Ravenglass saltmarshes showed varying concentrations of radionuclides. B. J. Howard and D. K. Lindley, also at the Institute of Terrestrial Ecology, found as much as a 50-fold concentration of caesium-137 in the liver and lambs compared with a control ewe and similar concentration factors in kidney and muscle. Up to a 100-fold concentration of plutonium was reported in the liver of ewes.

Because they are higher up the food chain, human beings are more vulnerable to radioactive contamination than are herbivores. Some members of the Cumbrian coastal community have been found to be avid eaters of fish and shellfish. Thus investigations by MAFF's Fisheries Radiobiological Laboratory (FRL) found that such individuals would consume up to 100 grams per day of fish, 18 grams per day of crustaceans and up to 45 grams per day of winkles and mussels. On the basis of the annual ICRP dose limit for members of the public, FRL calculated that in 1981 the critical group of fish and shellfish eaters would have received 69 per cent of the limit; in 1982, 54 per cent; and in 1983, 51.5 per cent, solely through diet. For consumers of seafood caught further away, by the commercial fisheries associated with Whitehaven, Fleetwood and the Morecambe Bay area, the dose to the critical group was calculated as 19 per cent of the ICRP dose limit in 1981 and as little as 13 per cent of that limit in 1982. For the average fish-eating member of the public consuming seafood caught by such fisheries, the estimated dose for 1983 was given as 1 per cent of the ICRP dose limit.

People spending any length of time on the Ravenglass estuary (for instance, salmon garth fishermen), or on board boats moored in

Whitehaven harbour will be exposed to external radiation. Taking the shielding afforded by the boat hull into account, FRL estimates that such a critical group may receive up to 8 per cent of the ICRP dose limit through that source, and another 4 per cent if they consume seafood. The alpha-emitters, including plutonium, coming ashore, will also add to the dose of those exposed, present calculations suggesting that it be 1 per cent of the ICRP limit.[4]

Discharges – As Bad as the Bomb

The potential for plutonium contamination of the environment through activities at Sellafield is considerable, and according to a report in the *Gazette Nucleaire*, a highly documented French newsletter, could exceed the entire global inventory of plutonium deposited from the atmosphere from all the test explosions that have ever taken place. To date, between four and five tonnes of plutonium has settled to earth from such tests, most falling out over the northern hemisphere. Nevertheless, Cumbrian coastal waters now show concentrations of plutonium and americium that are some 2,000 times higher than fall-out levels, while even the North Sea has double fall-out levels on account of discharge from Sellafield into the Irish Sea. In addition, Britain embarked on a sea-dumping programme in which solid wastes containing considerable quantities of plutonium were ditched in the Atlantic Ocean some 500 miles off Land's End.

Between 1949 and 1977 as much as 140 kilograms of plutonium may have been dumped with another third of a tonne contained in 3,000 cubic metres of waste awaiting similar disposal. Meanwhile, between 1968 and 1979, an additional 180 kilograms of plutonium were discharged into the Irish Sea from Sellafield. BNFL's present plans to reduce alpha discharges down to 200 curies and ultimately, when its 'enhanced actinide removal plant' is operational, to 20 curies per annum, will undoubtedly much reduce the alpha contamination build-up in the Irish Sea and Cumbrian silt. Nevertheless, the plutonium waste problem will not have been eliminated, merely transferred, and BNFL will undoubtedly seek that the sludges and plutonium contaminated resins from the actinide removal plant be dumped at sea after some form of 'waste conditioning'. There is no other place for it to go.

The Worst Reputation in the World

BNFL has not fared well in comparisons between its own operations at Sellafield and those of comparable plants elsewhere in the world. The closest in type of operation and in plans for the future is COGEMA's plant at Cap de la Hague on the Cotentin peninsula overlooking the English Channel in Normandy. Not only have discharges into the Channel from the French plant been lower by a factor of eight for beta-emitters and a factor of more than 200 for alpha-emitters – particularly during the mid-1970s – but worker exposure within the plant has been one-half or less than that registered at Sellafield. During 1980, for instance, the average dose to 2,671 workers at La Hague was 0.241 rem – thus approximately two times background radiation – while in 1982, one of BNFL's better years for radiation exposure the average dose for 5,223 workers was 0.64 rem.

Another way of indicating worker exposure is to relate it to the quantity of electricity generated from the fuel. Again the difference between the French and British reprocessing plants – the only two left in the western world with any commercial pretensions – is telling. Between 1971 and 1975 external irradiation alone at Sellafield amounted to 1.2 man–rem per megawatt (electrical) year and that for some 4,171 tonnes of magnox fuel reprocessed and 120 tonnes of thermal oxide fuel. The equivalent value for La Hague was 0.51 and that for 3,944 tonnes of magnox fuel and 356 tonnes of oxide fuel reprocessed.

Since 1977, La Compagnie Générale des Matières Nucléaires (COGEMA) has managed to bring down both individual doses to workers and the dose/energy relationship at La Hague, the latter being little over 0.22 man-rem per megawatt year in 1981. Meanwhile the man-rems per megawatt year at Sellafield for 1982 had been reduced to 0.93. The collective dose for the workforce at Sellafield has remained relatively steady over the years, the reduction in individual dose being achieved by increasing the workforce. In 1982 the collective dose was 3,370 man-rems, hence more than five times higher than the collective dose at La Hague. Per tonne of magnox fuel reprocessed BNFL's record at Sellafield has been at least 2.5 times worse than that of COGEMA.[5]

After considerable criticism, both during the Sizewell public inquiry and outside, BNFL has embarked on a costly programme to reduce its discharges. As much as £190 million was spent on control of beta-emitters and caesium in particular, the SIXEP ion-exchange treatment plant bringing total beta down to 30,000 curies when operational, supposedly during 1985. On 18 December 1984, BNFL announced that a further £150 million was to be spent on a new plant which

should cut total liquid discharges from Sellafield, including those from the THORP oxide reprocessing plant now under construction, to a target of less than 20 curies a year of long-lived alpha radiation emitters, compared with

383 curies in 1983. Annual discharges of mainly shorter-lived beta radiation emitters should be reduced to a target of 8,000 curies a year, compared with 67,000 in 1983.

The result of all the treatment plants working as planned will be, BNFL stated, to bring Sellafield discharges down to a level comparable with Cap de La Hague. Meanwhile criticisms are still voiced over discharges, such as they are, from La Hague, and over keeping the total down once attempts are made to keep abreast of the spent fuel coming from an ever expanding nuclear power programme, and thus of having to use a number of reprocessing plants in parallel. At La Hague, for instance, COGEMA has plans for two new thermal oxide reprocessing plants, UP 2–800 and UP 3A, with a nominal capacity of 800 tonnes of spent fuel per year. With regard to the emissions from those plants and their effect on the environment, the Castaing Commission – a body equivalent in France to the Royal Commission in Britain – stresses that:

(1) krypton-85 will be emitted in its entirety, approximately 11,000 curies per tonne for PWR fuel;
(2) liquid tritium discharges will increase from some 60,000 curies per year to one million curies, a factor increase of nearly 17;
(3) permissible liquid discharges of beta and gamma radionuclides will remain as they are, namely 45,000 curies per year (excluding tritium).

Given the expected increase in throughput of thermal oxide fuel to be reprocessed, COGEMA will have to improve its control of discharges by a factor of four. Just to keep abreast of COGEMA's expected performance BNFL would have to improve its containment of beta-emitters, including plutonium-241 which gradually transmutes into the highly toxic alpha-emitter, americium-241, by 30-fold, and of alpha-emitters by some 1,000-fold compared with its average over the mid-1970s.

Even if both companies achieve containment considerably better than has ever been achieved before they will be far from the zero discharges called for by the European parliament's Committee on the Environment, Public Health and Consumer Protection. In response to the unusually high incidence of cancer in the vicinity of Sellafield, the committee put before the European parliament a resolution including a demand that the British government impose a zero discharge of radioactive wastes into the Irish Sea, and that gaseous emissions of radioactive substances, including krypton-85, carbon-14 and tritium be controlled in accordance with available technology. The committee also called for a ban on the transport of spent fuel until discharges had been brought down to a technically achievable level and only when a final storage place for radioactive waste had been made available.

What to Do with the Waste

Much of the plutonium-bearing wastes remain in various forms on site at Sellafield and in need of conditioning. In 1974, as much as 580,000 curies of alpha wastes were contained in high-level waste storage tanks, in silos stocked with magnox cladding from stripped fuel elements and in various other radioactive sludges and wastes. The 500,000 curies of alpha wastes containers in the high-level waste storage tanks alone must amount to at least three tonnes of plutonium.

The aim is to solidify the high-level waste, through vitrification, and then dispose of it. One idea is to bury it in the ocean bottom, another on land. Concern has grown internationally over the use of the sea for the dumping of radioactive waste, and in February 1983 a consultative meeting of parties to the London Convention on Sea Dumping passed a resolution which called for a halt to the sea disposal of low-level waste pending further scientific investigation into the possible effects on man and on the environment.

The British government, although it agreed to a temporary ban on dumping, has made it clear that it sees few obstacles on scientific grounds to the resumption of dumping. On the contrary, if international opposition to dumping could be circumvented or ignored, then it is more than likely that Britain would use the Atlantic for getting rid of most, if not all, its nuclear waste, including pieces of derelict reactors. Even high-level wastes could legitimately be dumped if diluted sufficiently in the conditioning material, whether borosilicate glass or synroc, and if the release rates of radioactivity through leaching could be shown to be sufficiently low. Both the International Atomic Energy Agency (IAEA) and OECD's Nuclear Energy Agency are in favour of using the sea for dumping, singling nuclear waste out for special favour from all other potential pollutants of the marine environment. In its revised definition on 1978, the IAEA maintains that all waste from nuclear activities could be dumped provided that the release rates met certain criteria and that total quantities of waste did not exceed more than 100,000 tonnes per annum at any single dump site.

As the Political Ecology Research Group, an independent watchdog committee, points out:

> current regulations will allow a 100-fold increase in beta/gamma activity and a 50-fold increase in alpha. With suitable packaging, the UK could therefore

dispose of all its fuel element cladding, sludges and alpha-contaminated waste within a few years, with no effort being required to retard release of the activity at depth . . . High active liquid disposal would require demonstration of retarded release rates, otherwise 450 years would be required for wastes generated by the year 1985 alone.

In 1976, Grimwood and Webb of the National Radiological Protection Board advised the British government: 'No overriding reason connected with the radiological protection considerations has been identified which would preclude the disposal of suitably conditioned high-level waste on the ocean floor.' Meanwhile the government has been ordering new ships specially designed for dumping nuclear waste from the hull, hence overcoming the difficulties of dumping drums of waste over the side when being hassled by members of Greenpeace in rubber dinghies. To counter any such development, the National Union of Seamen (NUS) has declared its refusal to handle any nuclear waste destined for dumping in the Atlantic. The final straw for the British government has come with the majority decision at the September 1985 meeting of the London Convention to maintain indefinitely the ban on dumping.

Throughout the world nuclear wastes are now accumulating and no long-term solution for their safe disposal exists. The trend, especially among those countries determined to make a stand against nuclear weapons proliferation, is to keep spent nuclear fuel intact rather than reprocessing it. Temporary 'engineered' disposal sites for the storage of packaged spent fuel are now being constructed in countries such as Sweden. While such storage fails to provide a long-term solution, at least it offers a better alternative than reprocessing.

The contamination of the environment and living organisms continues, however, whether through routine discharges or through accidents. For the British public and the European community, the spate of accidents at Sellafield and its associated waste dump site at Drigg during the early part of 1986 were the last straw.

The incidents involved: the discharge of half a tonne of reprocessed uranium into the Irish Sea, an accident with much in common with the discharge of radioactive crud that contaminated 20 miles of beaches in November 1983; the escape of plutonium nitrate through a faulty valve into a building manned by reprocessing workers, a number of whom became internally contaminated; a fire at Drigg, apparently not involving the release of radioactivity; and, finally, a leak of radioactively contaminated drainage water from the Magnox storage and decanning plant.

The Irish in particular have been incensed by the continued polluting of the Irish Sea, Dr Garret FitzGerald, the Irish prime minister, calling for the plant to be shut down while the EEC carries out monitoring of BNFL's activities at Sellafield. Some of the strongest words against the plant came from the German member of the European parliament, Undine Bloch von Blottnitz. 'Europe is being polluted for the sake of a few jobs,' she told her follow members. 'The least we can do is tell the UK to close down its dump.'

As a result of the pressure on the government, the Health and Safety Executive is to carry out a six months' intense inspection of the Sellafield site. Simultaneously, studies will be carried out to test whether food in the locality has been contaminated. Nevertheless, the government has brushed aside appeals for the plant to be shut down until the Health and Safety Executive has made its findings public, as has been requested by some members of the European parliament and environmental groups such as Friends of the Earth.

As part of general policy, BNFL has consistently underplayed the nature and danger of its accidental discharges, claiming that the general public have not been put at risk. After the escape of plutonium nitrate which led to an 'amber alert', BNFL management made a statement that no more than two workers had been contaminated, one having received more than the annual maximum permissible dose according to ICRP standards. Later information revealed that as many as 15 workers had inhaled plutonium into their lungs. And with regard to the uranium discharge, BNFL's managing director, Con Allday, claimed that it was a mere drop in the ocean compared with the amount of uranium already in the sea. Again he was invoking the 'discharge, disperse and dilute' approach. Yet as ex-Harwell scientist M. E. J. Gilford pointed out, the truth was somewhat different, the localized dumping of radioactive uranium leading to far greater concentrations than naturally present either in seawater or in the top layer of the seabed. In a letter to the *Guardian*, Gilford stated:

According to Mr Allday 'The Irish Sea already contains many thousands of tonnes of naturally occurring uranium.' A BNFL spokesman also recently stated that the East Irish Sea contains about 1,000 tonnes of naturally occurring uranium in the sea water and 10,000 tonnes in the sea bed. My own calculations suggest that the sea water contains only 100 tonnes. Even if I grudgingly accept 1,000 tonnes in the sea water, I ask myself: isn't the East Irish Sea rather a big place? Surely, a large proportion of the uranium will stay within 100 square kilometres of the outlet for a considerable time. Now, according to BNFL figures, 100 square kilometres of Irish Sea contains only about 5 tonnes of naturally occurring uranium (my own figures suggest half a tonne). Compared with this, half a tonne dumped in a single day looks a little more alarming.

Ah, but I hear Con Allday say, perhaps not all of the uranium is in soluble form; some of it will fall to the sea bed where there is already 10,000 tonnes. Well, it depends what you mean by 'sea bed'. If you mean the top 200 metres of the sea floor, then yes I agree – 10,000 tonnes of naturally occurring uranium is about right. But if you mean the top 10 centimetres, where most of the living creatures reside, then I'm afraid the whole of the East Irish Sea contains only 5 tonnes of uranium, and the 100 square kilometres nearest the outlet contains a mere 25 kilograms. Compared with this, 440 kilograms dumped in a single day looks truly horrific.

We have now come to accept deceit and half-truths from the nuclear authorities. Even Sir Douglas Black in preparing his report on the spate of childhood leukaemia cases in west Cumbria was misled as to how much uranium had been discharged into the environment between 1952 and 1955. He and his team of advisers were told that the quantity discharged was 400 grams. Two scientists who worked at Windscale in the 1950s have now revealed that the quantity was at least 40 times greater. And what about other, far more dangerous radioactive pollutants such as americium and plutonium? Can we accept the official figures of the discharges? One would have to be exceedingly generous to give BNFL and other members of the nuclear establishment the benefit of the doubt.

The tragedy is that we have created a site that will remain dangerously polluted for all time. Gross mismanagement is simply compounding the problem. That is why the Sellafield plant must ultimately be closed down. However, the legacy of nuclear waste at the plant, together with the contamination of the soil from the various spills, dictate that Sellafield can only be shut down after a vigorous cleaning up campaign. A solution will have to be found for the waste on site, which will have to be disposed in an environmentally acceptable way. To date no one can agree on the safest form of disposal. The nuclear adventure at Sellafield must be brought to an end before an accident subjects us to yet another dousing of radioactive carcinogens, perhaps next time even worse than Chernobyl. A small, crowded island like Britain cannot accept the risk of nuclear power.

Notes

1. Officially only that plutonium extracted from the Calder Hall and Chapelcross reactors should be used for defence purposes, and only then, according to Dr Donald Avery of BFNL, when the original uranium fuel for those reactors had been derived from non-safeguarded sources obtained prior to the Non-Proliferation Treaty. Officially, too, the plutonium derived from the electricity boards' reactors is kept in storage on site until required for recycling either in thermal reactors or in the prototype fast reactor. The intention, too, is to win contracts for reprocessing spent fuel from abroad. A contract has

already been signed to import Japanese spent fuel and much of the rationale for building the Thermal Oxide Reprocessing Plant (THORP) was that it would largely be paid for by overseas customers.

2. On account of their occupations, leisure activities, eating habits or simply as a consequence of where they live, certain members of the public (known as 'critical groups') are more likely to be exposed to ionizing radiation from nuclear installations, whether of civil or military origin, than are other members of the public. Authorities such as the National Radiological Protection Board therefore assume that if radiation exposure of those in the critical groups can be kept below maximum permissible limits then automatically other members of the public will be even less exposed. As a result the likely exposure pathways of the various critical groups are followed and the maximum yearly dose averages calculated. And as long as critical group exposures are deemed to be below the maximum permissible doses, then, according to the NRPB, the safety of the general public will be assured. However, it must be appreciated that certain members of critical groups have had exposures close to the maximum permissible level based on the NRPB's assessment of risk. On account of elevated cancer rates among the public living close to nuclear installations, epidemiologists are no longer in agreement over the assessment of risk. Some would argue that members of critical groups have in many instances been exposed excessively to radiation.

3. In the *Marine Pollution Bulletin* (vol. 12, no. 5, pp. 149–53, 1981) S. R. Aston and D. A. Stanners from the Department of Environmental Studies report on similar results. As they point out, 45 per cent of all the 27,478 curies of alpha activity discharged via the pipeline into the Irish Sea between 1968 and 1978 is made up of americium-241, which then gradually comes ashore associated with fine silt sediments brought in with the tides. As a result, the entire range of the coastline from Maryport in the north to Wyre in the south has measurable activity, the greatest values being in the Ravenglass estuary where sediments from inner sections of the estuary reveal a sevenfold higher concentration of americium than has been reported for some Irish Sea sediments. Meanwhile plutonium levels 60 kilometres to the south of Sellafield in the Wyre estuary appear to be one-quarter of those for Ravenglass, the transport of the radionuclide taking six years from discharge to sweep down the coast. Much of the plutonium appears to be associated with organic matter in the sediment suggesting that movement through the food chain may be somewhat easier than if it were totally bound up with inorganic matter.

4. In the USA such exposures of members of the public to radiation from man-made sources would not be permitted on a routine basis. There the maximum allowable exposure from fuel cycle activities is 25 millirem, hence 20 times less than that allowed in west Europe. Meanwhile the most exposed members of the public in Cumbria will have received four or five times the US limit.

5. Peter Bunyard, evidence to the Sizewell inquiry, 1985.

24

Dumping Nuclear Waste at Sea

Jim Slater

Many people in the trade-union movement raised their eyebrows when in June 1983 the National Union of Seamen, along with the train drivers' union ASLEF and the TGWU, announced a ban on the movement of all nuclear waste to be dumped at sea. It was not the sort of issue that trade unions were expected to pursue, let alone take industrial action over. But within a few days of the announcement, the nine-man NUS crew of the *Atlantic Fisher* had walked down the gangway with their kit bags and the specially converted dump ship was left stranded at Barrow. Meanwhile, train drivers were refusing to take barrels of waste to Sharpness where the *Atlantic Fisher* was due to load its deadly cargo.

As well as preventing the dumping of nuclear waste in the Atlantic Ocean for the first time since 1949, our decision served one other important purpose. Along with the relentless campaigning of Greenpeace and others, our direct action helped alert the British public to Britain's appalling record of polluting the seas with radioactive rubbish. It was only when we began organizing a trade-union campaign against dumping that we discovered, for example, that Britain was responsible for dumping more than 80 per cent of the nuclear waste tipped into the oceans. Other countries, such as the USA, Japan and France had long ago opted to store their waste on land.

The union boycott certainly made an impact in the Trades Union Congress. Despite pressure from the pro-nuclear establishment within Congress House, an NUS motion supporting the dumping boycott was carried by seven million to two-and-a-half million votes at the 1983 annual conference. And within a few weeks the NUS had persuaded the congress of the International Transport Workers' Federation to announce a worldwide ban on ocean dumping. The controversy over nuclear waste had finally arrived on the agenda of trade-union concerns.

"You may be assured, Sir, that there is no cause for concern . . ."

It was after the TUC vote that the then environment secretary Patrick Jenkin was forced to suspend all further ocean waste disposal, pending the outcome of a joint Department of the Environment/TUC sponsored inquiry into dumping. There is no doubt that the government was gravely embarrassed by the initial NUS ban and was powerless to do anything, presumably because public opinion could have been mobilized very quickly on so sensitive an issue. That is the only reason to explain why the Royal Navy was not brought in to do the dirty work.

By 1985, the full impact of the NUS action was being felt, as the government began a frantic search for land sites to dispose of the low and intermediate radioactive waste piling up at nuclear establishments around the country. As a consequence, new and potentially politically damaging battles with local communities loomed on the horizon for 1986 and beyond.

The Reason for the Ban

How did the NUS come to take up this position on nuclear waste? We cannot claim to be scientific experts on nuclear technology or marine biology. What we do know is that if seamen fail to defend the seas, then no one else can be expected to do so.

Our position on nuclear waste is that dumping, whether at sea or underground at sites on land is grossly irresponsible. It is an 'out-of-sight-out-of-mind' solution. If Britain – or any other country – insists on maintaining military and civil nuclear industries, then a policy of storage must be developed in which the waste is kept in a monitorable and retrievable form. The nuclear industry would, of course, prefer to dump its waste, as that helps sustain the myth that nuclear power is 'clean'. Storage, apart from being safer than dumping, would remind people of the real price of nuclear power – a growing mountain of lethal waste on our own doorstep.

Events in the autumn and winter of 1983–4 surrounding the Sellafield (formerly Windscale) nuclear waste reprocessing plant inevitably overshadowed the union's achievement in preventing the dumping of 3,000 barrels of radioactive waste 500 miles south-west of Land's End. But the revelations concerning high childhood leukaemia rates and the contamination of a large stretch of the Cumbrian coastline again alerted the public, in the most dramatic and horrific fashion possible, to Britain's notorious record as the number one radioactive polluter of coastal and international waters. To many people – including, I suspect, most in the trade-union movement – disclosure of this shameful record has come as a nasty surprise. Perhaps naively, we always thought that Britain would behave more honourably.

On issues like nuclear waste disposal, the consequences of which could be felt for hundreds or thousands of years to come, all sane people would expect countries to heed world opinion and to abide by the decisions of internationally accredited watch-dogs. But when the London Dumping Convention (LDC) voted in February 1983 in favour of a two year moratorium on waste dumping, the British government immediately announced that dumping would continue. That, of course, is no way to treat the United Nations-sponsored agency charged with regulating the discharge of all potentially hazardous wastes into the oceans. The government's arrogance finally convinced the NUS that enough was enough and that the union had to try to organize direct action against dumping. When the LDC voted in September 1985 to extend its ban indefinitely, angry government officials again threatened to ignore the decision and even to withdraw from the organization itself – although more level-headed spokesmen later seemed to accept that sea dumping was no longer politically feasible.

Similar contempt is being shown towards the international community over the question of radioactive discharges from Sellafield. At the Paris Commission meeting in June 1984, the British delegation actually supported a resolution calling for the use of the best available technology to minimize the radioactive discharges as quickly as possible. This was hailed as a major change of heart by the British government. Most observers assumed that it meant that the discharges would end within three to four years. Yet within a few days, spokesmen for British Nuclear Fuels, presumably with official connivance, were insisting that the discharges would not be reduced for ten years. By that time the Magnox reactors which are the source of the contaminated waste will probably be decommissioned anyway.

Wider Issues

This scandalous disregard for public opinion both here and abroad on the question of waste is of course predictable from an industry and a government well known for their secretiveness and arrogance. We can at least take some comfort from the fact that these last few years have been a trying time for the pro-nuclear boffins who increasingly have had to account for their actions under the full glare of public attention.

The Sizewell inquiry helped. And Britain's whole approach to energy generation and conservation was put into sharp focus by the miners' dispute of 1984. It did not take long, for example, for many in the trade-union and Labour movement to discover that the power contributed to the National Grid by all 11 Magnox nuclear reactors could be produced by just two coal-fired stations, or the cut-back of four to six million tonnes of coal which Ian MacGregor wanted to impose on the mining industry.

Energy policy issues such as this, as well as the mounting anxiety over nuclear waste, played a decisive part in persuading the Labour Party's 1985 annual conference to call – for the first time ever – by 3.9 million to 2.4 million votes for a halt to the nuclear power programme and a phasing out of all existing plants. Let us hope that this decision will herald a new commitment to 'green' issues within the Labour movement. Concern for the environment should rightfully be the preserve of the Labour Party, with its tradition of care and concern for the welfare and health and safety of ordinary people and with its obligation to check the excesses of industry in the interest of greater social good.

What has to be overcome, however, is the natural inclination of trade unions to defend any industry or service if jobs are at stake, regardless of wider social or environmental considerations. This explains why the TUC has repeatedly endorsed a 'balanced' energy policy so as to reconcile the differing interests of unions in the coal-mining and nuclear industries. What this means in practice, however, is that the TUC is committed to the nuclear industry, simply because many powerful unions have members employed in it.

There is one hopeful sign that things might be changing for the better, however. Following the 1983 resolution to support a ban on sea dumping, the TUC set up a radioactive waste group, bringing together representatives from the established committees dealing with transport, health and safety and energy. Here at least is an example – probably unique – of the TUC recognizing that the implications and consequences of one industry's activities should be considered in a wider context than merely a self-interested defence of jobs and the status quo.

Despite the limited successes of recent years, the task for those groups campaigning against the folly of Britain's nuclear policies is still a daunting one. The next battleground is sure to be the plan by BNFL and the UKAEA to construct a fast reactor fuel reprocessing plant at Dounreay, Caithness. Disgracefully, the government is trying to gag opposition to the scheme by restricting the terms of reference for the public inquiry, which began in April 1986. Quite clearly, ministers are determined to avoid any discussion of the impact of such a project in terms of the increased quantities of nuclear waste which will be generated by reprocessing spent fuel not only from Britain, but also from France, West Germany, Italy and Belgium.

All this is hardly surprising. Controversy continues about the choice of land sites for nuclear waste disposal, and the last thing the government or nuclear industry want is the public to realize that Britain's waste problem could be cut at a stroke, in terms of volume, by a factor of between 12 and 20, if reprocessing at Sellafield – and soon Dounreay – were ended. In short, our nuclear waste crisis is almost entirely one of our own making.

One day the penny will drop. The events of the last few years, as nuclear waste and nuclear energy have become major issues of concern to the public, must surely have marked the turning point in the nuclear debate and have brought that day much nearer.

25

Radiation and Health

Peter Bunyard

The public has constantly been assured that emissions of low-level, that is, small, discharges of radioactive wastes from nuclear installations have caused negligible damage to health. After the November 1983 incident at Sellafield, when an unspecified amount of radioactive crud was sent into the Irish Sea, contaminating divers from the environmental action group, Greenpeace, who were then trying unsuccessfully to cap the effluent pipeline, the official response was that no harm would result – at least from the health point of view. Three months after the accident when radioactive debris was still coming ashore, Patrick Jenkin, then Secretary of State for the Environment, told the House of Commons: 'there is no evidence to suggest that this contamination, although very unsatisfactory, could cause significant damage to anyone's health.'

He added that the worst anyone might suffer would be 'localized irritation of the skin from prolonged contact with one of a number of pieces which have been found with much higher than usual levels of radioactivity.'

Radiation and Cancer

Even after the Chernobyl accident, which sprinkled fall-out over Europe and the United States at the end of April and May 1986, governments claimed there was no risk to health as long as certain precautions were taken in areas where the fall-out had been at its most intense. Equally no-one was supposed to have suffered any ill-effects as a result of the fire in the Windscale number one pile in October 1957, when according to official estimates, some 20,000 curies of radio-iodine were released into the atmosphere. The benign consequences of the Windscale fall-out were first effectively challenged by PERG – the Political Ecology Research Group based in Oxford. In looking through the health statistics for north west England, the area which received the highest fall-out from the Windscale fire, PERG found an excess of cancers in south west Cumbria, including evidence of a higher than normal incidence of thyroid cancer, a tell-tale sign that radioactive iodine had been released. Until then, the authorities, including the National Radiological Protection Board, had studiously failed to look for any excess cancers.

From the amount of radio-iodine expelled from the stricken pile, PERG assessed an excess of 250 thyroid cancers of which 5 per cent would probably have proved fatal. Even though the extra 12 thyroid cancer deaths would be virtually impossible to pick up against the mortality statistics of Britain as a whole – representing at most a third of a per cent increase – locally the excess might be apparent and in fact PERG found a doubling of thyroid cancers in Newcastle in 1977 compared to the national rate. Meanwhile Peter Taylor of PERG also found an above-average number of leukaemia cases in Barrow, but was refused permission to investigate official data on health statistics on the grounds that the information was 'politically sensitive'. In 1981 the Cumbria Area Authority claimed that 'deaths from leukaemia have not significantly altered from the national rates.' Yet whereas the leukaemia rates for 1951 to 1958 were 83 per cent of the national average, twenty years later, between 1971 and 1978, they had risen to 100 per cent.

Peter Taylor was initially baffled that radio-iodine fall-out from Windscale could have had such an effect. But he, like the rest of the British public, had been kept in the dark about the true nature of the emissions from the burning Windscale Pile. As John Urquhart, a statistician at the University of Newcastle-upon-Tyne, discovered, a major contribution to the fall-out was from polonium-210, a highly radiotoxic substance, which was being generated in the reactor so that it could be later used as a trigger in Britain's nuclear deterrent. According to Urquhart as much as 370 curies of polonium might have been released during the fire, giving a total dose to the UK of 5 million man-rems and a further dose to the continent of 0.5 million man-rems. 'We are therefore talking about more than a thousand deaths from the Windscale accident,' he says (*New Scientist* 31 March 1983), and puts the total toll up to 8,000.

The radiation releases from the Windscale Fire of 1957 were in fact far lower than those from Chernobyl from which between 100,000 and one million curies of iodine–131 escaped into the atmosphere together with some 5,000 to 10,000 curies of caesium–137. In parts of Cumbria, close to Windscale, the fall-out from that fire produced a concentration on the ground of up to 370,000 becquerels per square metre. Similar concentrations of radio-iodine were found after the Chernobyl explosion in parts of Sweden such as Gärle to the north of Stockholm and more than a thousand kilometres away from the reactor site in the Ukraine. Overall the Chernobyl accident will cause many more cancers and premature deaths than the Windscale fire. Dr Lambert, a radiobiologist at St Bartholomew's Hospital in London, believes that Chernobyl will result in an excess of some 500 cancers in Britain over the next few decades. Yet that number will be a tiny proportion of the total number of some 7 million cancer deaths taking place over the same period, and no-one will be able to prove categorically that a particular person's death will have been the result of Chernobyl.

An Ill Wind from Sellafield

The excess of cancer in south west Cumbria revealed by PERG's findings received national coverage with Yorkshire Television's *Nuclear Laundry* shown on 2 November 1983. This film revealed that dust contaminated with plutonium – like polonium an alpha emitter – was present not only in houses in the vicinity of the Sellafield reprocessing plant, but also 40 miles away in Scotland. In 1977, at the Windscale Public Inquiry into BNFL's application to build a new reprocessing plant to deal with thermal oxide fuel from Britain's AGRs as well as with spent fuel from overseas, the authorities, including BNFL and the NRPB, claimed that the discharge of radioactive wastes into the Irish Sea would be sufficient to disperse and dilute them to safe levels. Alpha emitters such as plutonium and americium would remain bound in the silt at the bottom of the sea and present no danger to those on shore. Dr Bowen of the Wood's Hole Oceanographic Institute in Massachusetts, among others, challenged that view, and described mechanisms by which plutonium could come ashore. Both Yorkshire Television and the Institute for Terrestrial Ecology have sufficient evidence to show that Bowen was essentially correct.

Undoubtedly the most disturbing information in the Yorkshire Television film was the revelation that the childhood cancer mortality rate was up at least tenfold in the tiny villages bordering Sellafield. Bootle and Waberwaite, which should have had no more than one case of cancer in children under 18 over a 30 year period, instead had had four in 20 years: whereas Seascale, 1.5 miles from Sellafield, had had 11 cases of childhood cancer since 1950. The chances of there being 15 childhood cancers in the three villages rather than the expected three, was no more than one in 50,000 – a very low probability indeed.

As a result of the public outcry which followed the film, the government asked Sir Douglas Black to head an investigation on the cancer cluster in Cumbria. In his report published in July 1984, he basically confirmed the excess cancer deaths, but played down any connection with radiation, on the basis that cancer clusters were not uncommon and that the levels of radiation in the environment would have to be higher by a factor of at least 40 to be responsible for the high cancer rate. His conclusion for the induction of cancer was based wholly on the NRPB's model of the dose/cancer relationship. At the same time, it was later revealed that he had not been given entirely correct information over the atmospheric discharges of uranium from the reprocessing plant. BNFL had told him that between 1952 and 1955 only 400 grams of uranium had been released whereas as the company now admits (*Guardian* 17 February 1986) 20 kg had been discharged, more than 40 times more. What other information was withheld from Sir Douglas Black and his team of investigators?

Cancer Deaths around Sizewell

The local press in East Anglia meanwhile discovered a high leukaemia rate among the population in the vicinity of the Sizewell A magnox power station, which had been in operation since 1966. The cancer rate was particularly high among CEGB workers at the station, and it was clearly embarrassing for the Board that the revelation should emerge in the middle of the Sizewell public inquiry, which was hearing evidence on the building of a second nuclear station at the Sizewell site. But again the authorities denied that radiation exposure, either within the plant or outside, could have been responsible for the excess disease.

Evidence at the Sizewell Inquiry, presented by witnesses for the *Stop Sizewell B Association* and *Ecoropa*, made it clear that leukaemia among the Sizewell A workers had reached a level of statistical significance. Although Dr Bonnell, medical adviser to the CEGB, kept on insisting that 'there is no CEGB station where the incidence of leukaemia or any other radiation induced diseases are statistically significant in excess of the anticipated incidence', he was corrected by Professor Harrington, chairman of the CEGB's Epidemiological Advisory Committee. Harrington put the expected incidence at 0.3 cases and stated that 'the expected incidence of leukaemia among the Sizewell A workforce would be rather less than one, whereas it is established that the actual incidence has been significantly in excess of this.'

Professor Blackith of Trinity College, Dublin, and his statistician colleague, Dr Michael Stuart, revealed that four cases of leukaemia among the Sizewell staff had been diagnosed, three of which had resulted in deaths. The CEGB tried to have one of the deaths ruled out from the statistics, because the man involved was an 'ex-employee' and his death had occurred one month outside the study period. Yet, as the Stop Sizewell B Association made clear, the ex-employee was only 'ex' because his disease had caused him to take premature retirement at 56, five months prior to his death.

Sellafield Deaths

A similar dispute went on over deaths among the BNFL workforce at Sellafield. Between 1948 and 1980, the total number of deaths from multiple myeloma was four, which as Dr Avery of BNFL pointed out was in accord with the expected number. Professor Blackith then discovered, however, that the four cases of multiple myeloma did not include any of the five additional deaths from the disease that occurred between 1980 and 1983. Those extra deaths altered the statistics from 4 expected to 9 observed. The extra cases pushed the statistics into the significance bracket – a point grudgingly conceded by Dr Avery.

The Hanford Data: The Healthy Worker Effect

The vast nuclear complex at Hanford in Washington State came into being with the Manhattan Project, its task being to produce plutonium for the United States' weapon programme. Hanford in fact produced the plutonium for the bomb that devastated Nagasaki. Because of its close association with research and development the workforce at Hanford has always had a considerable proportion of people with professional or technical qualifications. At the same time, prior to employment at Hanford – and the same applies to any nuclear establishment whether in the USA, Britain or elsewhere in Europe – there is rigorous health screening.

Health statistics for the general population show that on average those who have graduated from university or technical college have a much better life expectancy than those of the same age who go straight from school to work. Age for age college graduates also have much lower rates of cancer mortality compared with manual workers. If those who in general are most exposed to radiation during their work at a nuclear establishment such as Hanford come from the professional classes, then one might expect less of an effect on health than if the population at random were exposed to similar radiation.[1]

In fact, health statistics of those who have worked sometime or other at

Hanford from 1943 to 1977 indicate a lower mortality rate when individuals are standardized for age and sex with the general population than one would expect. Indeed, the standardized mortality ratio (SMR) for cancers of Hanford workers was at 86 some 14 per cent below average and at 78 lower still for other causes of death. Meanwhile those who remained employed at Hanford for more than two years had even lower ratios, 85 and 76 respectively. Those who stayed less than two years had SMRs of 88 and 86, and therefore 12 and 14 per cent lower for cancers and other causes of death than one would find in the general population.

The explanation that exposure to radiation must be good for health is not one readily accepted either within or outside the nuclear industry. Instead the concept of the 'healthy worker effect' is now generally accepted, it being recognized that health screening before acceptance for employment in demanding work combined with the requirement for those with professional qualifications leads to a better than average life expectancy.

Clearly if one is trying to find out what the effects of low doses of radiation are on health, the best place to look is in a nuclear establishment where careful records are kept of radiation exposure to individuals during the time of employment. Hanford has been scrupulous in keeping checks of both external and internal radiation exposure to its workers, and given that the social-security system in the United States makes it possible to follow up those who have died and the cause of death, some kind of evaluation of the effects of low-dose radiation are feasible. At the same time the relatively good health prevailing among the workforce at a nuclear establishment because of the healthy worker effect is likely to confuse any conclusion in terms of the effects of radiation on the general public.

Dr Alice Stewart has long been concerned with the effects of low doses of radiation. She came to eminence with her discovery that pelvic X-rays of pregnant women would substantially increase the cancer rate among those children who had been exposed *in utero*. She estimated that one rad of X-rays would double the incidence of childhood cancer deaths from its level of approximately 600 per million children under the age of 10. Although she was ridiculed at first by those setting the radiation safety standards – those for instance who comprise the International Commission on Radiological Protection – her data and conclusions are now widely accepted.

As a result of her work, Alice Stewart and her statistician George Kneale were asked to join an investigation into the occupational mortality risks for Hanford workers set up by Dr T. Mancuso under the auspices of what was then the Atomic Energy Commission, but which is now part of the US

Department of Energy. In essence Dr Stewart and her colleagues found evidence that those workers who had been most exposed to radiation at Hanford, always in the low dose range, had a 25 per cent increased risk of dying of cancer – even though the overall mortality statistics indicated that the chance of dying from all causes of death was significantly less than expected from standardized mortality rates. Indeed, when the cancers of the Hanford population of workers are classified according to the criteria laid down by the International Commission on Radiological Protection in its fourteenth report, there is a significant correlation between dose level and cancer mortality. Thus, whereas the observed to expected ratio is 0.79 at the lowest dose level of 0.1 to 0.25 rad (see figure 11, p. 256), at 4 rads and over the ratio climbs well above unity. For instance, for doses greater than 10 rads – accumulated over the years – the ratio goes to 1.32, thus indicating a mortality rate that is 30 per cent worse than would be expected in a normal population.

Stewart and her collagues divided the Hanford workforce into nine different groups, relating each to the type and nature of the job in the plant, and its ranking with regard to four levels of radiation exposure. The professional and technical staff exhibited a vigorous 'healthy worker effect', except for those exposed at the highest level when the 'healthy worker effect' all but vanished. Hence the increased exposure to radiation undermined the natural good health of the specially selected members of a high ranking class. On the other hand clerical staff as well as operatives and other manual workers not exposed to radiation – those working at the lower danger levels – showed positive scores with regard to differential mortality. They were thus naturally less healthy. In fact nearly half the dangerous work at Hanford measured in terms of radiation exposure was carried out by personnel with professional or technical qualifications. Indeed the levels of mortality were found to be considerably higher for the lower than the higher grades of work. The mortality risks only became similar for those specialists and non-specialists doing the most dangerous jobs.

Low Dose Radiation Causes Cancer

Through their interpretation of the Hanford data Alice Stewart and her colleagues have been able to show the likely shape of the dose/response curve – that is, the relationship between a given dose of radiation and the number of cancers which will result. The curve which fits best – the square root or half power law curve – is not the shape of curve selected as 'conservative' by such prestigious bodies as the ICRP or Britain's NRPB. (See Figure 12). Stewart's curve shows a non-linearity of dose response.

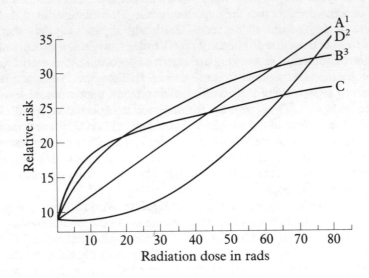

Figure 12 Theories of the relative cancer risk caused by ionizing radiation.

Notes:

1   The straight line hypothesis – according to which the relative risk rises proportionately with the radiation dose – is the one used by the ICRP and other standard setting bodies such as the NRPB on the assumption that it is 'conservative' and will, if anything, overestimate the relative risk at low doses.

2   This hypothesis assumes that very low radiation doses are less consequential per unit dose in causing cancer than are higher doses. Many health physicists believe curve D to be more representative of the facts.

3   Dr Alice Stewart believes that the evidence increasingly points to curves B or C being the true state of affairs. In both B and C very low doses of radiation are seen to be more effective in inducing cancer compared with higher doses. Indeed the point at which the cancer rate doubles in B and C is around 15 rads, whereas in A it is closer to 30 rads and in D to 50 rads.

Source: G. W. Kneale, T. F. Mancuso and Alice M. Stewart, 'Hanford Radiation Study III,' British Journal of Industrial Medicine, Vol 38, 1981, p. 162.

Whereas the linear model – a straight line relationship between dose and cancers – used by ICRP suggests that the cancer rate will double every 30 rads, Stewart's curve gives the overall doubling dose for low dose radiation as 15 rads.

Alice Stewart's findings indicate that the risk of cancer induction is far greater at very low doses than it is at higher doses. In fact, preliminary research both in Japan, and now in Britain, on the effects of background radiation suggests an increased incidence of cancer mortality among children in areas where background radiation is significantly higher, as for instance in areas overlying granite.

Some critics of Stewart have tried to argue that her contention of natural background radiation causing cancers is absurd, since (if she were right) the cancer rate would be higher than it actually is. But she points out that people in general, once they have survived birth, are relatively immune to low-dose radiation – whether of natural or artificial origin – and that susceptibility grows as a person ages. Furthermore, the long latency period for many cancers will only enable cancer to express itself in those who live long enough. Nor should one assume that cancer has only one cause. It must be appreciated, however, that all radiation carries dangers – and that some people are more vulnerable than others.

The Bomb Survivors

The rationale for using a linear dose/response curve is that higher doses are more likely to cause cancers than lower doses where a DNA chromosomal repair system is in operation and unlikely to be swamped by the damage caused. From this it is argued that extrapolation down from studies of populations such as the Hiroshima and Nagasaki bomb victims, if done in a linear fashion, is likely to overestimate, rather than underestimate, radiation risk. Stewart absolutely disputes such reasoning. She argues that where a repair system in the cell is effective, it is more likely than not to make mistakes, and at higher doses the cell is more likely to be killed outright. Furthermore, she does not consider the survivors of the atom bomb blast to be in any way representative of what prior to the blast was a normal population. Anyone who survived both the radiation and the trauma of the blast, let alone the aftermath of surviving without antibiotics in a devastated city with little shelter and rampant disease, must have had a particularly strong constitution. Stewart contends that those survivors who were closest to where the bomb went off must – by the very fact of their survival – have been innately the healthiest of all the original population. Stewart therefore invokes a 'healthy survivor effect' comparable in certain ways to the 'healthy worker effect', observed amongst nuclear workers.

Nevertheless those who survived radiation doses of 200 rads and more will not have escaped scot-free and, as experiments on animals have confirmed, will be likely to have residual bone marrow damage. That damage may not immediately show itself but, after some years, may suddenly appear in the form of aplastic anaemia, which expresses itself in a sudden, irreversible loss of the body's ability to make blood cells.

Stewart and Kneale have indeed found evidence in the A-bomb data for their postulated two 'silent forces'; namely, a 'healthy survivor effect' and residual bone marrow damage. By excluding both deaths from cancer and deaths from cardio-vascular disease from the mortality data on the survivors they are left with a large residual group which included all the deaths from infectious disease. The dose/response curve in that instance was found to be U-shaped, indicating a 'healthy survivor effect' acting more vigorously as the estimated dose approached 250 rads. From then on, that effect was counteracted by deaths from infectious disease, indicating that the survivors' immunological system had broken down.

In conclusion, Stewart points out that mortality rates, including cancer, among the A-bomb survivors are likely to give a false, optimistic picture of the dangers of low-dose radiation to a normal 'untraumatized' population. The cancer risk of low-dose radiation according to her model is likely to be 15 or more times greater than that assumed by both the ICRP and the NRPB. The available evidence supports her thesis.

The consequences of her findings are considerable. First, the refutation in the Black report that the high incidence of childhood cancer around Sellafield could be caused by radiation must be questioned. Secondly, the increased incidence of certain, radio-sensitive cancers among radiation workers is explained. Indeed, the nuclear industry, if it wishes to retain its image as a safe industry, will have to take all manner of precautions to keep radiation doses well below present levels. And that goes for discharges into the environment, which affect members of the public. Unfortunately, the nuclear industry and bodies such as the ICRP and NRPB which oversee its activities, are doing all in their power to dismiss Stewart's findings.

Conclusion

From the beginning, the nuclear industry has assumed, together with government watch-dogs such as the NRPB, that it understood the effects of low dose radiation on the public, and had a reasonably accurate knowledge of the likely pathways by which radionuclides would accumulate in the ecosystem. Facts have proved the industry wrong. It has had consistently to reduce its discharges and take economically crippling measures to improve safety – to the point where a sensible government might decide that the nuclear adventure should come to an end. The prestige associated with nuclear power, and the status in world affairs afforded to those with nuclear weapons, has meant, however, that the public is willy-nilly being driven into a technological cul de sac – one which threatens the very existence of the human species.

Notes

1. See G. W. Kneale, T. F. Mancuso and Alice M. Stewart, 'Identification of Occupational Mortality risks for Hanford workers', *British Journal of Industrial Medicine*, vol. 41, 1984, pp. 6–8.

26

Ignoring the True Cost of Nuclear Power

Peter Bunyard

Britain's civil nuclear power programme came into being because of the government's determination in the 1950s that Britain should continue to develop her own nuclear deterrent while keeping the cost down. The two Windscale piles, in operation during the first half of the 1950s, were expensive to run and produced no electricity to offset the costs; consequently after the 1957 fire in the number one pile, the government was quick to order the shut-down of the number two pile and leave the production of weapons grade plutonium to the Calder Hall and Chapelcross magnox reactors.

Although an intense public relations exercise had led the public to believe that nuclear power produced cheap electricity, the reality was otherwise and known as such both within the government and the newly formed electricity generating boards. Critics of the magnox programme, including Fritz Schumacher, who was then economic adviser to the Coal Board, claimed that by the late 1960s the civil nuclear power stations had cost the electricity consumers and taxpayers an extra £20 millions (in 1960 currency) per year over and above that which they would have had to pay if the generating plants had been operated on coal. In 1967, the chairman of the Coal Board, Alfred (now Lord) Robens told the House of Commons Select Committee that the magnox programme had led to the loss of 28,000 jobs in the mines due to the electricity boards' diminished demand for coal, and had cost £525 million more in capital costs alone than the £225 million a coal-generating capacity of an equivalent size would have needed.

Faulty Accounting

The civil nuclear industry has been heavily subsidized from the very beginning, not only through grants from the Treasury for research and development, but also through sharing with the Ministry of Defence such facilities as the Sellafield reprocessing plant and the Capenhurst uranium enrichment plant. The initial optimism over the likely costs of nuclear power was, moreover, based on faulty calculations. When planning a new coal- or oil-fired plant, for example, the CEGB always took account of site development and central engineering charges, incorporating them into the total capital costs. Yet, the nuclear planners completely overlooked such costs, which amounted to between 5 and 10 per cent of total station costs, when drawing up the designs for the magnox reactor programme in 1953. Interest charges on capital borrowed were also up to 6 per cent by 1961, 2 per cent higher than in 1954; such charges affected nuclear power plants, with their comparatively high construction costs, much more than they did conventional power stations. In fact, the capital costs of fossil fuel plant had been coming down spectacularly between 1955 and 1965 from £60 per kilowatt to £30 per kilowatt.

The government decided, however, to reduce the 'plutonium credit', based on the putative value of plutonium, as fissile material, extracted from spent fuel. Originally, the credit had been evaluated at 0.3 old pence per kilowatt-hour (p/kWh); now, it was to be more than 0.05 old pence – thus worsening, in one stroke, the overall cost of nuclear electricity by one-third. It is surely of some relevance that Britain was then negotiating to exchange plutonium for enriched uranium and tritium from the USA – for mutual defence purposes – and that it might have been embarrassing if the CEGB had received credit for such an exchange.

Rigging the Costs

The original undervaluation of the cost of nuclear electricity established a pattern of rigging the costs to show publicly that nuclear power was a valid competitor with other forms of electricity generation. For instance, during the mid and late 1970s, the CEGB announced that electricity from its magnox reactors was cheaper than that from other generators. 'The Board's nuclear power stations produced 11.4 per cent of the electricity supplied by CEGB power stations during the year, and the electricity they produced was the cheapest on the system,' the CEGB proclaimed in its 1979–80 *Annual Report*. In Appendix 3 of that same report, the CEGB indicated that electricity generated by the magnox stations had cost 1.3 p/kWh, while that from coal-fired and oil-fired stations respectively had cost 1.56 and 1.93 p/kWh. In the following *Annual Report*, that for 1980–1, the CEGB indicated that electricity from magnox stations was 0.2 p/kWh cheaper than electricity from coal-fired stations. Electricity from the new advanced gas-cooled reactors (AGRs) was said to be 0.4 p/kWh cheaper.

As Dr Colin Sweet shows in his book *The Price of Nuclear Power*, the costing of electricity from the CEGB's magnox stations had gone completely awry.[1] Conventional thinking on nuclear power believed that the high capital costs of nuclear power stations would be more than offset by cheap nuclear fuel cycle costs – the overall cost being cheaper than the capital and fuel cost components of coal-fired stations. For instance, in the mid-1970s, Sir John Hill, then chairman of the UKAEA, stated that while capital and operation charges for a nuclear power station were 0.41 p/kWh, fuel costs were just one-fifth of the total – 0.12 p/kWh. But during the latter part of the 1970s, nuclear fuel costs began to soar, primarily because of the sharp increases in reprocessing charges; by 1979–80, according to the CEGB's own *Statistical Yearbook*, they had reached more than 90 per cent of total generating costs. Thus, of the 1.3 p/kWh given for magnox generating costs, 1.19 were fuel costs. In effect, the published figures showed that while fuel costs had increased eight-fold, capital and operating costs had decreased by a factor of nearly three, falling from 0.28 p/kWh in 1971–2 to 0.11 p/kWh in 1979–80.

No Account Taken of Inflation

Such topsy-turvy figures, which make a complete nonsense of conventional thinking on the economics of nuclear power, were the result of the CEGB's statisticians massaging the figures to make nuclear power appear the best economic possibility for its thermal generating plant. The figures relied on 'historic cost' accounting: in effect, no account whatsoever was taken of the effects of inflation, since construction costs and fuel costs were not adjusted to present-day values. Indeed, a £100 per kilowatt difference in the capital cost of the two kinds of plant in the 1960s would appear insignificant in the money of 1980, unless adjusted to take account of inflation. Much of the uranium fuel for the initial loading of reactors was acquired and paid for when the magnox programme was launched; consequently its cost, which was generally incorporated into the capital cost component, would not be a true guide to the cost of purchasing that fuel today.

The CEGB is Taken to Task

The CEGB's attempt to misguide the public and government on the true cost of nuclear power was revealed in 1980, when the House of Commons Select Committee on Energy began investigating the Board's accounting procedures. After hearing evidence from the CEGB, the Committee commented: 'The historic cost method used by the Board to justify past investments distorts the effect of inflation on capital costs, rendering the resultant figures highly misleading as a guide to past investment decisions and entirely useless for appraising future ones.' The CEGB at that time was in the process of launching its campaign to promote the Sizewell B pressurized water reactor (PWR) and one of its main arguments was the significant savings to be had should the PWR be built 'in advance of need'.

Adjusting the Figures for Inflation

Undoubtedly one of the most devastating critiques of the CEGB's presentation of its comparative generating costs came from Professor Jim Jeffery, a crystallographer at London university. His analysis formed the basis of the report *Nuclear Energy – The Real Cost*, which was commissioned by *The Ecologist* in 1981.[2] Jeffery disintangled the jumble of different cost figures and then adjusted them for the effects of inflation using the retail price index. He was thus able to show that the cost of building the CEGB's magnox reactors (in constant money brought up to 1979–80 values) was four times greater than indicated in the CEGB's published figures (even without taking interest during construction into account). The adjusted figures brought the magnox generating costs up from 1.3 p/kWh to 2.25 and coal-fired costs up from 1.56 to 1.75 p/kWh – a considerable switch from the data in the 1979–80 *Annual Report*. The difference between coal and nuclear widened still further when Jeffery took account of the considerable increases in the cost of reprocessing magnox fuel which had taken place during the latter part of the 1970s, as a result of serious problems associated with spent fuel corrosion in the cooling ponds.

A similar exercise on the generating costs of the CEGB's most successful operating AGR – Hinkley Point B – indicated that far from giving cheaper electricity than Drax A coal-fired station, as the CEGB claimed, Hinkley Point B was some 40 per cent more expensive to have built and operate. The other AGRs – and especially Dungeness B, with their substantial cost overruns – were even less economic.

More than a year after *The Ecologist's* report, in February 1983, when the Sizewell public inquiry was already under way, the CEGB produced its *Analysis of Generation Costs*.[3] The CEGB analysis confirmed *The Ecologist's* figures. Indeed, the generating cost given for magnox, 3.37 p/kWh, and for coal, 2.28 p/kWh, were very close to those published a year earlier in *The Ecologist* when updated to March 1982 prices.

New Plants, New Costs

Clearly there comes a time when older plant must be replaced. For a station that has not yet been built, a number of variables and assumptions have to be considered before any conclusion can be drawn as to its economic viability. How much will the station cost? Will it be built to time, since delays will lead to extra interest charges and probably capital charges too? What will be its fuel cost, initially and over its lifetime, and how long will it operate? Will it operate to expectation? What rate of interest should be applied to the capital investment? And what rate of return should be expected? Equally, similar questions have to be asked of alternatives to the project in mind.

The Sizewell Inquiry

It does not necessarily follow that new plant, of whatever kind, will produce cheaper electricity. It may well be more expensive. But that is not how the CEGB presented its case for building a pressurized water reactor at Sizewell on the East Anglian coast.

In the statement it put foward to support the case for the Sizewell PWR, the CEGB gave three main reasons why new plant was required and why it should be a PWR. First, a considerable amount of generating plant would become 'time-expired' within the first decade of the new century and would have to be replaced. Secondly, an improvement in world economic growth in general, and in Britain in particular, with electricity gaining a greater share of the energy-use market, would mean that more plant would have to be built than was simply needed to replace decommissionings. And thirdly, the introduction of the Sizewell PWR into the system would lead to substantial savings. Not only would the PWR be the cheapest plant to operate within the system, the CEGB stated, but the savings to be made in its operation would make it worth while to construct the plant before it was needed – 'ahead of need' – so that more costly plant could be withdrawn from the system.

One of the main planks of the CEGB's economic case for the PWR was the cost of coal and the savings to be made through reducing the overall coal-burn of the generating system, in particular through the phasing out of thermally inefficient, older plant. Indeed the CEGB's argument leant heavily on the speculation that the cost of coal would rise sharply during the first few years of the PWR's operation during the 1990s, and would continue its rise throughout the 35 year lifetime given for the reactor. As Professor Jeffery pointed out in his evidence to the Sizewell inquiry, the effects of discounting would mean that a sharp rise in the cost of coal during the first few years of the Sizewell PWR's operation would prove far more effective in boosting the advantage of the nuclear plant in terms of coal saved than any rise which might come later. Thus a 40 per cent increase in the cost of coal coinciding with the commissioning of the PWR would be equivalent in terms of present value savings – hence discounted, annuitized savings – to a trebling of coal prices over the entire operating lifetime of the PWR. In that respect it surely cannot be idle coincidence that with each postponing of the forecast commissioning date of the PWR – from 1986 to 1994 – the CEGB has forecast that a sharp burst in the cost of coal – up to 40 per cent – will take place just before commissioning.

The economic viability of a new station does not depend solely on its own costs, but also on those affecting other stations in the system, and thus the CEGB has attempted to calculate the NEC or net effective cost of different plant in the overall generating system. An NEC of zero would effectively indicate that there would be neither advantage nor disadvantage from an economic point of view in building and operating the plant in question over its intended lifetime, given that all the assumptions fed into the analysis were reasonably correct. A negative NEC, on the other hand, indicates overall savings – the reason being that the assessed cost of construction, of interest, of fuel, of maintenence and of operation amount to less than the cost of the fuel needed to fire an alternative plant. Consequently the consumer would pay less for electricity than if the plant were not introduced into the system. A positive NEC would mean more expensive electricity than that generated by the existing system, and factors other than economics would have to weigh in the decision-making.

Coal: The Critical Fuel

In its evidence to the inquiry, the CEGB argued that a new coal-fired station would raise the cost of electricity by £1 billion over its lifetime, owing to the increasing cost of coal. By contrast, the proposed Sizewell B nuclear station would save Britain £3.5 billion because of nuclear power's avowed lower fuel costs.

Decommissioning and Reprocessing Costs Underestimated

Just as costs and savings soon after commissioning register strongly in the present value of the PWR, costs incurred after its final shut-down, put by the CEGB at 35 years, including decommissioning costs, reprocessing of spent fuel and nuclear waste disposal, are whittled away by discounting for 40 years and more. For instance, costs incurred 25 years after shut-down register as one-tenth of full costs, and Jeffery finds that all the post-shut-down nuclear costs comprise no more than one-seventh their full value. Since decommissioning and reprocessing of spent fuel are relatively expensive items, their real significance in the Sizewell NEC is much reduced.

One way around the discrepancy would be to evaluate all post-shut-down costs as if they took place during the plant's lifetime. In my evidence to the inquiry, I used such a methodology. I also suggested, in line with the government sponsored Castaing report on reprocessing in France, that reprocessing was likely to cost at least 40 per cent more than the figure provided by BNFL for its yet unbuilt thermal oxide reprocessing plant (THORP). Environmental considerations alone, which have become far more pressing since the 1977 Windscale public inquiry, were likely to raise costs.

- By far the biggest single factor in the claimed economic benefit of the PWR results from the savings to be made in the displacement of both coal and oil-fired generating capacity. Indeed, the combined fossil fuel savings are given by the CEGB as £221 /kW pa against a total expenditure, including capital, operational and nuclear fuel costs, of + £138 /kW pa.

  As Jeffery argued, the Board had not only weighted the savings by assuming substantial increases in the price of coal to coincide with the first years of the PWR's operation, it had used a 'marginal cost' of coal that was some 18 per cent higher than the cost of National Coal Board coal to a central coal-fired station. Meanwhile, as if it did not want to know what its left hand was doing, the CEGB had recently concluded a further 'understanding' with the NCB in which, in return for taking 68 million tonnes of coal per annum, the price would stay at the same level in real terms as it was in 1980. Furthermore, the NCB promised to make reductions in the price of coal taken above that amount. Rather than being more expensive as marginal cost theory would suggest, 'marginal coal' would be cheaper.

  By taking the CEGB's estimate for the price of NCB coal to a central-fired station rather than its putative higher cost, 'marginal coal' reduces the savings by £57 /kW pa. But even that conclusion is unreasonable, says Jeffery, because of the present 'understanding'. In addition, he argues, that by bringing on stream the Board's new coal and nuclear stations before the PWR begins operation, the need for any oil burning (other than that essential for firing coal stations), would be obviated. Hence no oil will be left in the system to be saved, and the CEGB cannot legitimately incorporate oil savings into NEC calculations. It so happens that the CEGB has postulated that high-cost oil will be 'saved' in those first early years after commissioning the PWR, when such savings will have maximum effect on the NEC.

With no oil left to be saved, and with coal prices stable until 1990, after which they increase by a linear 1 per cent per annum, Jeffery finds that the savings are reduced by another £65 /kW pa and the NEC of the Sizewell PWR swings from net savings of – £83 /kW pa to + £43, suggesting a net loss of £1.5 billion over the station's lifetime.

I also disputed the CEGB's use of a large credit for the uranium extracted from spent fuel, pointing out that the value of any reprocessed uranium was much diminished by the presence of the uranium isotopes, U-236 and U-234, both of which mop up neutrons and effectively poison the chain reaction. Recycled uranium also contains significant quantities of uranium-232, which is a potent emitter of gamma radiation. Extra precautions have therefore to be taken in handling reprocessed uranium and the question arises whether it should have any value ascribed to it.

When a higher cost of reprocessing is taken into account, when the uranium credit, which the Board had made equivalent in value to 40 per cent of the total cost of post-irradiated fuel management, is taken away, and when the process of discounting is carried out within the operational lifetime of the PWR, the effect on the fuel cost is to increase it from + £36 /kW pa to + £56. Decommissioning too increases from + £1 /kW pa to + £9 and the NEC for the Sizewell PWR becomes + £74 /kW pa, indicative that the extra lifetime cost to Britain of introducing the reactor will be some £3 billion.

Coal: The Subsidised Industry

At the Sizewell inquiry, the CEGB attempted to give reasons for its expectation that fossil fuel prices, mainly coal and oil, would rise at the rapid rates forecast. With regard to the present price of NCB coal, the CEGB remarked that it has been held down only 'with the help of significant and increasing deficit grants from the government'. It went on to argue that if such subsidies were removed, but the level of social grants maintained, the current pithead price would rise by more than 10 per cent. In its evidence to the Monopolies and Mergers Commission in 1980, the CEGB told of its expectation that the government grants to the coal industry would dry up during the 1980s, and that the NCB would therefore be forced to achieve a measure of profitability. Consequently the CEGB expected the pithead price of coal to rise by some 4 per cent per annum from 1980 to 1987, and subsequently by 2 per cent per annum until the end of the century.

The Miners' Strike

The crippling coal strike of 1984 over the question of pit closures and of what should be termed 'an uneconomic pit' certainly indicated the present government's intention to pare the coal industry down to suit the market conditions of today. The abandoning of older, 'uneconomic' pits would leave millions of tons of coal irretrievably underground; yet, if all worked out as the government intended, a coal industry would be generated that should appeal to the private sector. Breaking the power of the National Union of Miners would have to be a first, essential step in any plans for the future privatization of the coal industry, and the government clearly had every intention of succeeding in its aim – despite the enormous social and economic cost to the country. We hear of massive losses in the steel, railway and electricity supply industries because of the strike – £1.75 billion in the latter alone – and that does not take account of the policing of the strike, nor of the substantial loss of earnings of the miners themselves.

Walter (now Lord) Marshall, chairman of the CEGB, has stated that Arthur Scargill succeeded in promoting the case for nuclear power as he himself could never have done. That simplistic statement would perhaps have been better founded had the miners struck for more pay. Rather, they struck for their jobs and the saving of their pits from closure. Nor did Marshall add that one reason for the decline in the fortunes of the British coal industry has been the coming on stream of the AGR stations, each one of which displaces annually some three million tons of coal from total requirements. In addition to the five AGRs now working, two more are under construction. Indeed, if total demand for electricity in England and Wales stays at its present level of around 210 terawatt-hours per year, then with the Sizewell B PWR in operation, total coal requirements could be down to 55 million tonnes per annum, well below the record 80.6 million tonnes consumed during 1979.

Coal Imports

A prime reason for the deficit grants paid by the government to the NCB is the need to bring its average coal costs in line with cheaper imports so that the electricity boards will continue to restrict their consumption of overseas coal and maintain a high burn of NCB coal to support an indigenous industry. In its evidence at the Sizewell inquiry, the CEGB maintained that world demand for coal would put up the cost of imported coal to levels even higher than that of British coal – even with the deficit grant lopped off. Heavy oil would also show a massive price increase, according to the CEGB. Thus the CEGB projected that heavy oil would rise in price from its 1980 level of 237 p/gigajoule (gigajoule = 10^9 joules) to 570 p/GJ in 2000 – and to as high as 760 p/GJ in 2015. The price of internationally traded coal delivered to the Thames was projected to rise from 120 p/GJ in 1980 to 300 p/GJ in 2000 and to 450 p/GJ by 2030.

Even with demand for oil growing by 2.5 per cent per annum, Professor Peter Odell of Rotterdam university, giving evidence for the Town and Country Planning Association (TCPA), suggests that, over the next 30 years, oil need never reach or surpass the highest ever price of $30 per barrel paid in 1981. In his view: 'There is little more than a one in ten chance that oil prices will be as high even as the lowest oil prices which the CEGB uses as the basis for its calculations.' Mr Steenblik, also giving evidence for the TCPA, testified that international coal prices may fall in real terms by as much as one-sixth by 2000 from the 1980 figure of $60 per tonne. He expected steam coal prices 'to rise very little in real terms for at least the first 30 years of the 21st century'.

Should such energy experts be proved right – and they have given far more accurate forecasts of the oil and coal markets over the past decade than has the CEGB – then a main plank for nuclear power in Britain will effectively have been destroyed.

Capital Costs – Reality Rather Than Fiction

The CEGB's assessment of the capital costs and of the time taken for construction, as well as of the performance hoped for from the Sizewell B reactor, also came in for criticism, particularly from the Electricity Consumer Council, the TCPA and the Council for the Protection of Rural England (CPRE). With US experience of building Westinghouse reactors in mind, the CPRE claimed that the CEGB would be lucky to build the Sizewell PWR in less than 108 months, compared with the 90 months given as a central estimate in the CEGB's statement of its case. The CPRE also claimed that the station would probably cost just under £1,600 million (March 1982 prices) compared with the Board's central estimate of £1,147 million. The addition to capital costs alone would add £35 /kW pa to the NEC. With a load factor – hence the degree to which the plant is used over its lifetime – of 58 per cent, instead of the 64 per cent given by the CEGB, and a lifetime of 25 years rather than 35 years, another £14 /kW pa are added to the NEC.

By the time the criticisms of the various objectors are taken into account, including the increased 'back-end' nuclear fuel and decommissioning costs, the NEC of the Sizewell project swings from a net saving of − £83 /kW per annum given by the CEGB to a net loss of + £92 /kW pa; by comparison the NEC for coal deteriorates from the net loss of + £21 /kW pa given by the Board to + £65/kW pa. Hence, compared with the costs of keeping older plants going on the system the Sizewell PWR would lose more than £3 billion. By comparison, a new coal-fired station would lose close to half that amount.

In essence, objectors to the Sizewell B PWR argued that the project would be considerably more costly to the Board and to the electricity consumer than pursuit of other alternatives – including conservation, refurbishment of plant when it reached the end of its life, a proper planning for a non-growth situation, and the use of energy efficient methods such as combined heat and power. The belief that the kind of growth envisaged and hoped for by the government and the CEGB would be unlikely to materialize was expressed by all objectors; even were a station the size of the Sizewell PWR to be built, the very earliest it would be needed would be the end of the century.

Colin Sweet, a witness for the TCPA, pointed out the crippling financing that would be required to meet the expectations of the CEGB's 'middle-of-the-road' scenario. £1.75 billion would have to be spent on average each year for at least 17 years:

> If these capital expenditures were added to the non-nuclear capital expenditure currently being undertaken by the Board, then the capital requirement annually could be in excess of £3 billion by the middle 1990s (in 1982 prices) . . . In broad terms a requirement of this magnitude would equal about 30 to 40 per cent of the level of fixed capital investment for the entire UK manufacturing sector of the economy.

Nuclear Power, France's Waterloo

The French example is often cited as a major success story, whereby an aggressive nuclear power programme has not only brought down fossil fuel consumption, but has made electricity cheaper. The real facts tell a very different tale. To begin with, the main purpose of the nuclear power programme, which has led to more than 50 per cent of France's rapidly expanded electricity consumption being met by nuclear power in just over a decade, was to reduce substantially imports of crude oil. Yet whereas between 1973 and 1982, France succeeded in reducing its total oil consumption by 27 per cent, the reduction in Britain over the same period was 33 per cent, and in Denmark 35 per cent. And while electricity prices doubled between 1975 and 1984 – primarily because of inflation – in Britain and the Netherlands, and rose by a smaller amount in West Germany, in France they tripled. With its losses of up to 8 billion francs in 1982 – and substantial losses in all years since 1975 – Electricité de France (EdF) is the only electricity supply industry in western Europe and the USA to have consistently made a loss in its trading. The 200 billion francs borrowed by EdF for its nuclear power programme has also made France one of the heaviest borrowers of foreign exchange in the world.

Given that history, it is clear that economic catastrophe could be around the corner if Britain pushes ahead with its nuclear programme. Obsession with nuclear power, and determination to crush the National Union of Mineworkers, have clouded the government's judgement as to what is best for Britain. At the Sizewell inquiry, there was no proper appraisal of alternative strategies for supplying electricity. No one, for instance, tackled the question as to what would be the net effective cost of introducing small combined heat and power (CHP) coal-fired plants into the system; neither were the NECs of other energy supply alternatives considered.

The economic answers are in fact staring the CEGB and the government in the face. They include an active conservation policy with emphasis on improved end-use of energy; a refurbishing when necessary of older coal-fired plant with the possibility of introducing fluidized bed burning, or at any rate some method of controlling flue gases; investment in CHP schemes, and in alternative energy, including wind power. Meanwhile the CEGB should retain its oil-fired capacity, both as a stand-by and for use when the cost of low sulphur heavy oil falls to economic levels.

Notes

1. Colin Sweet, *The Price of Nuclear Power*, London: Heinemann, 1983.
2. Committee for the Study of the Economics of Nuclear Electricity, *Nuclear Energy – The Real Cost*, Report published as a special issue of *The Ecologist*, vol. II, no. 6, 1981.
3. Central Electricity Generating Board, *Analysis of Generation Costs*, London: CEGB, 1983.

27

Britain and Plutonium Exports

Peter Bunyard

Official sanction and support for the civil use of nuclear power came as a direct result of fears, particularly within the USA, that both knowledge and the actual manufacture and testing of nuclear weapons were proliferating ungovernably. By the early 1950s, the Soviet Union, Britain and France had developed nuclear weapons and China followed in the 1960s. The US lead, gained through the Manhattan Project, was rapidly being eroded and the US administration came to the conclusion that the promotion of a peaceful nuclear programme, under international surveillance, would legitimize nuclear power and make the concealment of bomb-making intentions less feasible. In his speech to the United Nations in December 1953, President Eisenhower called for: 'the fearful trend of military build-up to be reversed', and suggested that an International Atomic Energy Agency be set up to establish a bank of fissionable material which: 'would be allocated to serve the peaceful pursuits of mankind. A special purpose would be to provide abundant electrical energy in the power starved areas of the world. Thus the contributing powers would be dedicating some of their strength to serve the needs rather than the fears of mankind.'

With the coming into being of the International Atomic Energy Agency four years later in 1957, and a ready willingness on the part of US engineering manufacturers to win contracts for nuclear projects, the stage was set for the rapid promotion of a worldwide civil nuclear industry. The myth that the civil and military aspects of the atom could be kept apart, and, moreover, that the surveillance system would be sufficient in itself to pick up transgressors, had been engendered and with rare exception was to remain the accepted dogma until the early 1980s.

Nuclear Power's First Casualty

In fact, the boost given to a civil nuclear industry, both in terms of government subsidies and of technology, was the result of military developments in the first decade after the Second World War. In the USA, Westinghouse and General Electric set out to adapt light water reactors, that had been designed for powering nuclear submarines, for land based electrical power generators. Other types were also tested, including Walter Zinn's experimental breeder reactor – EBR 1 – which first generated some 45 kilowatts of electricity a few days before Christmas 1951. It did not take long for the hazards of civil nuclear power to be realized. Four years later the enriched uranium core of EBR 1 had melted into a solid mass, total destruction of the reactor having been averted in the nick of time.

The British government quickly perceived that the promotion of a civil nuclear industry would aid its commitment to nuclear weapons, both in terms of finance and of keeping public opinion quiet. For the most part that policy seems to have worked; indeed when the Campaign for Nuclear Disarmament (CND) was calling for unilateral disarmament during the 1960s and early 1970s, no one in the movement called into question the role of Britain's nuclear power stations in furbishing plutonium for defence. On the contrary, CND then accepted the official line that besides generating cheap electricity, nuclear power was thoroughly peaceful – truly the turning of swords into ploughshares.

Windscale's Costly Piles

Britain's two plutonium-producing piles at Windscale, both in operation during the early 1950s, were proving particularly costly to operate – the pumps to blow vast volumes of coolant air through the piles were alone consuming some £340,000 worth of electricity each, and that in money worth 10 times more then than it is today. The reprocessing plant, sited next door, was also an expensive item in the fuel cycle, and not surprisingly British scientists and engineers under the guidance of Sir Christopher (later Lord) Hinton were soon to come up with a reactor design – to be called 'magnox' because of the magnesium alloy used to encase the fuel – that would generate electricity and plutonium at the same time. For the British government the opening of Britain's first magnox station at Calder Hall in October 1956 could hardly have come at a more opportune time; one year later the Windscale number one pile caught fire, and both it and the number two pile were from then on permanently shut down – the full extent of the radiation leak only coming to light some 25 years later. The Queen described Calder Hall as a first in the production of electricity for civilian consumption. Yet the real importance of Calder Hall, and later of Chapelcross over the border in Scotland, was for the plutonium (now heavily subsidized by the electricity consumer) that would be generated for military purposes – at least until the electricity boards' magnox reactors were in operation during the 1960s and 1970s.

The Non-Proliferation Treaty

In 1970 the Nuclear Non-Proliferation Treaty (NPT) came into force, with Britain a signatory. The assumption ingenuously but vigorously sustained by each successive British government was that the Electricity Boards' nuclear reactors would come completely under the inspection of officials from IAEA. After 1978, Euratom took over the administration of peaceful uses of the atom for EEC countries. While some may have assumed that official surveillance of Britain's civil nuclear programme would be a sufficient guarantee of Britain's adherence to both the spirit and letter of the Non-Proliferation Treaty, reality has proved somewhat different. Thus, John Moore, Under-Secretary at the Department of Energy claimed on 24 January 1983 that a comprehensive system of international safeguard agreements had been set up to minimize the risks of the misuse of civil nuclear materials for military purposes. Yet, Dr Hans Blix, director-general of the IAEA, was more sanguine, stating in March 1982: 'My 140 IAEA Inspectors cannot possibly keep track of the 15,000 tons of nuclear material housed in 350 sites world-wide. You cannot stop proliferation by safeguards.' Clearly if a country wants to evade control, it will. Indeed, 11 days after the Israelis had destroyed the Iraqi research reactor at Tammuz, Dr Roger Richter, the IAEA inspector responsible for the Middle East, resigned because safeguards were 'not adequate' to detect violations by Iraq.

Britain, meanwhile, has proved to be a prime 'evader'. For more than a decade the British government has consistently denied allegations that plutonium from civil reactors could have found its way into nuclear weapons. On the other hand, it has admitted that over the years small quantities of plutonium have been exported – for peaceful purposes – to the USA under a bilateral agreement stretching back to the 1950s.

The first inkling that all might not be as the British government would wish us to believe came in a letter to *The Times* on 30 October 1981 from Dr Ross Hesketh, a senior nuclear scientist at the CEGB's research laboratories at Berkeley in Gloucestershire. Hesketh was voicing his concern over an agreement between Margaret Thatcher's government and the Reagan administration for the resumption of the export to the USA of British plutonium. According to official statements, the plutonium was to be used in the USA's fast reactor programme – seemingly a civil use of the material. The agreement for the export of plutonium to the USA coincided, however, with an announcement by the US Department of Energy – which designs and builds nuclear weapons for the Department of Defence – that there was a shortage of plutonium for weapon-building. The US Department of Energy not only announced its plans to use plutonium from the spent fuel of several research reactors for weapon manufacture, but in addition some 17.8 tonnes of plutonium destined for the breeder reactor at Clinch River, Tennessee, should that project ever get off the ground. The British plutonium would therefore go, the House of Commons was told, to replace some of the Clinch River fast reactor fuel.

Indeed, the British government insisted that there was no connection between the civil and military programmes in terms of plutonium export, nor had there ever been. For instance, in a parliamentary written answer on 1 April 1982, John Moore revealed that between 1964 and 1971 some plutonium from the electricity boards' reactors had been sent to the USA. But, he emphasized, 'The alleged linkage between the CEGB's nuclear power programme and nuclear weapons is wholly without foundation.' And in a statement to the press on 20 January 1983, his senior, Nigel Lawson, then energy secretary, stated: 'There is no more connection between the generation of power in a nuclear power nation and weapons than there is between a conventional power station and conventional weapons.'

While John Moore was still busy reiterating that no plutonium from the CEGB's nuclear power stations had ever been used 'for military purposes in this country . . . or exported for weapons' purposes either to the US or anywhere else', the nature of the specific agreement between Britain and the USA for the export of fissile material was coming to light. Indeed, within a few years of Eisenhower's speech to the United Nations calling for the peaceful use of the atom, Britain had entered into a bilateral agreement with the USA to exchange nuclear material for each country's respective weapons programme. The agreement permitted the two countries to share information and to transfer material derived from their reactors for the sole purpose of developing nuclear weapons.

Through parliamentary questions and through information released in the USA through the Freedom of Information Act, it has now become clear that Britain may have exported four tonnes of plutonium between 1964 and 1971 – enough for some 2,000 warheads – in return for enriched uranium and tritium, the former providing fuel for Britain's nuclear submarines and the latter for its thermonuclear bombs. If all the plutonium transferred to the USA from Britain came from Calder Hall and Chapelcross – the original magnox reactors for which a dual military/civil role had never been denied – then John Moore's statement would at least have a shred of honesty in it. Not surprisingly he was fervently backed up by the CEGB. John Baker, secretary to the Board, for instance, proclaimed:

> The current position is accordingly quite clear: no plutonium produced in CEGB reactors has been applied to weapons use either in the UK or elsewhere, and it is the policy of the government and of the CEGB that this situation should continue. The CEGB has no reason to believe that these policies will change in the future.

Where Has All the Plutonium Gone?

In *Science and Public Policy*, April 1982, Ross Hesketh, who was then in the midst of a struggle for his job with the CEGB, points out that Britain's civil reactors have produced between two and three times more plutonium than her military reactors at Calder Hall and Chapelcross.[1] On the basis that Calder Hall and Chapelcross have been operated on a mode that maximizes plutonium production, Hesketh suggests that they may have produced as much as 18 tonnes up until 1981 – or at the very least 11 tonnes – taking account of all plutonium isotopes. The civil reactors are likely to have produced as much as 35.4 tonnes, again taking account of all isotopes. Not all that plutonium will yet have been separated from the rest of the spent fuel through reprocessing.

If as much as four tonnes of plutonium were exported to the USA prior to 1974, and used for mutual defence purposes as ordained in the agreement, the question arises as to whether Britain would have exchanged such a large percentage of its military plutonium when it was building up its own weapons inventory. On the contrary, it seems particularly likely that civil plutonium from what should have been 'safeguarded' sources will have been used. Indeed, civil reactors are excellent generators of weapons grade plutonium during their first year of operation, and it must be tempting to divert that plutonium into the military stockpile.

Furthermore, whereas the Calder Hall and Chapelcross reactors have to be refuelled when shut down, the electricity boards' magnox reactors can be refuelled while still in operation. In theory at least, such on-load refuelling would allow fuel elements to be withdrawn when the plutonium-239 content is still proportionately high. In practice, such early withdrawal of fuel would be uneconomic – and therefore contrary to the electricity boards' obligation to generate electricity as cheaply as possible. On the other hand, as Ross Hesketh points out: 'In regard to load factor and in regard to consequent economics, it is preferable to use the civil system for brief irradiations such as would produce weapons-grade plutonium, and it is preferable to run the military system on the longest possible cycle, i.e. the civil cycle.'

Thus if economics are the yardstick of how best to operate all the magnox reactors – civil and military – to generate both electricity and sufficient weapons grade plutonium, the optimum mode of operation would be to maximize electricity production fronm Calder Hall and Chapelcross – and to use some of the civil reactors for weapons grade plutonium production. In 1958, it was revealed that the CEGB had been asked to modify the design of the refuelling machinery of three of the planned magnox stations to make it easier to obtain weapons grade plutonium from them. 'A most valuable insurance against future possible defence requirements,' was how one minister put it. One year later, the plan was altered and, according to the government, only Hinkley Point would be modified. The government has since claimed that Hinkley Point was never used to produce military plutonium.

A Slip of the Tongue

On 21 December 1981, John Moore stated in the House of Commons that 1,280 kilograms of plutonium had been exported from Britain to the USA since 1971 'all of it of civil origin'. Mr Moore made his announcement 'choosing my words with great care'. Three years later, in 1984, the US energy secretary, Donald Hodel, remarked: 'In summary, the United States received several shipments of plutonium, through barter, from the United Kingdom military stockpile between 1971 and the present, which were used for United States weapons.'

As a result a BNFL spokesman told the Sizewell inquiry that Mr Moore's words were 'a slip of the tongue', and that of the 1,280 kilograms, only 500 had been of civil origin while the rest had come from Calder Hall and Chapelcross.

Revelations at Sizewell

The drama associated with the accusation that the British government has since 1961 allowed civil plutonium to be used for weapon manufacture in the USA was enhanced by several revelations during the Sizewell inquiry. A few months before his death in 1983, Lord Hinton was interviewed by

David Lowry of the Open University, concerning his role in the launching of Britain's nuclear power programme. When asked about the CEGB's claim that no plutonium from its reactors had been diverted for weapon making, Hinton replied on tape: 'I am absolutely certain that the CEGB statement is incorrect . . . I don't know whether they should get permission for a PWR at Sizewell or not, but what is important is that they shouldn't tell bloody lies in their evidence.'

To compound the confusion the CEGB later revealed – on Day 193 of the inquiry – that the memory of what had happened to its plutonium stores up until 1977 had been erased 'by mistake' from its computer records. It then emerged, towards the end of the inquiry, that not only was BNFL's reprocessing plant at Sellafield out of bounds to Euratom inspectors, but that none of the CEGB's magnox reactors was designated for inspection. Indeed, as Hesketh and Lowry pointed out in the *Guardian*: 'No nuclear reactor in the UK is under IAEA safeguards.'

During cross-examination by CND at the inquiry, BNFL's Hugh Sturman, who is responsible for security at Sellafield, admitted that Euratom was 'unhappy' that its inspectors had no physical access to the magnox reprocessing line, where both safeguarded fuel from civil reactors and unsafeguarded fuel from the military reactors is reprocessed. Indeed, BNFL points out that the fuel is often coprocessed – a mixture of the two sources passing through the plant at the same time. Nonetheless, it claims that the extracted plutonium from such coprocessing is not used for military purposes. According to the British Department of Energy, the British government had already made it clear to Euratom in 1973 that coprocessing 'could make the negotiation of a mutually satisfactory inspection regime for the plant very difficult'. Finally, in their evidence for CND, three scientists, all members of SANA – Scientists against Nuclear Arms – claimed from their calculations based on published data that as much as three tonnes of plutonium were missing from official figures – and that the missing plutonium might well be in the USA, taking the total sent there from Britain to between six and seven tonnes.

Some Plutonium is Missing

Although secrecy and the destroying of records make it extremely difficult to know the whole truth about Britain's plutonium inventory, we can no longer doubt that the public has had the wool pulled over its eyes. Knowing what we now do about the ineffectiveness of safeguards and the way in which the British government has prevented the IAEA, through Euratom, from overseeing compliance with the Non-Proliferation Treaty, we can only wonder in astonishment at John Baker's remarks on behalf of the CEGB at the inquiry when he stated:

> The UK has voluntarily entered into safeguards arrangements which underline the government's and the CEGB's intention that plutonium produced in CEGB reactors will not be diverted for weapons purposes. The UK supports the aims of and is a signatory to the Non-Proliferation Treaty and as a result of this and our membership of the European Community, UK nuclear power stations and their operating records are subject to inspection by both the IAEA and Euratom to verify that there is no diversion of plutonium to weapons use.

The NPT is now up for review. Britain has always been quick to point out that she is a responsible nation. At the Windscale public inquiry of 1977, for example, it was suggested that only countries such as Britain should have reprocessing plants, in so far as Britain was committed to non-proliferation and could therefore be expected to carry out her duties responsibly. With Britain now seen to be flaunting the rules and obligations to non-proliferation how can she expect other nations to behave more honourably?

Notes

1. Ross Hesketh, 'The Export of Civil Plutonium', *Science and Public Policy*, April 1982, p. 70.

28

Overprescribed

Drugs and National Health

Charles Medawar

The greatest strength of Britain's policy on medicinal drugs is also its greatest weakness. Through the National Health Service (NHS), everyone has access to a relatively effective health care system and to the medicines they really need. On the other hand, we are a grossly over-medicated society, dangerously over-dependent on drugs and doctors for our health and well-being.

This is damaging to individual health and crippling for the NHS. Our obsession with health, together with loss of confidence in our own ability to attain it, 'is bad not only for the spirit of society; it will make any health-care system, no matter how large and efficient, unworkable.'[1] As C. B. Zealley, the chairman of Social Audit, points out:

> We are not sensible about drugs. We use too many drugs and expect too much of them. We have been dazzled by the relatively few really effective products that have emerged mostly in the last 50 years. But instead of accepting these gratefully and recognising their limitations, we use drugs as if they could provide an answer to all our ills . . . This expectation is foolish but understandable, since we all fear sickness and pain. But it has led us, our doctors, the drug manufacturers and our governments into a mess of which no-one can be proud. Our credulity as patients, and its legal exploitation by drug companies, are turning doctors into a conduit for drugs to patients – not of course all doctors, not all patients, nor all drug companies – nor all of the time. But the phenomenon is real, widespread and too well established.[2]

Few would disagree with D. M. Davies, editor of *The Textbook of Adverse Drug Reactions*, when he writes, 'There can be little doubt that much modern medicinal treatment is unnecessary.'[3] Moreover, over-medication directly contributes to an appreciable amount of serious illness:

- The Royal College of Physicians reported in 1984 that excessive prescribing 'constitutes a major cause of adverse reactions' – and notes that perhaps one in ten of all elderly patients are admitted to hospital solely or partly because of adverse drug reactions. An estimated 3–5 per cent of *all* hospital admissions are wholly or partly due to adverse experiences with drugs.[4]
- A study of antibiotic treatment in Scottish hospitals (the results broadly reflect practice wherever such drugs are used) reveals that treatment was of doubtful value in two-thirds of all cases. Since 11 per cent of all exposures to antibiotic drugs involve undesirable side-effects, the risks of treatment outweighed the benefit for many of the people treated.[5]

The prescribing of many other classes of drugs (for example, tranquillizers, cough medicines, and laxatives) is well documented as being profligate. This contributes to ill-health; and also involves substantial waste. The measure of this waste is not money, but human and health resources. We waste probably hundreds rather than tens of millions of pounds on unnecessary and undesirable medicines – and we need that money for other things. Thus:

- If doctors stopped prescribing the most expensive brands of drugs – and instead prescribed unbranded ('generic') versions of the identical drug – Britain could save at least £30 million a year. This is enough to give 1,500 people a three year training to qualify as state registered nurses.
- The NHS spends an estimated £30 million a year on the so-called 'peripheral vasodilator drugs', though there is no good evidence that they even work.[6] For this money, we could increase by about 15-fold the budget of the Health Education Council for campaigning on smoking and ill-health.
- Three-quarters of a million unwanted drug tablets were collected by chemists in Dudley, Hertford and Worcester, during a three week 'drug amnesty' held in 1982 – and this was nearly four times as many as the number collected in the 'drug amnesty' organized in these towns six years before. How many kidney machines could have been bought for that?

At the root of all this waste is a complex of relationships involving consumers, prescribers, producers and government. Our medicines policy reflects the needs and demands of each – as well as the abuse of power by some parties, and the failure of others to control it.

Consumers

In an average year, the 'typical' GP sees 600 people with coughs and colds, 325 with skin disorders, 300 suffering from depression and anxiety, 100 with chronic rheumatism, 50 with high blood pressure, 8 with heart attacks, 5 with acute appendicitis and 5 with strokes.[7] The estimates vary, but about one-quarter of all consultations with patients are thought by doctors to be 'trivial, inappropriate or unnecessary' – colds, cuts and dandruff being mainly to blame.[8]

Consumers constantly demand medication, but typically know very little about what medicines can and cannot do for them. About half of all patients do not take medicines as directed, including perhaps 20 per cent who never get their prescriptions dispensed at all. Today's consumer is, in short, the product of yesterday's views on health education – the underlying concept being that 'only a few people knew certain facts and that the majority of the population knew little or held the wrong views.'[9] Even today, health education programmes seem more concerned with telling people what is good for them – rather than helping them find out for themselves.

At the same time, consumers have been exposed to a flow of partial information, encouraging them to put their trust in doctors, drug manufacturers and government – and to have full confidence in drugs. Thus, present British policy on medicinal drugs is pursued in the name of consumers and at their expense, but it is premised on the view that people cannot control their own health – and need not (and may not) question even how it is controlled.

Doctors

Doctors have traditionally blamed patients for the overprescribing of drugs. Many probably still share the view of the Southport GP who challenged his colleagues: 'Try listening to some dopey witterer and send him off with a catechism of advice and no FP10 in his hot little hand.' This particular doctor and many of his colleagues prescribe drugs as a matter of course: 'In 30 years of general practice I have only on three occasions managed to see patients out of my surgery without a scrip – the talisman, the reason for the visit and the evidence of the recognition of the seriousness of their complaint.'[10] On average, doctors prescribe in about four out of every five consultations though, in a few practices, the proportion of consultations ending with a prescription may be as low as 15 per cent. There is little evidence to go on, but nothing to indicate an association between low prescribing levels and ill-health.

What doctors prescribe, and how they prescribe, is adversely affected by two main factors. First, they get little training in effective prescribing; and secondly, doctors are exposed to a formidable amount of promotional activity:

- Medical students are still taught very little about the principles of drug use and drug effects; and learn little about assessing drug effectiveness and safety in clinical trials. The most recent official report on effective prescribing concluded that much more basic training in drug use was needed by medical students and post-graduates alike.[11]
- The drug industry spends, on average, £5,000 per GP, per year, on promoting its products – about 40 times more than is spent (mainly by government) on providing doctors with independent information about the *comparative* value of different drugs, and the use on non-drug treatments.

Prescribing errors are commonplace; and the level of prescribing of inferior or inappropriate treatments is high.[12] At the same time, the detection and reporting rates of adverse drug reactions (ADRs) are very low.[13]

The real failure of the medical profession lies in its refusal to recognize its limitations. In particular, the profession denies that the drug industry's promotional activities have any significant adverse effect on doctors' prescribing decisions – in the face of substantial evidence to the contrary. The profession still vigorously defends the 'clinical freedom' of doctors, their right to prescribe what they want in the manner they think best – without having to explain or justify what they do. 'Clinical freedom' implies that all prescribing decisions are above criticism, unless they are plainly bad. There has been no shortage of obituaries for 'clinical freedom' but in general practice, in particular, it remains a bedrock of prescribing practice.

The Drug Industry

The drug industry vigorously supports the medical profession's defence of 'clinical freedom', because restrictions on prescribing threaten to limit drastically the market for drugs. There are now of the order of 6,000 prescribable products on the NHS, (in a variety of strengths and formulations), though this is over ten times the number of drugs that any teaching hospital or individual practitioner would actually ever need to use.

The drug industry defends having a free market in drugs on the grounds that any restriction would threaten future drug innovation. Though there is some justice in this argument, it is absurd to suggest that all innovation should be encouraged and rewarded – because most of it involves no real therapeutic gain. The US Food and Drug Administration, for example, estimates that two-thirds of all new chemical entities (and 80 per cent of all new products) offer 'little or no therapeutic gain' over existing products.[14] Similar assessments by DHSS officials have concluded that drug innovation in Britain has led to the introduction of 'an abundance of analagous drugs . . . not rarely with exaggerated claims for efficacy'.[15]

Equally, the industry defends its very high expenditures on promotion on the grounds that promotion allows companies to differentiate between products, and encourages the expansion of the drug market – or at least allows the market to be maintained at its present size. In general, the industry suggests that promotional activity keeps doctors informed – and that if some of its activities seem excessive, perhaps doctors themselves are largely to blame. For example:

- A senior spokesman for the industry recently admitted that 'there is apparent extravagance in spending by certain companies. The industry's strict code of practice is particularly hard to administer. Doctors are flattered by generous and unusual entertainments provided to attract their attention . . . The recipients of lavish hospitality do not complain. The uninitiated do.'[16]
- Joe Eagle, Director of Marketing at Ciba-Geigy told a London conference of drug industry executives that a study of why doctors attended medical meetings at which 'hospitality' was provided had 'found that 30 per cent of doctors qualified as genuinely interested in the topics, while the rest were just professional eaters'.[17]
- The combination of 'clinical freedom' and a free market in drugs has made effective price control impossible. Since 1950, every enquiry by the all-party Parliamentary Public Accounts Committee has found serious weaknesses in the voluntary price control arrangements between industry and government, and has concluded that drug industry profits have been excessive. The most recent PAC enquiry (1983) found that the return on capital for the pharmaceutical companies was over 5 per cent higher than for industry generally – and increasing in spite of the recession.[18]

There has been increasing criticism, in Britain and overseas, about the cost of drugs and about the proliferation of superfluous or undesirable products – and about the promotional and lobbying activity which underpins the market in its present shape. The industry responds to this in three main ways. First, and in spite of overwhelming evidence to the contrary, the industry stoutly maintains that everything is in order, everywhere. The industry insists that all its products 'have full regard to the needs of public' and that they are promoted with 'scientific information with objectivity and good taste [and] with scrupulous regard for truth'.[19]

Secondly, the industry keeps up a barrage of self-congratulatory propaganda (much of it orchestrated through the medical profession) and lobbies intensively to protect its own. The defensiveness of the industry is legendary, and not infrequently suggests delusions of persecution. Thus, a senior spokesman for the British industry recently warned, 'There are those working in the pharmaceutical industry who are so obsessed by the attacks on it that they doubt whether it can survive into the 21st Century.'[20]

Thirdly, the industry continually emphasizes the (undisputed) fact that it is a very successful part of British industry, which makes substantial contributions to the country's balance of trade. This line of argument has proved very persuasive to successive governments, which have largely allowed the industry its own way.

Government

In 1985, the Conservative government proposed the introduction of a 'limited list' of drugs that would be reimbursable under the NHS. Almost all of the blacklisted drugs were either acknowledged by the medical profession to be 'less suitable for prescribing',[21] or very expensive minor variants of well-established alternatives. Nevertheless, the decision provoked a tremendous row. What went on behind the scenes is unclear. Publicly, the government was smeared in a £1 million industry advertising campaign; while the British Medical Association forbade any of its members to co-operate and also tried to sue the government for violation of EEC law.

This was effectively the first occasion on which any government had opposed the pharmaceutical industry and the medical profession combined – and had it not been for an unusually powerful and obstinate government, it is unlikely that it would have succeeded. Previous governments tended to cringe before both the medical profession and the drug industry. Though concerned about drug prices and about irrational and uneconomic prescribing, British governments have typically indulged both parties, not least because of the value placed on the industry's business performance.

The DHSS acts as the industry's 'sponsoring department' and is formally charged with responsibility for helping it to expand its business here and overseas. This conflict of interest has meant that the NHS has been required to pay excessively high prices for drugs, in order to boost trade.

The relative weakness of government in the face of the drug industry and the medical profession is underlined by the inadequacy of the main drug laws, and the ineffectiveness of their implementation. The main hallmarks of this include chronic secrecy; lack of resources and expertise (particularly in the regulation of industry accounting practices); an ingrained disinclination to regulate (for instance, over standards of promotion and drug advertising); and the exclusion of patient/consumer participation in decision-making.[22]

In general, the criticisms apply also to the main advisory body to the Department of Health – the Medicines Commission and its seven subsidiary committees, including the Committee on Safety of Medicines (CSM) and the Committee on Review of Medicines (CRM). In his excellent appraisal of the licensing and provision of medicines in Britain, Dr Joe Collier particularly criticizes the secrecy in which these bodies operate: this prevents, not least, disclosure of information about the relationship between the commission and committee members and the pharmaceutical industry. Overall, Collier has concluded that the time has come for a comprehensive review of the Medicines Commission and the drug-licensing system:

> The Medicines Commission, the CSM and the CRM now seem to need changes in their legal framework and their methods of operating. I believe it is time to amend the terms of reference relating to safety and efficacy so that the requirements of the committee more closely coincide with those that prescribers need.[23]

The main drug law (the Medicines Act, 1968) has an overwhelming disadvantage in that it requires the registration and licensing authority (the DHSS) to approve the introduction of new drugs *regardless of their relative merit and price*. Pharmaceutical companies wishing to introduce a new drug are required only to demonstrate: (a) that the new drugs are not obviously less safe than other drugs, and (b) that they are 'efficacious'.

In most tests and trials on new drugs, the sample sizes are too small and the length of treatment too short to gain aything other than a general impression of the safety of a drug in use. The DHSS continues to permit claims for the safety of new drugs to be made on the basis of very modest and often inadequate data, though it fully accepts that the safety of the drug can only be proved or disproved in normal use. The recent withdrawal of several anti-arthritic drugs (Opren, Flosint and Osmosin, for example) shortly after their launch, underlines the point that unrealistic claims for the safety of new drugs are made all too often.

To prove that a new drug is 'efficacious', a company need only show that the drug does what it claims to do – though this need not be of benefit to patients. An efficacious drug need not be as effective as alternatives, nor cheaper, nor easier to use. Efficacy falls short of 'effectiveness' (which implies that patients benefit from treatment) and falls far short of meeting the standard of real 'medical need' (implying the best or optimal treatment available).

What Is To Be Done?

Two main reforms are prerequisite to the proper use of drugs: (a) a reduction in the numbers of drugs available for prescribing, and (b) the active involvement of consumers and their representatives in the planning and implementation of drug policies.[24]

There is an overwhelming case for reducing the numbers of prescribable drugs. There is no therapeutic advantage to be gained from having several thousand products on the national drug list when several hundred would meet all real medical need. On the contrary, the use of limited lists is associated with more rational and economic prescribing, and with greatly simplified purchasing, distribution and stocking arrangements.

Reducing the number of drugs is a prerequisite for controlling the abuse of clinical freedom; also for controlling the volume and quality of drug promotion which nourishes it. Clearly, reducing the number of drugs is the first step: there is no point in trying to control the promotion of, or provision of information about, drugs that are not needed in the first place. We should be aiming to reduce the number of drugs to the number we can use well – a fraction of the number we use now. Thus the World Health Organisation has identified just 250-odd 'Essential Drugs', and most of the Nordic countries use one-quarter of the drugs used in Britain. It is worth adding that the drug formulary (limited list) used in the average teaching hospital would contain around 600 drugs – while the average GP prescribes from a limited list of probably no more than 100 drugs in all.

Appropriate arrangements can and should be made at the same time to encourage *useful* innovation. This could be achieved, for example, by extending the patent life on drugs which offer a significant therapeutic gain. Beyond this, consumers (that is, patients) need to become actively involved in the planning and processes of the health care system. As the

World Heath Organisation notes:

> It is realised today that science and technology can contribute to the improvement of health standards *only* if the people themselves become full partners of the health care providers in safeguarding and promoting health . . . people have not only the *right* to participate individually and collectively in the planning and implementation of health care programmes, but also a *duty* to do so.[25]

These high ideals were adopted by Britain and other signatories of the Declaration of Alma-Ata in 1978 – but little seems to have changed since then. The point is made by A. Herxheimer and C. Davies in a recent article in the *Journal of the Royal College of General Practitioners*. Patients, they write,

> need to grasp a few basic concepts relating to medicines. They will understand the physician, the pharmacists and the PPI[26] much better if they understand the notion of a benefit/risk relationship, and that this is influenced by the nature and seriousness of illness, that the body gets rid of drugs, and that the speed with which this happens affects the frequency of doses; that a drug can have effects on the body that are not intended; and that medicines do not keep for ever. We have to find ways of teaching patients these and other elementary ideas, *but have hardly begun to do so* [emphasis added].[28]

We have a very long way to go. Nothing much is likely to be achieved until people are able to understand what their health resources rightfully are:

> We are, in real life, a reasonably healthy people. Far from being ineptly put together, we are amazingly tough, durable organisms, full of health, ready for most contingencies. The new danger to our well-being if we continue to listen to all the talk, is in becoming a nation of healthy hypochondriacs, living gingerly, worrying ourselves half to death.
>
> And we do not have time for this sort of thing anymore, nor can we afford such a distraction from our other, considerably more urgent problems. Indeed, we should be worrying that our preoccupation with personal health may be a symptom of copping out, an excuse for running upstairs to lie on a couch, sniffing the air for contaminants, spraying the room with deoderants, while just outside, the whole of society is coming undone.[28]

Notes

1. L. Thomas, *The Medusa and the Snail: more notes of a biology watcher*, New York: Bantam, 1979, p. 39.

2. C. B. Zealley, personal communication, 1983. The extract quoted was salvaged from notes scribbled on an early draft of C. Medawar, *Drugs and World Health*, The Hague: IOCU, 1984.
3. D. M. Davies (ed.), *Textbook of Adverse Drug Reactions*, Oxford: Oxford University Press, 1981.
4. Royal College of General Practitioners, 'Medication for the Elderly' (report of the working party under the chairmanship of Sir Douglas Black), *Journal of the Royal College of Physicians*, vol. 18, no. 1, 1984.
5. D. C. Moir, L. J. Christopher and D. H. Lawson, 'Data Collection in Scottish Hospitals' in World Health Organisation, *Studies in Drug Utilisation*, Copenhagen: WHO Regional Office for Europe, 1979, pp. 93–102.
6. J. Avorn, M. Chen, R. Hartley, 'Scientific versus Commercial Sources of Influence on the Prescribing Behaviour of Physicians', *The American Journal of Medicine*, no. 73, July 1982, pp. 4–8.
7. Royal College of General Practitioners, *Trends in General Practice*, 1977.
8. F. Fitton and H. W. K. Acheson, *Doctor/Patient Relationship: A Study in General Practice*, Department of Health and Social Security (DHSS), London: HMSO, 1979, p. 53.
9. World Health Organisation, *New Approaches to Health Education in Primary Health Care*, Geneva: Report of a WHO Expert Committee, Technical Report Series, no. 690, 1983, p. 18.
10. C. Swan, 'Doctor knows best', *General Practitioner*, 22 November 1985, p. 33.
11. P. R. Greenfield (chairman), *Report to the Secretary of State for Social Services of an Informal Working Group on Effective Prescribing*, London: DHSS, 1983.
12. C. Medawar/Social Audit, *The Wrong Kind of Medicine?*, Consumers Association, London: Hodder and Stoughton, 1984.
13. J. P. Griffin, 'Is better feedback a major stimulus to spontaneous adverse reaction monitoring?', *Lancet*, 10 November 1985; W. H. M. Inman, 'Study of fatal bone marrow depression with special reference to phenylbutazone and oxyphenbutazone', *British Medical Journal*, no. 1, 1977, pp. 1500–5.
14. *Scrip*, no. 763, 26 January 1983, p. 10. *Scrip* (World Pharmaceutical News) is published bi-weekly by PJB Publications Ltd, 18/20 Hill Rise, Richmond, Surrey, TW10 6UA.
15. J. P. Griffin and G. E. Diggle, 'A Survey of Products Licensed in the United Kingdom from 1971–1981', *British Journal of Clinical Pharmacology* no. 12, 1981, pp. 453–63.
16. G. Teeling-Smith, *The Future for Pharmaceuticals: The Potential, the Pattern and the Problems*, London: Office of Health Economics, 1983.
17. *Scrip*, no. 913, 11 July 1984, p. 1.
18. Public Accounts Committee, *Dispensing of Drugs in the NHS*, London: HMSO, 1983. See also: *Appropriation Accounts*, vol. 8: classes XL and XII, 1981–82, London: HMSO, 1983.
19. *International Code of Pharmaceutical Marketing Practice*, Geneva: International Federation of Pharmaceutical Manufacturers' Associations, 1982.
20. Teeling-Smith, *The Future for Pharmaceuticals*.
21. *British National Formulary*, London: British Medical Association and the Pharmaceutical Society of Great Britain, 1982 (published twice yearly since 1981).
22. J. Collier, 'Licensing and provision of medicines in the United Kingdom', *Lancet*, 17 August 1985, pp. 377–81; Medawar, *The Wrong Kind of Medicine?*.
23. Collier, 'Licensing and provision of medicines'.
24. Medawar, *The Wrong Kind of Medicine?*; C. Medawar/Social Audit, *Drugs and World Health*, The Hague: IOCU, 1984; C Medawar/Social Audit, 'International Regulation of the supply and use of pharmaceuticals', *Development Dialogue*, no. 2, Uppsala, Sweden: Dag Hammarskjold Foundation, 1985.
25. World Health Organisation, *New Approaches to Health Education*, pp. 10–13.
26. Patient Package Insert: a leaflet provided by some medicines, giving patients instruc-

tions for use, etc.

27. A. Herxheimer and C. Davies, 'Drug information for patients: bringing together the messages from prescriber, pharmacist and manufacturer', *Journal of the Royal College of General Practitioners*, no. 32, 1982, pp. 93–7.

28. Thomas, *The Medusa and the Snail*.

29

Fighting Back against Cancer

The Need for a Broad Campaign

Alan Irwin and Doogie Russell

In the early part of this century, infectious diseases were the major killers in our society. For the urban poor, tuberculosis and influenza were a particular cause of fear and suffering. Today, infectious diseases are still the main cause of death worldwide, but in the 'developed' nations they have been superseded by the twin evils of circulatory disease and cancer. Cancer alone is responsible for 20 per cent of British deaths (about 150,000 deaths per year). In just the same way that tuberculosis symbolized the unhealthy living conditions of an earlier age, cancer has been represented as a 'disease of affluence'; just as with tuberculosis, a sense of fatalism has grown about the likelihood of contracting the disease and about one's chances of subsequent survival. We would argue that it *is* possible to 'fight back' against cancer – but only when the political and economic roots of cancer causation are recognized. In particular, a broadly based cancer prevention campaign is urgently needed to counteract the secrecy and complacency that has characterized the state's approach to regulation.

The starting-point for any broad campaign must be the recognition that cancer is *not* just a disease of old age ('everyone must die of something' say the fatalists). Cancer is the single most common cause of death for people in the 35–54 year age group and is second only to accidental death in the 5–34 year group. Further clues to the roots of cancer can be found if we examine the statistics in terms of social class. You are far more likely to die of cancer at any age if you belong to social class IV or V than to the 'higher' classes of I and II. What must also be recognized is that, despite all the publicity and funding devoted to a 'cure for cancer', survival rates for most of the common cancers have not improved greatly over the last 30 years. Cancer is not a simple disease for which a single 'magic bullet' is likely to be found. Rather, there are many different types of cancer most of which appear to be caused by the interaction of several different factors. We know that cancer rates vary according to occupation and geographical location (as well as social class). What this implies is that at least some of the underlying causes are environmental.

Environmental Factors

So what are these environmental factors and how can they be controlled? Broadly, views on the major environmental factors fall into two camps. The life-style approach argues that the main cancer-causing agents are those over which the individual has a choice (smoking, diet, alcohol, exposure to sunlight, sexual/reproductive behaviour). Prevention, therefore, is best approached by persuading individuals to adopt a 'lower risk' life-style (don't smoke, eat less cholesterol and fat, cut down on alcohol consumption, and so on). For instance, in 1984 the US National Cancer Institute (NCI) announced a plan to halve the number of cancer deaths by the year 2000. Almost 50 per cent of that reduction is to be achieved – according to the NCI – by persuading people to stop smoking and to change to a lower fat diet. This broad life-style argument has won many supporters in recent years through its basic insistence on the individual's 'choice' as to what cancer risks should be run.

In opposition to this life-style approach, a more radical approach to cancer prevention argues that chemical and radioactive agents in the environment cause a far higher proportion of deaths than supporters of the life-style argument presently claim. Trade unions in Britain point to research suggesting that occupational carcinogens are responsible for up to 30 per cent of cancers – examples such as asbestos, 2,4,5–T, vinyl chloride monomer, benzene, BCME, low-level radiation and aromatic amines can be put forward as cases in point where a carcinogenic risk to workers is known. Equally many of these chemicals *also* pose a risk to the general environment due to their presence in consumer products – as with asbestos or 2,4,5–T – or to atmospheric pollution – as with low-level radiation. Not surprisingly, the chemical industry opposes this argument and prefers to quote figures which suggest that 30 per cent of cancers are caused by smoking and only 5 per cent by exposure to occupational and environmental carcinogens.

There are very good grounds for arguing that the 5 per cent figure for workplace and environmental pollution is a severe underestimation.[1] But what is more important than an academic debate over percentage points is that we should tackle the pressing question of what we actually *do* about cancer. The 5 per cent figure for occupational cancers may indeed be an underestimation – but even at this unrealistically low percentage we are still assuming that some 7,000 people are dying each year as a consequence of workplace exposure. Looking more closely at the life-style argument, we need also to consider just why factors such as smoking and diet are conventionally taken to be a matter purely for individual choice and re-education. What about the powerful social forces which push tobacco so relentlessly or which shape the pattern of agricultural production and consumption in this country? As the examples below demonstrate, fighting back against cancer means tackling carcinogenic substances in the environment but *also* opposing the powerful lobbies in this country which act against good health.

Asbestos

The case of asbestos production in Britain demonstrates well the industrial pressures against external control and the reluctance of official bodies to take firm regulatory action. The processing of raw asbestos in this country actually began in 1879 and the first recorded case of asbestosis (lung fibrosis) came in 1900. As early as 1906, a Home Office report pointed out the real dangers of asbestos but it was not until 1931 that a set of regulations were introduced to control levels of asbestos dust in a limited number of factories. By 1934, the link between asbestos and *cancer* was pointed out – but it was only in 1969 that new asbestos controls were enacted. Throughout the 1970s and early 1980s, evidence of the severe health effects of asbestos exposure has continued to emerge – notably through the 1976 Ombudsman's report on an asbestos plant in Hebden Bridge and the 1982 Yorkshire Television programme *Alice – a Fight For Life*. The epidemiologist, Richard Peto, has now concluded that some 50–70,000 deaths from asbestos exposure can be expected over the next 30 years.

Further controls on asbestos exposure in the workplace were introduced in 1979 and again in 1984. Increasingly, however, the major concern has been over the myriad of locations in the community where asbestos can be found. From hospitals to housing estates, from schools to health centres, asbestos has been extensively used as an insulating material. The public may be put at special risk following fires, explosions or building demolition – yet the assistance given by central government to local authorities so that they can deal with this problem safely has been minimal. Even after a century of harsh experience, the story of asbestos is still not over.

Food Additives

A similar story of 'too little, too late' can be found in the case of food additives. According to one recent approximation, British food and drink manufacturers have spent between £160 and £180 million on additives – a category which includes preservatives, colourings, sweeteners and flavourings. Certainly, there has been a rapid expansion in the use of these substances since the Second World War. Whilst a number of additives are regulated in this country, however, the vast majority – including the ubiquitous 'flavourings' – are not. Accordingly, the level of monitoring for health effects is extremely low. Even more crucially, commercial secrecy is so dominant that consumers know very little about the hazards involved. As with other official bodies responsible for toxic substance control, the Food Advisory Committee of the Ministry of Agriculture, Fisheries and Food conducts all its deliberations in secret. It is thus impossible to uncover the real evidence on which commercial clearance is allowed – or, more accurately, it is impossible to discover which substances have even been raised in the advisory committee.

Fighting Secrecy

The first step, therefore, in this fight-back against cancer must be to examine the official bodies which at present make decisions about carcinogenic hazards. From pesticides to drug assessment, from environmental to workplace hazards, the blanket of secrecy is almost total. This absence of freedom of information is all the more crucial because of the 'balancing act' between physical cost and economic benefit which is at the core of any decision over carcinogenicity. It seems indefensible for small and unrepresentative bodies of 'experts' to make such decisions without the knowledge or participation of those actually suffering the costs.

The political arguments for a basic democracy in decision-making need also to consider the technical uncertainties which exist – and are likely to continue – over carcinogenic risk assessment. Epidemiology, animal testing and 'short-term' tests (such as the 'Ames' technique) all attempt to provide evidence of carcinogenicity – but each has its deficiencies. Witness the protracted debates about the link between tobacco and lung cancer – despite the undoubted potency of the effects and the long period over which data have accumulated. Where a clear-cut cause-effect relationship may *never* be developed, the 'burden of proof' becomes crucial as does the manner in which this is decided. Closed committees where an easy consensus can develop between regulatory authority and regulated industry seem an especially inappropriate forum for establishing the carcinogenic guilt or innocence of any substance.

Trade-union Campaigns

Of course, it would be wrong to suggest that the British system for controlling carcinogenic hazards has been totally resistant to change. In the workplace, for example, a reasonably well-organized trade-union movement has been successful in launching both a general anti-carcinogen campaign and also a number of single-issue initiatives – for example, on asbestos, radiation, and benzidine based dyes. Some success has also been gained through the tripartite structure of the Health and Safety Commission, which gives trade unions at least notional representation on advisory committees relating to occupational hazards. But despite this progress, trade-union demands for 'No Carcinogen' agreements in the workplace or for tighter controls on specific substances have met with considerable opposition from industry. If these difficulties are apparent in the workplace where the existing trade unions provide a well-established basis for campaigning, the problems for effective lobbying *outside* the workplace seem all the more immense. 'Constituencies' of those especially exposed to a hazard are difficult to form. Very often, no background of negotiation and bargaining skills can be drawn upon. Individual citizens feel remote from the centralized decision-making bodies and their style of operation. The apparently technical nature of the arguments serves as a further barrier to communication. In a system based upon the restriction of information to a small technocratic elite, weakly organized groups of citizens are unlikely to have a substantial impact.

An Alliance Against Cancer

These problems need to be tackled before any truly preventive cancer campaign can begin. What is needed is an *alliance* between those groups within and outside the workplace who are concerned to attack the problem of cancer. Trade unions have already made some efforts in this direction. For example, GMBATU has been active in alerting local authorities, tenant organizations and the general public to the hazards of asbestos in the community. The need for such concerted action is clear when risks are not restricted to one group alone but also involve consumers, local communities, environmental groups, health workers, political associations, indeed *every* section of the population at one time or another.

These groups cannot be successful alone but need to recognize and strengthen the links between their activities.

What such a broad cancer prevention campaign must also tackle is the politically awkward life-style factors – notably smoking. It needs to be remembered that the 'victim-blaming' approach often adopted by industry diverts attention away from the cancer risks of other substances such as those discussed above. Present patterns of taxation and industrial subsidy also foster the growth of industries which contribute little to the good health of any nation. Equally, the propaganda machine which encourages present food and tobacco habits hardly serves to perpetuate a free choice – especially when so much advertising and sponsorship is directed towards young children.

Two central points must be made about any broadly based campaign. First, alternative employment must be created for those sectors which would be progressively squeezed by a prevention campaign – otherwise these is no possibility of winning the support of workers in the tobacco or agricultural industries. Secondly, the Third World implications of improved policies for health must be monitored carefully – better cancer prevention strategies in one nation should not lead to hazards merely being exported elsewhere.

What would a broad campaign against cancer actually consist of? The details could only be finalized after discussion with concerned groups – a single blueprint would be idealistic and anti-democratic. The key elements, however, might well be as follows:

- Much greater emphasis should be given to cancer prevention – both in terms of scientific research effort and national policy-making. An adequately funded institute – independent of corporate control – is needed for the testing and evaluation of toxic substances.
- The present procedures for evaluating carcinogenic risk need urgent – and public – re-evaluation in terms of the technical consensus on which they operate, the openness of their activities and the freedom of information which they practice.
- It is generally accepted that over 50,000 people die prematurely in Britain each year due to tobacco-smoking. It is simply not good enough to continue blaming the victims. A stronger anti-smoking campaign is needed which would ban the advertising and promotion of all smoking materials, restrict the economic development of the tobacco industry and facilitate the redeployment of those who depend – directly or indirectly – upon tobacco for their living. At the same time, we should not allow a concern with tobacco to play down *other* carcinogenic risks in the workplace or environment.

- A major feature of such a campaign must be its ability to unite groups who would otherwise focus only on the specific issues which affect them most directly. The practical experience of those, such as environmentalists and trade unionists, who have already fought individual battles over cancer hazards must be brought to the assistance of those who have only recently become involved.

The difficulties of sustaining such a campaign should not be ignored – powerful economic and political opposition can be anticipated. The alternative, however, is to condone a system in which the bulk of the population is excluded from any say in such vital matters and in which vested interests and profit carry more weight than human needs.

Notes

1. See L. Doyal, K. Green, A. Irwin, D. Russell, F. Steward, R. Williams, D. Gee and S. S. Epstein, *Cancer in Britain; the Politics of Prevention*, London: Pluto Press, 1983.

30

Britain and the Third World

John Madeley

Soon after the end of the United Nations Conference on New and Renewable Sources of Energy (UNCNRSE), held in Nairobi during August 1981, Britain's former prime minister Edward Heath found himself with an overnight stop in the Kenyan capital city. As befits a former prime minister he was duly taken care of by the British High Commission. And Mr Heath was anxious to enquire of the Commission what had happened at UNCNRSE, which according to press reports had been about as little use as a spent light bulb. A highly placed official did not agree. 'The conference was a remarkable success for the British government,' he told a bemused Mr Heath, 'we managed to stop anything from happening at all.'

As bizarre as that remark is, it is nonetheless an accurate summing-up of Britain's policy on Third World issues over the last nine years. During that time, Britain has had a 'do nothing, block everything, damage every initiative' government that has totally failed to recognize the legitimate interests of the developing world and barely begun to conceive where Britain's own long-term interests lie.

The decline in Britain's relationship with the Third World did not start with the coming to power of a Conservative Party government in May 1979. Long before that Britain's government was conservative in its approach to the issues, its chief aim being to conserve what Britain had and give nothing away, despite the worsening problems from food and energy shortages to resource depletion and mounting debts that developing countries were facing. From early 1976, however, the relationship began to go badly wrong.

In 1975, Britain had a relatively good year; in May, the then Labour government had proposed new ideas to stabilize the commodity prices on which many developing countries are heavily dependent.[1] This was followed by the publication of a white paper that was subtitled 'More Aid to the Poorest'.[2] The British government, it seemed, was committed both to giving the poorest of the poor more aid, and to helping poor countries earn more themselves from trade.

Why did it all turn sour? From 1976 onwards, government ministers seem to have cared so little about the issues that they allowed civil servants to take the upper hand. So British policy to the Third World came to be dominated and virtually dictated by Whitehall. Despite the 1975 white paper, Britain's aid to poor people did not grow under the Labour government of the late 1970s. The Treasury successfully blocked any attempts to give money to anyone who might be caught spending it on something other than British goods. Aid for a rural health clinic or for a water pump that was run by a village co-operative, which could buy locally or make materials for itself, was never likely to be passed by Britain's Treasury. Backed by the Department of Trade, the Treasury wanted aid recipients to buy British and so help the British economy.

In Aid of Trade

Aid therefore by-passed the small local projects and went to the big grandiose schemes. Thus it was the same Labour government that compiled 'More Aid for the Poorest' which looked favourably on a request from the Sri Lankan government for £100 million – more than Britain had ever given before to one project – to be given to build a huge dam across the Mahaweli river to provide irrigation and hydropower. Because of inflation, the project was to provide some £150 million worth of orders for British companies.

The Mahaweli project, which was sanctioned by the incoming Conservative government, is costing the Sri Lankan government so much money that it is the poor of that country who are ending up paying for much of it. The 'aid' that has so handsomely helped British firms is impoverishing, not helping, the Sri Lankan poor. In the late 1970s, the environmental consequences of the scheme – including the displacement of some 50,000 people, the clearing of trees that needed to be cleared, the loss of thousands of acres of forest, the destruction of wildlife, and the likely silting of the river bed – were pushed aside. Yet Mahaweli is typical of the type of aid project which Britain backs. The real effect on the poor and the ecological damage that might be caused – such longer term considerations lag far behind promoting the interests of British companies.

In 1980, the British government was at least honest about where its aid programme was going. It said that when deciding aid allocations, more emphasis would in future go to 'political, industrial and commercial considerations'.[3] Between 1980 and 1984, the government siphoned off increasing amounts of money from the aid budget for the Aid-Trade Provision – a device introduced by Judith Hart, Minister of Overseas Development in 1978, under which aid is openly given to secure export orders. Inevitably, there is less for the poor; the last few years have seen Britain reducing its aid to the starving peoples of Africa so that there is more in the kitty for better off countries such as Turkey and Mexico. The result is that as far as the poor of the earth are concerned, Britain today scarcely has an aid programme. It has been slowly bled to death.

Yet even within the existing programme, elements of decency are struggling to break out. For despite, rather than because of the British government's approach to aid, some excellent work continues to be done by the scientific research units. In 1983 the government ignored protests and merged two of the units, the Tropical Products Institute and the Centre for Overseas Pest Research, into a single unit, the Tropical Development and Research Institute. With a limited budget (under £10 million a year, compared with overall British aid in excess of £1,100 million) TDRI is involved in projects which are helping developing countries to tackle some key ecological problems. In Egypt, for example, the unit has demonstrated that a virus suspension applied to cotton can control leafworm as effectively as chemical insecticide, whilst not affecting the beneficial insect population.

The Scandal of Pesticide Exports: Britain's Role

Although the British government may not yet have realized it, the work of the TDRI in Egypt could mean that less insecticide is purchased from Britain. British pesticide exports have grown massively in recent years – by 211 per cent in value over the 1975–9 period.[4] Britain is today the world's third largest exporter of chemical pesticides, selling some £350 million worth in 1983, over half of which went to the Third World. The government has shown little inclination to curb the exports of pesticides, even when they have been shown to be dangerous. Indeed, Britain has continued to use the Third World as a dump for pesticides, with British firms selling formulations that contain active ingredients (such as disulfoton and terbufos), which are either banned or severely restricted on health or environmental grounds in Britain and other western countries. Pesticides containing such ingredients are judged harmful to British farmers, even though those farmers have protective clothing and can read labels. Yet they are apparently deemed 'safe' for use in the Third World, where it is usually too hot for people to wear protective clothing (even if it can be afforded) and where many people cannot read labels – especially if they are only printed in English.

The British companies that export such lethal goods are under no obligation even to warn prospective buyers about their dangers. The British government maintains that developing countries should pass legislation to restrict imports if they wish. But even if legislation exists in theory, it is often weak in practice. Much of the pesticide trade takes place between companies – Third World governments are frequently unaware which pesticides are coming into their countries, let alone which ones may be causing sickness and death. According to David Bull, director of the Environmental Liaison Centre in Nairobi, at least 10,000 people die each year in the Third World because of occupational pesticide poisoning – whilst a further 400,000 become sick.[5] In the isolated fields and communities of the Third World no one knows what havoc and suffering is caused by the chemicals that leave Britain's shores.

In October 1983, the European parliament called on EEC countries to control the export of pesticides that are banned or restricted in their own countries. The UN General Assembly and OECD have made similar appeals. The British government has ignored them. It announced in June 1984 that it was drawing up legislation on pesticide use in Britain to ensure that unsafe pesticides are not used by British farmers. A consultative document was circulated; it contained not a word about exports to the Third World.

Campaigners from Friends of the Earth and Oxfam (the two British organizations who jointly co-ordinate the Pesticides Action Network, set up in February 1984), protested about the omission. The government, however, showed few signs of being moved to do anything.

An Abysmal Record

Aid and pesticide exports are but two of the issues over which the British government has done nothing positive in the last decade. Having made the running on commodity trade in 1975, the government's record has since been abysmal. It has been my own grave misfortune to have attended many of the North–South conferences of recent years. Such a ringside seat has been harrowing but intensely illuminating. Not only has the British government had nothing positive to offer, it has generally blocked every useful initiative that other countries have made.

Thus the Common Fund for Commodities was stalled for years by a negative British approach and has still not started. In 1980, a special

"But if we don't sell our food, how can we earn the money we need to buy it back again?"

session of the United Nations, called to try to launch global talks on key North–South issues, was effectively torpedoed by Britain's unwillingness even to come to the negotiating table. Attempts to reform the world monetary system – which is now so shaky that there is speculation as to whether Britain's high street banks are at risk – lie impaled on the Treasury hook.

Britain has also been deaf to suggestions that it might curb sales of armaments to the Third World. Such exports of death now run to over £1,200 million a year, more than Britain's entire 'aid' programme.

The past decade has seen the British government abandon leadership and give up thinking about the long-term consequences of its policies for the Third World – but then government ministers have shown they care little about these issues. 'There are no votes in Aid,' they chorus smugly. And so they leave it to their civil servants – many of whom are unsurpassed in their narrow, utilitarian approach to Third World problems. Britain's relationship with the Third World is in urgent need of being rescued by the democratic process.

There could yet be votes for politicians who show people that they understand why and how Britain's relationship with the Third World does matter – why it is not in our interests that Third World people are poisoned by our chemicals, go hungry because we refuse to pay a fair price for their products and give aid to build giant white elephants, like the Mahaweli project, which can only serve to impoverish still further the poor of the Third World.

Notes

1. *World Economic Interdependence and Trade in Commodities*, Cmnd. 6061, London: HMSO, May 1975.
2. *The Changing Emphasis in British Aid Policies: More Help for the Poorest.* Cmnd. 6270, London: HMSO, October 1975.
3. Statement by the Minister for Overseas Development, House of Commons, 20 February 1980.
4. David Bull, *A Growing Problem: Pesticides and the Third World Poor*, Oxford: Oxfam, 1982.
5. Bull, *A Growing Problem.*

31

Pesticide Exports

Britain's Record

Chris Rose

Worldwide, some of the most important markets for pesticides are now in tropical countries – especially where crops such as peanuts and cotton are grown. Pesticides sold in the markets of Europe are in the main designed to meet the needs of markets in tropical areas. Such markets only exist where traditional subsistence agriculture has been replaced by a cash economy and where intensive agricultural practices have been introduced, creating both the need and the necessary cash to buy pesticides. It goes without saying that such markets did not exist a few decades ago.

In 1982, Oxfam responded to the growing reports of environmental and social upheaval created by pesticide use in the Third World, by publishing the now classic study by David Bull, *A Growing Problem: Pesticides and the Third World Poor*. Bull describes how lack of training, illiteracy, and the near complete absence of environmental controls has led to the gross abuse of pesticides in the Third World. In Mexico, for example, it was found that 50 per cent of all containers were incorrectly labelled. In Ethiopia, market stall-holders were using old pesticide cans to sell beer in. In Ghana, fishermen were employing toxic insecticides as a cheaper method of collecting fish from streams: whole villages were poisoned as a result. In Nepal peasants frequently spent several months' wages on one can of fungicide, with instructions written in a foreign language, and forwent their pest resistant crops for higher yielding but vulnerable foreign varieties.

In 1972, the World Health Organization estimated that there were 500,000 cases of pesticide poisoning a year, with over 9,000 deaths. Since that time, pesticide use has increased around 50 per cent.[1] It is estimated that although people in developing countries use only a fifth of the world's pesticides, they suffer 50 per cent of all poisonings and 75 per cent of the deaths (375,000 and 10,000 respectively).

Britain: A Major Exporter of Banned Products

In 1985, shortly after opposition MPs and peers and Conservative back-benchers had failed in their attempts to persuade the government that Britain needed to control her exports, Catherine Caufield published an article, entitled 'Exporting Death' in *New Scientist*. Caufield pointed out that Britain is one of the world's most active exporters of pesticides. Indeed, she is the third largest. In all, 15 per cent of all international trade in pesticides comes from Britain. And, contrary to industry propaganda, the chemicals include those banned at home because of their environmental or health hazards. DNOC (whose hazards have been known since the 1940s) and DDT are two examples. As DDT uses were gradually disallowed under the government's Pesticide Safety Precautions Scheme in the 1980s, so exports were stepped up. A parliamentary answer to Dr David Clark MP (the opposition spokesman on the natural environment) revealed that while in 1980 DDT exports stood at 1,840 kg, by 1984 they had leapt to 125,503 kg, with a value that had increased 20-fold. Officially DDT was banned in Britain in October 1984: in practice, it is still used (see p. 151).

Caufield found in her research that although records of exports are kept a government secret, Britain was sending abroad at least a dozen highly hazardous pesticides. These included dinoseb, disulfuton, DNOC, endosulfan and phorate: all part II substances under the Poisons Rules that require a full protective set of clothing with respirator and hood if the concentrate is handled in Britain. Part III substances need a face shield, rubber gloves and other protection in similar circumstances: exports of part III pesticides include demeton-S-methyl, draxolon and pirimphos methyl as well as trebufos (which must not be used in Britain). Paraquat, which is covered by the Poisons Act, is also exported.

In Britain, substances covered by the Poison Rules require the buyer's signature and are restricted to certain retailers; all are subject to legal Health and Safety Regulations. There is no guarantee – indeed it is very unlikely – that Third World governments will apply similar constraints, however inadequate those constraints might be. Indeed the governments involved may not even know that the chemicals have been sent into their country.

Prior Informed Consent to Pesticide Exports

For these reasons environment and development groups have pressed governments to adopt the principle of 'Prior Informed Consent'. Proposals for 'Prior Informed Consent' were first put forward by the

United Nation's Food and Agriculture Organization. Under the proposed FAO scheme, exporting governments would prohibit shipments of a pesticide until they received notice from the government of the importing state that it had not only received full information of the ecological and health implications of the chemical, and of the regulations governing its use in the country of manufacture, but that it still wanted the shipment to proceed. This would simply ensure full exchange of information, making sure that the receiving government would be aware of the shipment, and its potential risks.

'Prior Informed Consent' was made a cornerstone of the Pesticides Action Network's campaign to eliminate double standards between 'North' and 'South'. British members of the network – known as PAN – were active in lobbying for 'Prior Informed Consent' during the debates over the Food and Environment Protection Bill. Amendments inspired by their evidence were put forward by peers such as Lord Melchett (Labour) and supported by Lord Stanley of Alderly (Conservative), Lord North-bourne (cross benches) and Lord Mackie of Benshie (Liberal). There was also all-party support in the Commons for making 'Prior Informed Consent' a legal requirement for British pesticide export. Dale Campbell-Savours (Labour), Simon Hughes (Liberal) and Richard Body (Conservative) were among the MPs who called on Britain to follow the example of the Dutch government which had just extended its own pesticide legislation to embrace the principle. But faced with heavy pressure from British and American multinationals, channelled through the British Agrochemicals Association, the government adopted the Foreign Office approach and simply took powers to ask companies for information in order to comply with international obligations if such arose. And it made sure that they did not arise by joining other exporting nations in lobbying the FAO to see that Prior Informed Consent did not become a legal requirement of its proposed Code of Pesticide Exports. Consequently when the government published its consultative document on the Food and Environment Protection Act in November 1985, it simply stated that the use of powers 'will be considered if the voluntary approach does not prove satisfactory'.[2]

Britain Gets Her Way

The FAO code, finally agreed in December 1985, only sets guidelines on exchange of information, childproof containers, and the language of labelling. It *has no legal force*. In a suitable twist of irony, as reports of the FAO decision reached the press, another pesticide story was in the news. *New Scientist* reported on 5 December 1985:

Four hundred scientists and staff employed by Brazil's Institute for Geography and Statistics have lost their jobs, because of a pesticide accident . . . [They] were fumigated with a mixture of parathion, aldrin, BHC and DDT – pesticides that are forbidden or restricted in Europe. The chemicals should have been diluted before use but the fumigators sprayed them at full strength. As a result, workers in the building started sweating and salivating, vomited and had headaches and diarrhoea. Within a month, four women had miscarriages. Government authorities in Rio de Janiero responsible for the building, cleaned the furniture and told workers it was safe to return. When they did, the symptoms recurred.

And who benefits from the pesticide export trade? In 1985, ICI's plant protection division (that is, its pesticide division) received the Queen's Award for Industry for Export Achievement with sales of £635 million in 1984, up 32 per cent. Trading profit for 1984 was £82 million, up 51 per cent. One of ICI's most successful promotions was Paraquat, a herbicide now widely used as a post-harvest dessicant as well as a simple weed killer but banned in West Germany and controversial in Malaysia and elsewhere. In their 1983–4 *Annual Report*, the British Agrochemicals Association reported, 'Once again the industry's outstanding growth was export-led with overall sales increasing by 26 per cent to £690.8 million. Following 27 per cent growth in 1982, exports increased by a further 31 per cent to £361.5 million . . . '

Notes

1. In fact, from 1964 to 1974, pesticide use in Africa grew 500 per cent. Between 1972 and 1978, the Philippines saw imports increase by a factor of four. And in 1979, 25 per cent of all pesticides exported from the USA were either banned or not registered there.
2. Postscript: Some good news is that in 1986, the Commission of the European Communities proposed a Regulation (stronger than a Directive) to the Council of Ministers, which would have the effect of making Prior Informed Consent mandatory throughout the EEC. This came about as a result of lobbying by environmental, development and health groups. The Commission also asked for a mandate to pursue the question of Prior Informed Consent with UNEP (The United Nations Environment Programme) and OECD (Organisation for Economic Co-operation and Development). The end result may be that Britain is forced to reform by European pressure, once again.

32

Environmental Secrecy

The Government's Record

Maurice Frankel

In May 1985, more than 50 people died after fire swept through the stands at Bradford City Football Club. The fire could have been prevented. Within 48 hours, West Yorkshire County Council released the text of letters it had sent to the football club ten months earlier identifying the 'unacceptable' hazard and calling for improvements 'as soon as possible'.[1] Later, it was disclosed that as long as five years earlier, the Health and Safety Executive had known there was a 'substantial fire hazard' at the ground – but had not informed the fire brigade.[2] Had these letters been released when they were written, public pressure may well have forced the football club to take action. But the letters – like all correspondence between environmental or public health authorities and those they regulate – had been secret, and the public left ignorant of the hazard they faced.

Where such a hazard is caused by a chemical with gradual, long-term effects on health, the consequences of such secrecy may be equally severe, but very much less obvious.

The IBT Scandal

In 1984, senior executives from a US company, Industrial Bio-Test Laboratories (IBT) Inc., were convicted of fraud for improperly performing – and in some cases falsifying – crucial safety tests on pesticides and other chemical products. In a bravura display of incompetence, the laboratory had kept its mice in cages built for rats, and the animals had simply escaped and bred wild, in the laboratory. To make up the numbers, missing mice were replaced with new animals – and these were falsely described as having been present throughout the test. Periodically technicians would attempt to hunt down the escaped animals with chloroform sprays; but chloroform itself is a carcinogen, and the spraying will have wrecked any toxicity tests carried out in the vicinity.[3]

IBT had carried out some 800 vital tests on 140 pesticides. The US Environmental Protection Agency later discovered that three-quarters of these tests were so badly performed as to be useless.[4] In Britain, as elsewhere, IBT data had been accepted and regarded as reliable. *Unlike other countries, however, Britain responded to news of the scandal in total secrecy.* In the USA, the EPA has published a complete list of the pesticides which had in whole or part been tested by IBT. In Britain, this information has always been withheld, despite requests for it in the House of Commons.[5]

Canada and Sweden banned a number of pesticides after their reliance on IBT data was recognized. No pesticide was removed from the British market for this reason. The Ministry of Agriculture, Fisheries and Food has claimed that IBT data were never of central importance in its decision to approve pesticides – but has refused to produce any specific evidence to substantiate this assessment. In Canada, several pesticides were required to carry warnings on their labels to point out that safety data could not be regarded as complete until IBT studies had been replaced.[6] MAFF also sought replacements for work done by IBT on British pesticides – but users were never warned to treat any pesticide with special caution in the meantime.

The Official Secrets Act

It is often difficult to know whether such secrecy represents a deliberate attempt to conceal the indefensible – or whether it is merely the bureaucrat's instinctive response to any awkward question. On entering the civil service, every government employee is required to sign a declaration acknowledging that 'I understand . . . that I am liable to be prosecuted if I publish without official sanction any information I may acquire in the course of my tenure of an official appointment.' The prosecution, of course, would be for breach of section 2 of the Official Secrets Act, which makes the unauthorized disclosure of any official information an offence. The effect is to condition officials into believing that answering the public's questions is, potentially, a criminal activity.

The consequences can be truly bizarre. In 1978, the organizer of a conference on waste recycling was threatened with prosecution under section 2 if he distributed to members of the audience the text of a talk they had just heard a civil servant deliver. The official had himself provided a copy of his talk (entitled 'The Importance of Recycling') for distribution – but presumably had not first obtained the necessary written authorization from his department. On realizing this, he telephoned the conference organizer insisting on its immediate return, and warning of legal consequences if this was not done. Under protest, the conference organizer returned his paper. In the event, not a single member of the audience even asked for it.[7]

The Health and Safety at Work Act: Keeping the Public in the Dark

In addition to the general effect of the Official Secrets Act, several pollution statutes contain specific restrictions on the disclosure of information. The Health and Safety Executive, for example, has always maintained that section 28 of the Health and Safety at Work Act 1974 prevents it from releasing information to the public that it has obtained from firms about such matters as asbestos stripping, nuclear waste, pesticide spraying and the location of factories storing dangerous quantities of flammable or explosive chemicals.

In 1983, asbestos stripping began at the disused Fulham power station in London. Monitoring by the local authority showed that asbestos contamination was occurring outside the power station boundary. A resident then asked the Health and Safety Executive for details about the way in which the work was being carried out, and the results of any monitoring within the power station boundary. The following reply was received:

> I would re-emphasize that the contractors have not given permission for internal monitoring results to be released to outside bodies. In this connection your attention is once again drawn to the provisions of section 28 of the Health and Safety at Work etc Act 1974 . . . no relevant information shall be disclosed without the consent of the person by whom it was furnished . . .
>
> You may not have a copy of the prohibition notice (served on the contractor for unsafe working) nor the written method statement supplied (by the contractor) . . . In addition to any restrictions imposed by Section 28 mentioned above there is also the question of the public debating of every prohibition notice or improvement notice issued by the Executive. I am sure you will agree that this is undesirable and impracticable . . .
>
> Bulk sampling of lagging on external pipes was carried out by the London School of Hygiene and Tropical Medicine and by scientists from this department. I am afraid you may not have the results of this bulk sampling for your files . . .

Lifting Restrictions, Closing Doors

If secrecy is imposed by – or at least justified by reference to – the law, what happens when such legal restrictions are relaxed? Regulations made under the 1974 Control of Pollution Act allow local authorites to obtain and publish information which previously could not be disclosed.[8] By issuing notices, local authorities can obtain details of industrial air pollution which must then be entered on a publicly available register. These powers came into force in 1977. But a 1982 survey by the Institution of Environmental Health Officers found that only five authorities had used these powers in the preceding year. When these five were later contacted, only one acknowledged having used the new powers – the others suggested the Institution had included them in error.[9]

Another illustration of how authorities exercise their discretion on openness occurred in 1983 when the water authorities were freed from their existing legal obligation to hold their meetings in public.[10] The Water Act 1983 reorganized the boards of water authorities, and while it allowed them to meet in public, it gave them the option of excluding press and public from their deliberations. Only the Welsh Water Authority carried on meeting in public: all nine English authorities went into permanent closed session.[11]

Openness after 1983 was, if anything, even more important than before, for the new boards contained a very high proportion of directors drawn from local industry – an obvious opportunity for conflict of interest, since one of the boards' functions is to control industrial pollution in their areas. Questioned about the need for openness, the government's spokesman in the House of Lords, Lord Skelmersdale, explained that the authorities should now be thought of as *private* companies, adding 'there is no reason why their board meetings should be any different from those of a company in the private sector.'[12]

One step towards greater openness did, however, follow. The 1974 Control of Pollution Act provided for the opening of public registers containing details of discharges made to rivers and estuaries. Over a ten year period, successive governments had avoided implementing these provisions. But in 1985, the necessary regulations were finally introduced, and registers opened by all water authorities.[13] Nonetheless, the public's right to examine monitoring data is subject to a number of caveats. The most important of these are as follows:

- The Secretary of State for the Environment may exempt certain information from disclosure in order to protect trade secrets or because its release would be against 'the public interest'.

- The public only has access to the results of sampling undertaken by the water authorities themselves – and then only if the samples were taken 'in pursuit of section 173 (1) of the Water Resources Act 1963'. If a company performs its own monitoring (and many do) or if the water authority does not invoke the 1963 act, then there is no requirement to enter the monitoring data on the public register. Indeed, were an authority to so do without obtaining the permission of the discharger, it would be breaking the law.

- Although the water authorities can require a company to meter the volume and rate of its discharges, that information does not have to appear on the register: nor does the company have to reveal data on the nature, composition or temperature of its discharges. Yet such information may be of critical importance to watch-dog groups who wish to assess the environmental impact of a discharge.[14]

Secrecy: Standing in the Way of Environmental Improvements

A constant critic of unnecessary secrecy on such matters has been the Royal Commission on Environmental Pollution. In a series of reports, starting in 1972, it has argued that secrecy stands in the way of environmental improvements, and often produces unjustified public suspicion of both pollution control authorities and industry. In 1984, it reviewed the progress made on the issue and found that, while there had been some improvements, its recommendations had not 'been tackled systematically enough or with sufficient urgency'. It restated its fundamental belief in the damage done by unnecessary secrecy and recommended that all future legislative and administrative measures to deal with pollution should be based on: 'a presumption in favour of unrestricted access for the public to information which the pollution control authorities obtain or receive by virtue of their statutory powers, with provision for secrecy only in those circumstances where a genuine case for it can be substantiated'.[15]

Perhaps because the government perceived itself as being electorally vulnerable on 'green issues', the recommendation struck home. In December 1983 the government had rejected out of hand a modest recommendation from the Royal Commission on pesticide secrecy.[16] In 1984, just a year later, it accepted the Royal Commission's far more wide-ranging recommendation for openness as the basis for its future policy.[17]

In future, a presumption in favour of openness would exist 'in relation to every sector of the environment'. An interdepartmental working party was set up and asked to report by the end of 1985 on ways of establishing 'a uniform regime' for implementing the Royal Commission's recommendation for openness, provided it could avoid both 'red tape' and an 'unacceptable increase in costs for either industry or pollution control authorities'. At the end of 1985, the working party's report had not been completed, but a consultation paper suggested that it favoured a system of publishing on registers information which was routinely collected in a standard form by pollution authorities.[18] There was no indication that a general 'unrestricted' right of access – as proposed by the Royal Commission and accepted by the government – was to be recommended.

Signs of Hope?

In two areas, however, a more specific indication of future developments is available. The first relates to pesticide safety. At about the time it accepted the Royal Commission's recommendation, the government introduced a new bill to give statutory backing to the existing voluntary system of controlling pesticides. One feature of the proposals for a 'Food and Environment Protection Bill' that immediately attracted criticism was the absence of any provisions for ending secrecy about pesticide safety.

The pressure for change was given further impetus by a ruling in June 1984 from the US Supreme Court that disclosure of pesticide data under earlier US legislation could now go ahead.[19] Within months, Friends of the Earth in London had obtained from Washington a complete set of safety studies on the Monsanto herbicide Roundup, and had lodged requests for similar data on other pesticides also used in Britain. MAFF's attempts to keep such data secret were immediately rendered futile.

Under all-party pressure in parliament, the government added a clause to its bill which allowed (though did not require) the disclosure of safety information, and promised that a substantial degree of disclosure would result. It has since proposed that, as new pesticides are approved in future, its own evaluation of the safety data received from manufacturers will be made public. The full studies, however, will not normally be released. Where an evaluation is challenged on grounds that have 'scientific justification', MAFF has proposed that 'exceptionally' access may be given to the full data.[20]

This is a major step forward in principle. But in practice there are two serious shortcomings:

- Access to the raw data is entirely at the discretion of ministers or their advisers. Though members of the British public can obtain raw data from the USA *as of right*, they may still be refused the equivalent information by the British authorities.

- The new arrangements will apply only to newly approved pesticides – not to the many hundreds already in use. Data on these will become available only after their approvals has been reviewed, which may not be for many years. So even if the new regulations are implemented, data on the vast majority of pesticides will still not be available.

The Health and Safety Commission – Still Wed to Secrecy

The other area where specific disclosure proposals have been made relates to information about public health or safety hazards held by the Health and Safety Executive. A discussion paper issued in March 1985 proposed to open up just two, narrowly defined, categories of information. It proposed that lists of prosecutions or enforcement notices served might be made public in future, and that premises subject to various notification or licensing requirements might be publicly identified. This would mean, for example, that the location of sites notified to the HSE because they stored large amounts of flammable or toxic substances on their premises – and which could thus pose a threat to the community in case of accident – would for the first time be released.[21]

But that would be all. For any further information, the public would be expected to turn to employers. In effect, the Health and Safety Commission (HSC) had rejected the government's new policy of favouring 'unrestricted' public access. Rather, it was proposing that the public should only be allowed access to limited items of information, whose disclosure would be administratively convenient.

Workers already have substantially greater rights of access, under section 28(8) of the Health and Safety at Work Act. Under this section, factory inspectors routinely supply them with copies of any letter sent to an employer after an inspection asking for safety improvements. They also receive the results of any monitoring carried out by inspectors. The HSC's proposals would continue to deny equivalent information to the public, and would give them no access to any assessment of the hazard made by inspectors. Such information, it was proposed, should come from employers. They might be required to publish a one-off assessment of the implications of their activities for the public – but any other disclosure would be entirely at their discretion. As the HSC put it, employers 'will have control over the information to be disclosed . . . Each employer would need to decide in relation to his own circumstances how much information needed to be divulged.'

Employers would also be free to decide whose questions to answer – and who could be refused information. The HSC explained that their knowledge of the local population would enable them to decide 'who might "need to know"'. But perhaps the most objectionable aspect of the proposal was the suggestion that the public would not in any case receive the full facts. Employers would only be expected to disclose 'some relatively simple account of the type of hazard which might occur . . . What would be required by way of public information would therefore be the conclusions of an analysis, not the analysis itself.' The public would be given the employer's overall conclusion – and denied access to anything that might allow it to be independently verified or questioned. There would be no access to raw data, no information about assumptions used, and no disclosure of the HSE's own opinion on the issue.

At the time of writing these remain draft proposals. If they are implemented, the government's new openness policy will have little public credibility.

Notes

1. *Guardian*, 15 May 1985.
2. *Guardian*, 6 June 1985.
3. See, for example, *The Miami Herald*, 6 April 1983; *The New York Times*, 1 May 1983 and 9 May 1983.
4. Environmental Protection Agency, Office of Pesticide Programs, *Summary of the IBT Review Program*, July 1983.
5. *Hansard*, 5 June 1984, col. 74.
6. *Chemical Marketing Reporter*, 27 June 1983, p. 5.
7. Campaign for Freedom of Information, *Secrets* (newspaper), no. 5.
8. The Control of Atmospheric Pollution (Research and Publicity) Regulations, 1977.
9. Campaign for Freedom of Information, *Secrets File*, no. 2, January 1984.
10. The Public Bodies (Admission to Meetings) Act, 1960.
11. Campaign for Freedom of Information, *Secrets* (newspaper), no. 4.
12. *Hansard*, 2 April 1984 (Lords), col. 473.
13. The Control of Pollution (Registers) Regulations, 1985 (SI 1985/813).
14. J. McLoughlin and M. J. Forster, *The Law and Practice Relating to Pollution Control in the United Kingdom*, London: Graham and Trotman, 1982, p. 138.
15. Royal Commission on Environmental Pollution, Tenth report, 'Tackling Pollution – Experience and Prospects', London: HMSO, 1984.
16. Department of the Environment, *Pollution Paper*, no. 21, December 1983.
17. Department of the Environment, *Pollution Paper*, no. 22, December 1984.
18. Department of the Environment, *Public Access to Information Held By Pollution Control Authorities: A Discussion Paper*, 1985.
19. Supreme Court of the United States, Ruckelshaus vs. Monsanto Co., no. 83–196, decided 26 June 1984.
20. Ministry of Agriculture, Fisheries and Food, Pesticides and Infestation Control Division, *Pesticides: Implementing Part III of the Food and Environment Protection Act 1985. A Consultative Document*, November 1985.
21. Health and Safety Commission, *Access to Health and Safety Information By Members of the Public. Discussion Document*, London: HMSO, 1985.

33

Beyond Environmentalism

Jonathon Porritt

The environmental movement in the second half of the 1980s is very much at a watershed. It would seem to be going through one of its periodic peaks, where the issues in which its protagonists are involved coincide with people's everyday concerns to force politicians and 'decision-makers' into a more open and responsible pattern of activity.

The visible evidence of this is to be found in the consistent and unequivocal findings of one opinion poll after another, namely that a clear majority of people consider the environment to be an area of major, *mainstream* concern. The less visible evidence of this is to be found in a whole array of often subtle political and life-style changes, by which people give substance to their theoretical concerns. Attitudes towards health, or towards diet and nutrition; an increase in the number of bicyclists and those who choose to walk; a greater readiness not to waste things, but to see them recycled and energy conserved: the cumulative impact of such changes is difficult to quantify, but together with a growing sense of apprehension about nuclear power, acid rain, toxic wastes, the pollution of both air and water, and the continuing destruction of our countryside, it is enough to put the pressure on politicians who for the most part are still indifferent, or whose hearts are only in it when there are votes at stake.

There has been considerable talk over the last couple of years of the 'greening of British politics', and the extent to which all the major parties have begun to assimilate the ideas and, on occasions, even the specific policies of the environmental movement.[1] Frankly, much of this talk is wishful thinking, coming either from those within the political parties who are trying to bring the rest of their curmudgeonly colleagues with them, or from those within the environmental movement whose success is partly measured by their ability to green the contemporary political scene. More

". . . And, Prime Minister, I am pleased to announce that the economy has never done better"

dispassionate observers remain sceptical about both the depth and the durability of the groundswell of change that is said to be going on.

True enough, certain individuals in all parties are now openly identified with the green cause; officially or unofficially, they provide for their party the 'green-speak' that is now de rigueur. As a consequence, particularly within the opposition parties, one may discern changes of emphasis in their stated goals and, much more rarely, substantive policy decisions that should ultimately benefit the environment. But the political limits to such changes are made quite clear, and unhesitatingly called upon whenever the adoption of green ideas threatens to dilute or contradict the fundamental consumer-orientated, growth-driven principles that still hold sway at the heart of our industrial society. Politicians have become so accustomed to appealing to people as *consumers*, that elections have become nothing more than rhetorical jousts between those who promise jam today and those who promise yet more jam tomorrow if we only forgo the jam today. One might almost suppose that without economic growth, there would be no politics, so obsessively are politicians engaged in the pursuit of it. Their fate is entirely contingent upon their ability to generate notional increases in gross domestic product, regardless of the side-effects, in terms of human costs and environmental impact, of such growth and regardless even of the number of people who genuinely benefit.

The limits to green thinking are manifested in different ways in the different parties: in the endless skirmishing between the leadership of the Liberal Party and its rank and file; in the debate over the apparent mismatch between 'middle-class environmental concerns' and the nitty-gritty of working-class politics that still perplexes so many socialists; and in the sterile sophistication of the SDP's attempt to resolve fundamental political dilemmas by coining largely meaningless phrases such as 'green growth'.

Confronted with the obligations of government, any clashes within the Conservative Party caused by the first stirrings of green awareness have been resolved rather more brutally. On one occasion after another, the Department of the Environment, and its excellent if temporarily hapless minister, William Waldegrave, have been on the receiving end of interdepartmental thuggery; initiatives to do more about acid rain, the balance between farming and conservation, nuclear waste and recycling have all been thwarted by the heavyweights in the Treasury, the Department of Energy, the Ministry of Agriculture, and the Department of Trade and Industry. The priorities of the Conservative Party have thus been unambiguously revealed: environmental concerns are at best marginal, at worst totally irrelevant.

From this more dispassionate perspective, and confronted with the staggering range of problems covered in this book, it would have to be said that any greening of British politics going on beyond the Green Party is as yet a patchy if not paltry affair. Though the environmental movement has indeed been growing in strength over the last few years, so that its influence is now greater than it has been since the early 1970s, this has not brought about the kind of fundamental shift that one might have anticipated. Despite the occasional victory, despite the overwhelming support of the opinion polls, despite much more sympathetic coverage in the media, it would be an illusion to suppose otherwise. To what extent, it must then be asked, are the causes for this relative lack of success to be found within the environmental movement itself?

Reformation or Radical Overhaul?

This is not an easy question to answer, since the growing confidence of today's environmental movement has not been accompanied by any significant or profound heart-searching about its actual *goals*. It could be said, of course, that the problems are so severe, and the need for action in so many areas so pressing, that this provides sufficient rationale in itself. When the Earth is being assaulted and poisoned at every turn, need we really bother to work out why we seek to stop such devastation? I would argue very strongly that we should and, even more strongly, that it is our failure to do so that accounts for our inability to mobilize the undoubted public support to bring about real change. The reactive, defensive, often arbitrary interventions of environmentalists are simply not up to the job that now confronts us. For on what grounds do we intervene? Do we merely seek to stem the extent of the damage (i.e., to reform the system), or to establish alternative principles to prevent the damage taking place altogether (i.e., to replace the system)?

Up until now, the reformists have been dominant with the environmental movement. Realizing that people cannot cope with too much change all at once, that single issues can be dealt with far more effectively than whole value systems or fundamental political platforms, and empowered by a sense of urgency and irrefutable moral rectitude, the reformists established a code of environmental conduct that has achieved much over the years: there were indeed positive advantages to be derived from *not* addressing oneself to the root cause of the problems one confronted. There were also disadvantages. Band-aid environmentalism of this sort can only be vindicated by the touching but undoubtedly erroneous belief that industrialism as a system is somehow perfectible – despite all the evidence to the contrary.

By 'industrialism', I mean adherence to the belief that human needs can only be met through the *permanent* expansion of the process of production and consumption – regardless of the damage done to the planet, to the rights of future generations, to the human spirit and to the living standards of all those who end up as the losers in this global, all-encompassing human race. The often unspoken values of industrialism are premised on the notion that material gain is quite simply more important to more people than anything else, so that higher productivity, sustained expansion and faster economic growth become the dominant, if not the exclusive, criteria for determining 'progress'. Industrialism leaves little breathing space for those who see progress in less materialistic, gentler terms as something having more to do with the satisfaction of a whole range of human needs than with the satiation of individual greed.

The superficial sound and fury of many environmental battles conceals the fact that the real struggle is between these different value systems, a deeper confrontation which is only marginally influenced by the outcome of any one specific issue. It has therefore been deeply frustrating to have to re-fight the same battle in different places at different times, as if nothing had been learnt from previous clashes. At the same time, it has been depressing to see many individuals and organizations, knowingly or unknowingly, co-opted by that dominant value system, and therefore no longer defining their own area of concern or even determining their own practice and policy.

A more profound consequence of the reformists' approach has been its effect on the *image* of the environmental movement. We have predominantly campaigned *against* things, reacting defensively to a sequence of threats so as to hold steady the thin green line of environmental sanity. There have been correspondingly few opportunities to present more positive alternatives. The bunker mentality resulting from this approach has allowed our opponents to caricature us as retrogressive reactionaries intent on blocking progress itself. The single-issue, reformist perspective put blinkers on our vision, and rendered us incapable of presenting more than the barest outline of that vision to anybody else.

Not everyone went along with the tactics of the reformers. There have been many, aware of the dangers of compromising with the system, who have preached a different, more visionary faith, and from them the environmental movement in the second half of the 1980s has much to learn. There is also much to beware of, for their preaching has not always been as constructive as it might have been. Some have welcomed and even sought to accelerate the collapse of industrialism, hoping that the green alternative would arise like some phoenix from the ashes, despite the dislocation and intense suffering that would be caused thereby. They often seem unconscious that the only phoenix likely to emerge from such a collapse would be iron-clad and baton-wielding, the avatar of global totalitarianism, and thus alienate themselves from many still working for change within the system.

And some, of an eco-anarchist persuasion, have found it easier to opt out of the system altogether, privatizing their vision of a new Jerusalem, scorning those who remain to work within, and, having supposedly seen the light, nurturing it narrow-mindedly under their own moribund bushel. Many have fallen foul of that particular streak of chronically unrealistic escapism over the last couple of decades.

Taking on the 'Super-ideology' of Industrialism

I am deliberately raising the negative features of some of those who have sought to transcend industrialism rather than reform it, for their visions have often been rather elitist and their reluctance to dirty their hands in the muck and grime of contemporary politics is regrettable, if understandable; but this is not to deny the power both of certain individuals, outside of any organized movement, and a small number of organizations, to provide the vision and inspiration without which our movements simply would not move. Such people and organizations are normally rooted deep in everyday life, and whilst seeking to transcend contemporary industrial reality, are well aware of its capacity both to compel and seduce its victims.

During the whole dramatic, traumatic history of the industrial revolution, alternative movements have for the most part fought shy of taking on industrialism itself, or even perceiving it to be the problem. The identification and critical analysis of this 'super-ideology' (embracing as it does capitalism, communism, most variations of socialism and practically every other -ism in every part of the world beyond the range of so-called 'primitive societies'), therefore constitutes an exceptionally important advance. In essence, the critique we make, though genuinely radical, is in fact relatively simple: that the pursuit of industrial expansion and economic growth simply cannot produce the kind of sustainable, just and self-reliant way of life that we should now be seeking.

Sustainability is of course the key to the green alternative. The prophets of doom and gloom in the early 1970s may have been a little premature in their prognostications, but their basic argument remains unchallengeable. The simple fact is that if we go on using up the Earth's non-renewable resources (its oil, coal, minerals, etc.) at the rate we do now, and at the same time misusing the Earth's renewable resources (its fertile soil, clean water and forests) at the rate we do now, then at some stage in the future, the whole system is going to fall apart. Those who claim that the fact it has not done so *yet* is proof that it never will, are dangerous idiots. Technological innovation may well postpone the final crunch, but it cannot obviate it altogether.

Moreover, only the most deliberate self-deception would allow one to suppose that the prospects for a just and fair society, both nationally and internationally, are improving. The ecological crisis is getting worse, particularly in many Third World countries. In the developed world, technological progress is already narrowing the options available to people for finding satisfying work; in a society where the Protestant work ethic still remains dominant, this tends to render superfluous more and more people. By the same token, our centralized, homogenized economy makes it harder and harder for people to create space for themselves or to withstand the pressures of consumerism.

Unequivocal opposition to industrialism is therefore fundamental to the environmental movement's prospects of success. That is why it must move beyond the reformist model that has held sway up until now, and unapologetically get stuck into the business of bringing about changes in people's values by promoting positive and constructive alternatives. What we need now is a new generation of pragmatic visionaries!

Many would argue that in the writings of people like Fritz Schumacher and Barbara Ward, a suitable model has been around for many years – it is just that we have failed fully to appreciate the enduring radicalism that marks out these two for our particular attention. It is clear that the best of the visionaries have always been those with their feet on the ground rather than their head in the clouds, and that the best of the reformists are those whose work is inspired by a very different vision of the future from that normally on offer. The question for Fritz Schumacher and Barbara Ward was not an either/or question (reform *or* replace industrialism?) but how best to combine the two (how to reform *and* replace industrialism at the same time).

Many would now argue that only by finding common ground with those in the peace movement and the development movement, with human rights activists and those whose work is primarily spiritual rather than political, and with all those whose concerns lead them in different ways, but at the same time, to fight free of the suicidal imperatives of industrialism, will it become possible to achieve what at the moment remains unattainable.

Synthesis through the Green Movement

Only the green movement, unformed and untried as it yet may be, has the capacity to achieve that synthesis, not only because of its radical analysis of industrialism and visionary yet realistic presentation of alternatives, but also because the holistic philosophy that underpins the movement is alone likely to overcome the competitiveness, separatism and crude territoriality that characterize so many other alternative movements.

For some, this will mean involvement in the mainstream parties, bringing pressure to bear on their colleagues not just by way of specific policies, but in terms of reconsidering the often redundant assumptions and values that lie at the heart of these parties. Others will prefer to join the Green Party, not with any expectations of immediate power, but with a clear understanding of the need to present the whole picture and not just its more superficially attractive aspects. And the influence of pressure groups, such as Friends of the Earth and Greenpeace, will undoubtedly increase – just so long as their single-issue campaigning is set within a broader, more radical context, and they succeed in co-operating more closely with other potentially 'green' organizations.

Consider the informal coalition of groups that came together around the issue of the Treaty on the Non-Proliferation of Nuclear Weapons, and its Third Review Conference in September 1985. Anti-nuclear activists, peace campaigners and representatives from the development movement all found common cause in their concern to keep the treaty alive, to bring maximum pressure on the super-powers to start negotiating arms reductions 'in good faith', to promote non-nuclear energy strategies and to set up a Fund for Energy Development to ensure that developing countries receive adequate and appropriate assistance. Some of the groups involved might have thought of themselves as 'green', and some might not; but the coming together and the joint campaigning was *definitely* green.

There is much work to be done here and time is not on our side. Some years ago, the environmental crisis seemed to threaten the end of industrialism; few of us appreciated the astonishing resilience of the system and the ingenuity of those whose affluence was most directly threatened by such a prospect. In the last few years, we have witnessed an overhauling and restructuring of the productive apparatus of industrialism, and of the methods of social organization that go with it, with a view to eliminating its more destructive aspects and promoting a new, improved model as the answer to all our problems.

The rise of new eco-industries has been particularly significant. The very process of generating wastes and pollution has brought into being many 'end-of-the-pipe' technologies: desulphurization equipment for power stations, water purification plants, catalytic converters for cars, recycling for domestic and industrial waste. In an astonishing spasm of technological 'rearmament', industrialists have found ways of creating new markets and have increased profits out of cleaning up their own mess. One person's pollution becomes the next person's profit – and all done in the name of sound environmental practice.

It is an illusion, of course, to suppose that this resource-intensive clean-up operation can do the trick: the effectiveness of the repair measures (in terms of energy saving, reduced pollution and so on) will inevitably be neutralized by the continuing expansion of industrial activity. And there will remain some forms of pollution, particularly those of chemicals on individuals and communities, entirely beyond the redress of the eco-industries' would-be do-gooders.

A second variant of this new improved industrial model is that promoted by today's high technology enthusiasts, who hope to rescue industrialism from its smoke-stack image in a new age of enhanced efficiency. Microelectronics and biotechnology have become the buzz words of a new generation of green growthists; 'small is beautiful' is a phrase that trips effortlessly and yet sincerely off their tongues. Having at last realized that the glorious highway of industrialism has become nothing more than a dirty, dangerous cul-de-sac, they now seek ingenious ways of lengthening it, tidying it up and providing plenty of deck chairs for people to sit in along their way. But for all their efforts, it remains a cul-de-sac.

Reducing the scale and the external costs of the production process, and the energy intensity of individual products, does little to solve the crisis caused by the dominance of consumerist values and a ruthlessly materialist philosophy. It does even less to resolve the dilemma of mass unemployment, for the high-tech society would seem to depend for its success on encouraging millions of people to become the passive recipients of the fruits of automation. Without some radical redistribution of work, we shall inevitably end up with a dual economy, a sophisticated form of apartheid, with a minority of people in secure, well-paid jobs in the capital-intensive sector, and the majority either jobless or partially employed and increasingly dependent on state handouts.

But even in raising these warnings, one is conscious of reinforcing the image of environmentalists as backward-looking progress blockers. After all, the new industrialists have taken on board many of our arguments and concerns, and would no doubt argue that they had come more than half way to meet us. It thus becomes even more important for the environmental movement to start positively promoting its ideas of a sustainable, just and self-reliant society, and to weigh in the balance the advantages of these alternative models as well as the disadvantages of industrialism. Let that be the task of our next book!

This is not a challenge to be taken on lightly, for those who hold the strings of power are unlikely to relinquish them without a fight. But as

John Maynard Keynes was fond of pointing out, the power of vested interests is vastly exaggerated compared to the power of ideas. In calling on the environmental movement to re-evaluate its goals, to find common ground with others in a broader green movement, and to move from being primarily reactive and negative to a more constructive role, one cannot help but be aware that what is lacking at the moment is a true sense of vision. As Barbara Ward said:

> We learn from the visionaries, we do not learn from the practical men of affairs. They are marvellous, once the direction is set, but you will not find them in the forefront. Our visionary perspective is the true realism and that is what we have got to pursue.

Notes

1. There is inevitable confusion here, to be dealt with later, between those who see the 'environmental movement' as synonymous with the 'green movement', and those who see the environment as just one part of a far broader green movement. For the moment, let us stick with the former interpretation.

34

The Case for Direct Action

Nick Gallie

In 1952, when the British wanted a place to test their nuclear weapons, they drove the aboriginal peoples of central Australia from their secured 'dreaming lands', confined them to reservations many hundreds of miles away and proceeded to blow the 'place of dreams' apart. In so doing, they not only laid waste and contaminated a land whose worth they could hardly measure; they had uprooted and destroyed a fragile culture which had evolved over 50,000 years. Similar nuclear colonization has taken place all over the South Pacific region, by the 'great' western nations. As a consequence, island peoples have been bullied and evicted from their homelands, some have been deliberately exposed to radiation and most only allowed parasitic roles in the new westernized economies, often as prostitutes or neo-slave labourers.

In 1972, sickened by the continuation of atmospheric nuclear tests by the French at Mururoa Atoll, the newly formed environmental group Greenpeace sailed their tiny protest vessel into the Mururoa test site, deliberately to provoke a French reaction and thereby focus the world's attention on that lonely spot, a thousand miles from 'civilization'. The French did react, by savagely beating the Greenpeace protesters – who offered no physical resistance – but they were also obliged to react to an outraged world opinion. Two years later, they stopped their atmospheric nuclear tests and aligned themselves with the signatories of the Partial Nuclear Test Ban Treaty which forbids nuclear tests in the atmosphere.

Why did Greenpeace – a handful of unarmed protesters pitted against the military might of a fully fledged nuclear nation – win, when thousands of Polynesian and Australian aboriginals had lost, not only their livelihoods and traditional dwellings, but their whole worlds, past, present, future and transcendent as they perceived and understood them? The answer must lie in the extraordinarily persuasive powers of non-violent

"Who are the Real Conservatives?"

direct action. Greenpeace had taken the 'passive resistance' of Gandhi and Martin Luther King and turned it into an active vehicle of confrontation – a fascinating cultural and economic twist. This was not the protest of the underprivileged or downtrodden. It was an outraged but successful white western businessman who, with his friends, sailed his yacht – itself a symbol of accomplishment and respectability – into a nuclear test zone. They knew how to use the media. They had all been brought up on diets of pop-corn and television. They understood the need to make their protest visible to the world before it could become effective. And later, when they adopted the regalia of the American Indian and called their flagship *Rainbow Warrior*, they took on board the guilty conscience of modern America (and by implication all the conquering western nations) and pitched that as well against the powers that be.

The Power of Direct Action

It is important to try to grasp the invisible constituents of direct action so that we can begin to understand why and how it can work effectively as a means of persuasion.

First, protesters must establish a clear moral high ground. They must gain and retain this initiative in order to push the opponent into a moral dilemma; this of course will only work where protagonists share a moral perspective. It would be difficult to imagine a successful direct action campaign being waged against a wholly guiltless and amoral opponent; it is therefore imperative to understand where and how such opponents derive their own moral strength, how and to whom they justify their actions. It is to this weak point that the lever of direct action will be applied.

Secondly, and in full view of his or her moral peers, the protesters must place themselves at the opponent's mercy and provoke a response by actively interfering in the opponent's business. The direct action must be peaceful and must not damage the opponent's property. The point of this is to intensify the opponent's moral dilemma. It is, as it were, the fulcrum of the lever.

Thirdly, the protester must endure and refuse to move. In the fullness of time opponents will be toppled. They will be literally unhinged and forced to shift their ground to find a new equilibrium where they can regain their self-esteem.

From this it can be seen that successful direct action must form part of a holistic campaign. The establishment and retention of the moral high ground, the correct assessment of the opponent's strengths and weaknesses, the willingness and ability of the protesters to endure must be brought together in a coherent strategy.

> Planet Earth is 4,600 million years old. If we condense this inconceivable time-span into an understandable concept, we can liken Earth to a person of 46 years of age. Nothing is known about the first 7 years of this person's life, and whilst only scattered information exists about the middle span, we know that only at the age of 42 did the Earth begin to flower. Dinosaurs and the great reptiles did not appear until one year ago, when the planet was 45. Mammals arrived only 8 months ago and in the middle of last week man-like apes evolved into ape-like men and at the weekend, the last ice age enveloped the Earth.
>   Modern Man has been around for 4 hours. During the last hour, Man discovered agriculture. The industrial revolution began a minute ago. During those sixty seconds of biological time, Modern Man has made a rubbish tip of Paradise. He has multiplied his numbers to plague proportions, caused the extinction of 500 species of animals, ransacked the planet for fuels and now stands like a brutish infant, gloating over his meteoric rise to ascendency, on the brink of a war to end all wars and of effectively destroying this oasis of life in the solar system.

Thus Greenpeace states its claim for the moral high ground, and preludes its environmental campaigns. Defence of the environment is of course a 'natural' for direct action campaigns; seals and whales can hardly speak up for themselves, while the educated urban population, cut off from nature, is highly susceptible to protectionist arguments. But good campaigners are wary of public opinion. What is obviously 'right' must be scientifically demonstrated and presentations must be made to the appropriate fora. Getting the science right is as necessary as whipping up public support; only then can the legislators be moved, goaded by the need to appease their electorate and provided with the appropriate rationalizations to give objective backing to their 'concerns'.

Whaling and Nuclear Waste

In this political matrix, direct action serves two purposes. First, it highlights issues which are inherently complicated and simplifies them to a point where they become visible and tangible to a wide public. Secondly, direct action cuts through jargon, esoterica and bureaucracy and demands a straight answer. In media terms, a political and scientific wrangle that may have been going on for years is suddenly reduced to a simple headline and picture. It becomes news. A whaling ship, an explosive harpoon, a fleeing whale and between them a tiny, manned inflatable with the word

'Greenpeace' emblazoned on its side – it says it all. The image is a
'God-send' for television news, and instantly hundreds of millions of
people have shared an experience of 'Save the Whales'. How many years
of petitions and arguing over quotas in the International Whaling
Commission could equal that, and yet when Greenpeace entered that very
forum carrying with it the power of its television image, it could demand
and eventually win an international moratorium on all commercial
whaling.

Britain quietly began dumping nuclear waste at sea in 1948. Joined by
Belgium, Netherlands and Switzerland, close to 100,000 tonnes of low-
level waste were dumped at an agreed point in the Atlantic Ocean some
500 miles south-west of Lands End. No one had ever heard of it. Then in
1982, and again in 1983, Greenpeace appeared on the world's television
screens, this time placing their tiny inflatables directly beneath the barrels
of waste intended for the deep. Behind the scenes, Greenpeace had made
scientific presentations to the London Dumping Convention and the trade
union movement. The Convention called for a moratorium on nuclear
waste dumping in 1983, pending further scientific studies, but the British
government refused to acknowledge the moratorium, so the transport
unions 'blacked' the waste shipments. Now, proof of long-term safety has
to be provided to the LDC before any further dumping of nuclear waste
can take place – the practice is effectively stopped.

Greenpeace and Windscale

In 1978, Greenpeace began a campaign designed to draw attention to the
consequences of the radioactive waste discharged from the giant nuclear
reprocessing plant at Windscale (Sellafield) in west Cumbria. Greenpeace
were also concerned that the hazards of shipping spent nuclear fuel to the
plant were either misunderstood or being deliberately ignored. Greenpeace employed an independent team of researchers to calculate the likely
consequences of an accident at sea involving a ship carrying spent nuclear
fuel. The research indicated a possible outcome of thousands of prompt
and delayed cancer deaths. After presenting these findings, Greenpeace
blocked the harbour entrance at Barrow in Furness to highlight their
claim. The protesters were given a small fine and obliged to promise not to
do it again.

Greenpeace then switched tactics and began to focus attention on the
discharges from the Windscale pipeline. Through this pipe, two million
gallons of radioactive waste-contaminated water pass directly into the Irish
Sea every day. It has done so for over 20 years. Again Greenpeace did its

homework, which this time revealed that an estimated quarter of a ton of plutonium had been released to the sea bed. Further research showed that far from being carried out to sea and safely diluted, the radioactivity was being remobilized from the sea bed and blown back onto the land. Greenpeace then briefed a team of documentary film makers, who carried out their own research. They found that the incidence of leukaemia amongst children living in Seascale, a village close to the Windscale plant, was ten times higher than the national average. They also discovered plutonium in the house dust of a local resident. The film, when broadcast in November 1983, pointedly accused the nuclear authorities of threatening Cumbrian children with myeloid leukaemia (the only known cause of which is exposure to radiation).

The stage was set for Greenpeace divers to attempt to block the Windscale pipeline. In actual practice they failed, but soon the environmentalists found themselves back in the High Court once again being ordered, as men and women of good conscience, not to do it again. Greenpeace could not give that understanding without compromising the moral basis of their campaign. Their decision cost them £50,000 in fines, but won them public sympathy.[1] The government was obliged to react. Its first move was to commission its own research, which confirmed the statistics but fell short of drawing a causal connection between the dead children and the nuclear plant. Later, the government had to defend its position at the Paris Commission, the international body concerned with ocean pollution from land-based sources. Today, under international pressure plus the threat of further antics by Greenpeace, Britain has finally agreed to work towards a zero discharge policy from the Windscale plant.

Back to Muroroa

Peaceful direct action by a group of people internationally organized and well versed in both the theory and practice of peaceful direct action makes a formidable opponent to national governments who inevitably are politically tainted, bound by compromise, limited by nationalism, weighed down by bureaucracy and who can only profess peace whilst condoning their own and therefore others' military forces. When Greenpeace set out to return to Mururoa in the South Pacific for the twelfth year of its campaign against nuclear tests, French secret agents detonated two limpet mines below the waterline of the protest ship *Rainbow Warrior* as she lay at anchor in Auckland harbour. One crew member, Fernando Pereira, was killed and the ship sank in four minutes. That outrage focused attention on the activities of Greenpeace, which became a household word around the world. Greenpeace met violence with silence, and sent a replacement ship halfway round the world in 42 days. As she arrived, the nuclear test zone itself was lit up by a magnificent rainbow. Meanwhile, a government had been discredited, and the whole notion of the right to explode nuclear devices in the South Pacific had been called into question. When Greenpeace returns again in peace to continue the campaign, its power will have been strengthened immeasurably.

Luxulyan

We can take a certain amount of hope from the examples set by Greenpeace, particularly if that organization continues to amass its moral power and apply it with flair and integrity. But perhaps the greatest hope rests with the many spontaneous uprisings of ordinary people who find themselves confronted with a seemingly totalitarian threat to their communities or fundamental values by the agents of big business and central government.

The story of a handful of determined villagers from Cornwall who successfully thwarted the combined attempts of a government monopoly, the Central Electricity Generating Board, the High Court and the Court of Appeal to inflict a nuclear power plant upon them, serves to illustrate the point.

The CEGB began searching Cornwall in 1978 for possible sites for a new nuclear power plant in the south west of England. By 1981, three sites had been determined, and some Cornish people had formed a loose alliance known as CANA – the Cornish Anti-Nuclear Alliance. At Nancekuke they threatened to put their lives at risk to protect a colony of Grey Atlantic seals from the potentially devastating effects of seismic and hydrographic surveys to be made by the CEGB. They would, if necessary, go to sea in small boats and encircle the survey vessels and enter the explosion zones. The Board withdrew their plan. At Luxulyan, they barricaded a farm commandeered by the Board for a drilling site. Women chained them-selves to rigs, protesters took up residence and maintained a vigil for 172 days. During this time they refused to react to violence of any kind and won the respect of both the local police and the contractors. The Board was granted 32 writs of injunction against the protesters, who immediately disbanded and were replaced by 32 new protesters and there were plenty more in the queue. The whole community experienced a sense of unity in the face of adversity that began to create a formidable moral pressure. The entire community was polled and 92 per cent were opposed to the siting of the nuclear plant. Local MPs and dignitaries publicly voiced the fears and implacable opposition of the community. A strong case was made that the plant was neither beneficial nor even necessary to the immediate locale or the wider community. The motives for siting it in Cornwall were called into question. Even the Cornish Chief Constable was hauled up before the full bench of the Court of Appeal after he had repeatedly refused to identify any breach of the peace by the protesters, who were acting peacefully on private land with the consent of the owner. The police had likewise refused to aid the Board's men to evict the protesters. But the Appeal Court redefined a breach of the peace. 'There is a breach of the peace whenever a person who is lawfully carrying out his work is unlawfully and physically prevented by another from doing it', said Lord Denning in the Court of Appeal on 21 October 1981. The protesters finally withdrew after six months of blockade.

Not long after, the CEGB announced that there was no site in Cornwall suitable at that time for the construction of their new nuclear plant. The prize went elsewhere.

Is it Right to Break the Law?

Elected governments, because they claim that a parliamentary majority automatically confers moral authority; business, because it is just going about its 'lawful business'; and the law itself, as arbiter and codifier of the public morality, rightly distrust and fear the usurptive powers of direct action. Those people resorting to direct action would therefore be naive if they were surprised to find big business, the government and the law ranged against them in a solid phalanx of opposition. Appreciating this, it is essential to present one's case clearly, and with a view to scotching some popular misconceptions. Our aims are not political; we are not a revolutionary body, we are reformists. We respect our governing institutions but we believe that the law itself is in need of change; the issue at stake is of much fundamental significance and we have no other way to make our voice heard. We have presented irrefutable evidence time and again and no one has listened. These are the arguments by which a moral initiative can be taken and put to work as the ground for subsequent direct action.

Non-Violent Direct Action: the Weapon of a New Age

Some people take the view that the ecological critique of latter-day twentieth-century society represents as profound a shift in consciousness as that caused by Copernicus when he elevated the status of the sun. By casting humankind from its central position of worth amidst creation and by redefining peace, in its broader sense, to include peace with nature as a necessary constituent, a new ethic has been born. In its struggle to attain social ascendancy, this ethic will be met by violence of every conceivable kind, as the dominant establishment increasingly finds itself with its back to a wall whose foundations are already eroded. Non-violent direct action ranks high in the armoury of that new ethic and to whatever extent it helps to establish it, it will make its own case, on its own terms.

If this really is what is happening, then these are early days for a fundamentally new, last resort technique of persuasion. One would hope that the very simple expressions of it that have been made, not only by Greenpeace but by countless small groups of caring and intelligent people, will be greatly amplified and modulated through the coming years. If humankind can develop the variety and sophistication in peaceful techniques of persuasion that it has in techniques of brutality, there is some real hope that humankind will survive the crisis of its evolution through which it is currently passing.

Notes

1. One year and a half later, after being taken to court by the public prosecutor and being found guilty on a number of charges, including negligence, BNFL was fined £10,000 for the accidental discharge. During the court hearing, BNFL claimed that the bulk of some 4,500 curies in the sea tank had remained on site and had been pumped to a treatment plant. Sir Samuel Edwards, from Cambridge university and Chief Scientific Adviser to the Department of Energy, disputed BNFL's contention, and concluded that the majority of the 4,500 curies had been discharged to sea.

35

Green Hopes

Lindy Williams

When the Green Party was formed in 1973 it was dismissed as being a party of freaks, drop-outs and dreamers. For a political party to be concerned with environmental issues was considered to be quaint, unrealistic and certanly not seriously political. But,now, although the process has seemed at times to be unbelievably slow, the Green Party has established itself as a serious and radical party, one which cannot be ignored, the fourth political voice in the UK. In addition the words 'green' and 'environmental' have now been absorbed into everyday political vocabulary, so strong have been both the rise of pressure groups and the surge of concern from the public about issues such as acid rain, nuclear power and food quality.

The transition from being considered a freak fringe party to a minority party with a strong and important voice has not been easy, and with the present unrepresentative electoral system there is little chance of an influx of Greens to Westminster! However the support shown at the polls, especially in general elections, by no means reflects the true support that exists amongst voters. Time and again Green Party candidates throughout the country hear people express total support for all of our policies, but regretfully say that they must vote for one of the main parties in order to keep X or Y out, but maybe next time. . . . It is very likely that if there were a system of proportional representation there would by now be some Green MPs.

From the main parties there are constant requests for Green Party candidates to stand down either to give the Alliance a chance and 'wait' for proportional representation, to give Labour a clear path to unilateral nuclear disarmament or even to join the Liberals who claim always to have been green. Additionally we are frequently accused of stealing votes from the other parties, the implication being that voters exist to give power to the parties, whereas the Greens take the opposite view – that the political parties exist to give the voter a voice. There is no chance of coalition or compromise on the part of the Greens, for not only would our engulfment into a large establishment party take the pressure to change off that party, but it would take away from the voters the opportunity to vote positively for a complete Green package.

It has been, and to a large extent still is, convenient and less threatening for established parties and the establishment media to perceive and portray the Green Party as one solely concerned with environmental pollution, whether it be radioactivity, sulphur emissions, pesticides or food additives. Nowhere was this more ironically portrayed than during the passage through parliamant of the Representation of the People Bill in 1984. The Greens (then still called the Ecology Party) were mounting what proved to be a successful campaign against raising the parliamentary deposit to £1,000, when an MP said during the debate, '. . . I cannot speak for organisations such as the Ecology Party, but I imagine that they hope to move from being single issue organisations, which attract a small but respectable following because of public feeling on those single issues, to becoming broader parties with full manifestos.' There was nothing empty about the Green Party's manifesto in 1983.

The Green philosophy is firmly grounded in the principle that the planet and its resources are finite, and its inhabitants interdependent, and this global perspective is shared by Green parties all over the world. All our policies stem from this, and although it can be described as an environmental base, it is important to stress that Greens do not see 'the environment' as existing in a vacuum – it involves all living things, especially since exploitation of the planet inevitably involves exploitation of people. The wide range of Green policies includes proposals for a post-industrial economy, including a basic income payable to all, co-operative, small-scale employment initiatives, locally-based agriculture, safe energy, massive conservation measures and moves towards non-violent social defence. The fact that all political parties are now attempting to look green indicates that at last they are starting to recognize that the Green Party does cover the whole spectrum of politics, and is therefore a threat to them on all issues. But none of the established parties has even begun to address the issues of *why* environmental degradation, social injustice and violence are still accelerating at such an alarming rate; or if they have, they are constrained by their own traditions. It is perhaps beyond their self-imposed remit to take a closer look, for they are locked into a system that prevents them from seeing further ahead than the next opportunity to be elected. That opportunity is now threatened by the Green Party from a radical stance which challenges the very roots of industrial society, of the institutions and the hierarchies which we see as the cause of the global crisis. Without this particular challenge the results of industrial expansion and obsession with economic growth will continue to manifest themselves in exploitation of people and planet alike.

Whilst conventional politics, with its commitment to economic growth as the panacea for all the ills of the world, represents the static political positions in society, the Green Party represents a broadly based movement which does not seek power in order to rule, but which calls for the change which is essential if there is to be any justice left on the earth and for the earth – or indeed if there is to be any earth at all. But this change will not come solely through parliamentary channels. Although it is part of the role of the Green Party to work through those channels, radical change also needs to be initiated at grassroots level. For too long people have waited for change to come from above and all too often they are numbed by the cold rationality that is churned out by paternalistic experts and politicians, or drown themselves in a sea of consumerism to avoid having to contemplate the crises that threaten our existence and that of our children. Our education and conditioning encourage us to do little else – such is the power of established institutions. The Greens' role is to empower people, to encourage every person to take part in the shaping of a post-industrial Britain, to convince them that no contribution is too small and that every contribution is vital. This entails minimizing exploitation and domination by the breaking down of hierarchies and replacing them with co-operative networks to maximize communication and concensus decision-making. In this way the Green's commitment to non-violence extends beyond our view of the fragility and interdependence of the eco-system to our relationships with each other and our ways of working. The social and the environmental cannot be separated.

The human race faces a broad challenge, and the Green Party with its clear, uncompromised philosophy has a vital part to play in the 'greening' of Britain. Our role is beautifully summed up by Petra Kelly, when she says about the German Greens.

We aim for a party system that is truly representative, with no entry restrictions. Right now there is a very real need for an anti-party party, which will genuinely espouse the cause of the weaker members of society. The aspirations of the peace and ecology movements should be represented within a political forum, in addition to their expression outside parliament. As Greens, it is no part of our understanding of politics to find a place in the sun alongside the established parties, nor to help maintain power and privilege in concert with them. . . . Society and the economy now pose such a threat to survival that they can only be resolved by structural change, not by crisis management and cosmetic adjustments. The Greens can make no compromises on the fundamental questions of the environment, peace, sexual equality and the economy.[2]

Notes

1. *Hansard* 10 December 1984 col. 789.
2. Petra Kelly, *Fighting for Hope*, trans. Marianne Howarth, London: Chatto & Windus, 1984.

List of Addresses

British Association of Nature Conservationists, Rectory Farm, Stanton St John, Oxford

Campaign for Freedom of Information, 3 Endsleigh Street, London WC1H 0DD

Conservation Society, 12a Guildford Street, Chertsey, Surrey

Earth Resources Research Ltd, 258 Pentonville Road, London N1 9JY

Earthlife, 10 Belgrave Square, London SWIX 8PH

The Ecologist, White Hay, Withiel, Bodmin, Cornwall

Ecoropa (UK), Crickhowell, Powys, Wales

Farm and Food Society, 4 Willifield Way, London, NW11 7XT

Friends of the Earth, 377 City Road, London EC1

Greenpeace, 36 Graham Street, London N1 8LL

Henry Doubleday Research Association, Ryton on Dunsmore, Coventry CV8 3LG

Nature Conservancy Council, Northminster House, Peterborough

National Society for Clean Air, 136 North Street, Brighton

National Union of Seamen, Maritime House, Old Town, Clapham, London SW4 0JP

Save Britain's Heritage, 68 Battersea High Street, London SW11

Soil Association, Walnut Tree Manor, Haughly, Stowmarket, Suffolk

Vegetarian Society, 53 Marloes Road, London, W8 6LA

Index